T0331179

Math Problems in Water and Wastewater

This book covers the fundamental concepts required to solve typical problems in water and wastewater engineering. Water professionals working in the industry require a license to work in water plants, and *Math Problems in Water and Wastewater* aids readers in preparing for the mathematics portion of these exams. It lays a sound foundation that not only helps with the certification examination but also helps water operators in performing their daily activities. The basic concepts and volumes of various unit devices followed by specific problems in water and water treatment are presented through solved example problems.

- Includes examples both in Imperial and SI units throughout.
- Covers common and specific topics both for water and wastewater operations.
- All calculations shown with unit cancellation.
- All example problems are followed by practice problems.
- Examples include problems suitable for all level of certification.
- A brief description of the water and wastewater treatment is given.

Math Problems in Water and Wastewater

Subhash Verma

CRC Press
Taylor & Francis Group
Boca Raton London New York

CRC Press is an imprint of the
Taylor & Francis Group, an **informa** business

Designed cover image: Shutterstock

First edition published 2025
by CRC Press
2385 NW Executive Center Drive, Suite 320, Boca Raton FL 33431

and by CRC Press
4 Park Square, Milton Park, Abingdon, Oxon, OX14 4RN

CRC Press is an imprint of Taylor & Francis Group, LLC

ISBN: 978-1-032-74045-4 (hbk)
ISBN: 978-1-032-74328-8 (pbk)
ISBN: 978-1-003-46874-5 (ebk)

DOI: 10.1201/9781003468745

Typeset in Times
by Apex CoVantage, LLC

Contents

About the Author

Before leaving for Canada in 1978, Subhash was an assistant professor in the Department of Soil and Water Engineering of Punjab Agricultural University Ludhiana, India. After completing his master of engineering in water resources engineering from the University of Guelph, he stayed with the school of engineering and researched watershed modelling. In 1982, Subhash joined the Sault College of Applied Arts and Technology to head a new program in water resources engineering technology. In addition to teaching, Subhash developed training programs in many areas of water and wastewater engineering technology. These training programs were delivered to Ministry of Environment personnel in the province of Ontario. During his time at Sault College, Subhash developed course manuals to supplement courses in hydraulics, hydrology, water supply and wastewater engineering technology. In 2007, Subhash was appointed chair of Sault Ste. Source Protection Committee by the Minister of Environment. After his retirement, Subhash's pastime is developing and teaching online courses in water and wastewater technology on Ontario Learn, a provincial initiative. Subhash has been the lead author of two books: *Water Supply Engineering* and *Environmental Engineering, Fundamentals and Applications*. He has recently published a book titled *Water and Wastewater Engineering Technology*. He has published two books *Applied Hydraulics* and *Engineering Hydrology* and is working on groundwater and wells as part of a series in water resources engineering. He has started working on another book titled *Farm Irrigation Engineering*.

List of Figures

List of Tables

1 Introduction

As plant operators of engineered water and wastewater treatment systems, you need to evaluate the plant and individual processes. As part of your job, you monitor and control the processes with the objective of optimizing the operations. Do you think it is possible to accomplish your job without the use of the language of math? The information you collect as part of operations is processed by using the tools of mathematics, and based on these, decisions are made about changes, if any. You can very well appreciate the importance of communicating the right information. The values of process variables are not mere numbers; these numbers speak about the processes and hence the health of the plant. It is not sufficient to say that the plant is operating at good efficiency. It needs to be substantiated with factual data. One person's good may be another person's best or worse. To communicate in the scientific sense, it requires that the operator be familiar with basic mathematical concepts.

The objective of this book is to help the operator to determine the process efficiency through the use of mathematical calculations rather than trial and error or the guessing method. This book will not dwell on math per se but rather use it as a tool to process the information and hence control the process. There are some basic concepts which, once mastered, will make your calculations seem simple. The equations or formulas may appear to be complex, but once you understand what process or relationship they represent, you will find them interesting and useful. Good luck and have fun with math.

1.1 USE OF THIS MANUAL

This book is designed with the assumption that the student possesses basic math skills. If you are weak or it has been long time since you have had the opportunity to use or apply these skills, it is advisable that you first brush up on the basic concepts. This may include the units, priority of mathematical operations and algebraic manipulation of formulas. An equivalent of high school mathematics is considered sufficient to move you along smoothly.

The contents of this book are divided into three sections. The general section includes those topics which are common to the water and wastewater fields. The topics discussed in the general section cover a wide range of applied math problems. The topics are arranged in a logical sequence. It is strongly recommended that students follow this sequence with as little deviation as possible. Students with a strong math background might find the problems too easy in the beginning. It is still worthwhile to do those problems just to get used to the methodology of solving applied mathematical problems. This will also allow you to build strength and confidence and get rid of the math "scare".

Sections two and three are specifically for water and wastewater treatment. Topics in these two sections are grouped by unit processes. The topics are arranged in order

DOI: 10.1201/9781003468745-1

of the flow through a plant. You must know the treatment processes. Those who do not have prior background in plant operation are advised to learn about treatment methods before attempting the math problems. These manuals may be useful in preparing for certification examinations. Depending on the level of certification, the complexity of the problems and topics may vary. In such cases, it is advisable that you first determine the topics and type of problems to be covered in a particular examination and then focus on the topics.

1.2   BASIC APPROACH

Half the solution lies in understanding the question right. So be patient and do not attempt any calculations until you have read the question thoroughly. A logical sequence of steps in solving a mathematical problem is as follows:

1.2.1   KNOWN AND UNKNOWNS

First write down the information provided and what is to be found in a given question. You must put down the units with each quantity. Quite often the questions are 'word problems', and some information may be hidden somewhere in the question, perhaps in another form. That is why it is important to read the statement of the problem carefully.

1.2.2   DIAGRAM

Draw a diagram and mentally visualize the problem. This step may be done as part of step one, where you can label the diagram with knowns and unknowns. Remember the diagram should be schematic rather than artistic. The time spent drawing is worthwhile, as it helps to maintain a focus and to organise the data.

1.2.3   CONCEPT

What concept is involved? Is the question about area, hydraulic loading or chemical feed rate? Identifying the concept will also allow you to select the appropriate formula/equation required to solve the problem.

1.2.4   APPLICABLE FORMULA

Familiarize yourself with all the formulas in the formula sheet. It is suggested that you use one equation for each basic concept. Regardless of the unknown, always write the formula in the same original form. Using the same form over and over helps you to remember the formula and become more comfortable with its use.

1.2.5   ALGEBRAIC MANIPULATION

Rearrange the formula from its original form to solve for the unknown. Write the applicable equation and manipulate the terms such that the unknown (to be found) is on the left-hand side (LHS) and all other terms are on the right-hand side (RHS) of the equality sign (=). Follow the rules of algebra when transposing the terms left or right of the equals sign. For example, one of the common formulas is discharge rate Q is equal to flow velocity v times flow area A, as shown:

$$Q = v \times A$$

This form is okay if you want to calculate discharge or flow rate. However, if you would like to find flow velocity, then you will need to manipulate. One of the important things to know is that when you perform an operation in an algebraic equation; you must do the same to both sides of the equation. To separate v, we can divide both sides by the area.

$$\frac{Q}{A} = \frac{v \times A}{A} \quad or \quad \frac{Q}{A} = v \quad or \quad v = \frac{Q}{A} \ similarly \ A = \frac{Q}{v}$$

So, when terms on both sides of the equation do not involve plus and minus terms, then if a variable is part of the numerator (above the bar), you can move it to other side and place it in the denominator (below the bar). Similarly, when the equation involves more than one term on either side, then change the sign of the term to move it to the other side.

1.2.6 SCIENTIFIC NOTATION

Scientific notation helps us to represent the numbers which are very huge or very tiny in a form of multiplication of single-digit numbers and 10 raised to the power of the respective exponent. It is a form of presenting very large numbers or very small numbers in a simpler form. As we know, whole numbers can be extended to infinity, but we cannot write such huge numbers on a piece of paper. Also, the numbers which are present at the millions place after the decimal needed to be represented in a simpler form. Thus, it is difficult to represent a few numbers in their expanded form. Hence, we use scientific notations. For example, 100000000 can be written as 10^8, which is the scientific notation. Here the exponent is positive. Similarly, 0.0000001 is a very small number which can be represented as 10^{-8}, where the exponent is negative.

1.2.7 DIMENSIONAL ANALYSIS

Substitute the values and make the necessary unit conversions to get the desired answer. If after cancellations the units do not tally up with the unknowns, stop. Stay cool and check all the previous steps to figure out your mistake. The unit cancellation procedure is discussed in the next section.

1.2.8 ANSWER

Most important of all is checking the answer critically after performing the calculations. For example, if the detention time for a clarifier comes in hundreds of hours or the units after cancellation are that of velocity (distance/time), something is wrong. After you have straightened out the problem, *weigh the value of the answer in terms of unknowns and original data*. Does it make sense? Is it reasonable?

1.2.9 ROUNDING OFF

The final step is to present the final answer. Based on the number of minimum significant figures in the original data, round off the answer. It is a good practice to highlight or underline your answer. If the calculation only involves addition and subtraction, produce the final answer to the minimum number of decimal places.

1.2.10 PRACTICE

There is no substitute for this. When you read a solved problem, it may appear easy. However, the real test is to do yourself. More and more practice will build your confidence and help you get rid of math apprehension.

1.3 UNIT CANCELLATION PROCEDURE

A simple straightforward procedure called unit cancellation will ensure proper units in any kind of calculation. The steps of the procedure are described subsequently. This procedure when properly executed will work for any equation or formula. It is really very simple, but some practice may be required to use it. The more you practice it, the more you will feel comfortable with its use. Make this a habit. This will go a long way in getting the correct answer.

1.3.1 STEPS FOR UNIT CANCELLATION

1. After solving the equation for the unknown, substitute values including the units.
2. Decide on the proper units for the result. Most of the time the units desired for the results are specified in the statement of the problem.
3. Cancel units which appear in both the numerator and denominator of the expression.
4. Use the conversion factor to eliminate unwanted units and obtain the proper units, as decided in step 2. To cancel out, the conversion factor should appear opposite to the unwanted unit. *Make sure the units are given the same exponents as the variable in the equation.*
5. In step 4, if you fail to get the desired unit, it may be due to a mistake made in the preceding steps. Increase your concentration by a notch or two and find your mistake. Successfully getting the right units for the unknown is a good indication that you are on the right track.
6. Perform the calculation and check the answer.
7. Weigh your answer, and if satisfied round up to the appropriate number of significant places. As a rule of thumb about significant figures, do not

produce your final answer in more than three significant figures. If the original data is expressed in two significant figures, the final answer can not only be correct to two significant figures.

Let us say we want to express a velocity of 1.5 m/s as m/min. Should we multiply by 60 s/min or min/60 s? You guessed it right; 60 s/min is the correct way, since it will allow us to cancel out the seconds.

Sol:

Velocity expressed as m/min

$$v = \frac{1.5\,m}{\cancel{s}} \times \frac{60\,\cancel{s}}{min} = 90.0 = \underline{90\,m/min}$$

Velocity expressed as per km/h

$$v = \frac{1.5\,\cancel{m}}{s} \times \frac{60\,\cancel{s}}{\cancel{min}} \times \frac{60\,\cancel{min}}{h} \times \frac{km}{1000\,\cancel{m}} = 5.40 = \underline{5.4\,km/h}$$

In the previous case, conversion process was done in three steps, seconds to minutes, minutes to hours and finally metres to kilometres. We could have made the same conversion by converting seconds to hours using the conversion factor of 3600 s/h. The following example illustrates conversions from English to metric units, converting flow velocity of 1.5 m/s to ft/s. If you do not remember the conversion factor between ft and m, refer to a conversion table. Conversion tables are available when you write certification examinations.

Sol:

Velocity expressed per ft/s

$$v = \frac{1.5\,\cancel{m}}{s} \times \frac{3.28\,ft}{\cancel{m}} = 4.92 = \underline{4.9\,ft/s}$$

1.4   BASIC MATH EQUATIONS/FORMULAS

1.4.1   DENSITY

Density indicates the denseness or relative lightness or heaviness of a given liquid or solid. Mass density is mass per unit volume. The mass density of water is 1 kg/L or 1.0 g/mL. **Weight density** is weight per unit volume. It is important here to make the distinction between mass and weight. That is, weight is force due to gravity. **Specific gravity** is the ratio of the density of liquid or solids as compared to the density of water. This parameter, being the ratio of two densities, is dimensionless or is just a number.

Expressions for density

$$Mass\ density,\ \rho = \frac{m}{V} \quad Weight\ density,\ \gamma = \frac{w}{V} \quad Specific\ gravity,\ SG = \frac{\rho}{\rho_w} = \frac{\gamma}{\gamma_w}$$

m = mass; w = weight; V = volume; ρ_w = mass density of water

1.4.2 FLOW RATES

Flow rate is quantity of liquid (volume, mass, weight) flowing per unit time. *Volume flow rate* is the volume of liquid carried per unit time, mass flow rate is mass per unit time and weight flow rate is weight per unit time.

Expressions for various types of flow rates

$$Volume\ rate,\ Q = \frac{V}{t} = v \times A \quad Mass\ rate,\ M = \frac{m}{t} = \rho \times Q$$

$$Weight\ rate,\ W = \frac{w}{t} = \gamma \times Q$$

1.4.3 FLOW VELOCITY

Flow velocity is essentially the average speed in the direction of flow. Another way of saying it is flow rate per unit flow area.

$$Flow\ rate,\ Q = v \times A \quad or \quad v = \frac{Q}{A}$$

1.4.4 CONTINUITY EQUATION

Based on the continuity of flow, flow rate at various sections of a conduit must be the same in a steady flow situation. This leads us to the following relationship, more commonly called the continuity equation.

$$v_1 \times A_1 = v_2 \times A_2$$

A = area, v = velocity, and sub 1 and 2 indicate the flow section

1.4.5 CONCENTRATION

Concentration indicates the strength of a given liquid chemical in terms of the amount of substance in a given volume of liquid or solution. When expressing concentrations of highly concentrated solutions, it is usually indicated per unit mass of the liquid. The mass to volume relationship is very commonly used, and the most common unit is mg/L or g/m^3.

Expression for concentration

$$Mass\ conc.,\ C_{m/v} = \frac{m}{V} = \frac{M}{Q} \quad Mass\ to\ mass,\ C_{m/m} = \frac{m}{m_{sol}}$$

1.4.6 DILUTION FORMULA

Diluting solution

$$V_1 \times C_1 = V_2 \times C_2$$

Simple equations like this one can be easily manipulated algebraically to solve for the unknown. For example, continuity equation shown previously, if velocity in the section 2 is the unknown, equation can be modified as follows:

Flow velocity and flow area

$$v_1 \times A_1 = v_2 \times A_2 \quad or \quad v_2 = \frac{A_1 \times v_1}{A_2}$$

Chemical feeding rates

$$Q_1 \times C_1 = Q_2 \times C_2$$

1.5 STEPWISE SOLUTION

Problem solving becomes easy by following logical steps. A little bit of extra time spent on putting things down and understanding the known and unknowns is the key to success. These steps are illustrated by showing the stepwise solution to the following problem.

EXAMPLE PROBLEM 1.1

An operator chlorinates well water using a hypochlorinator (Figure 1.1). The well pumping rate is 45 L/s, the hypochlorite solution concentration is 20% (200 g/L)

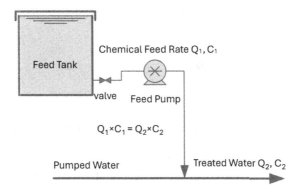

FIGURE 1.1 Liquid chemical feeding.

available chlorine and the dosage is 2.5 mg/L. Calculate the feed rate of the hypo-chlorite solution in L/d.

STEP 1:

Draw a sketch of the system and label the diagram with known and unknowns. Make sure all the units are indicated.

GIVEN:

$$C_1 = 200 \text{ g/L} \qquad C_2 = 2.5 \text{ mg/L} \qquad Q_2 = 45 \text{ L/s} \qquad Q_1 = ?$$

STEP 2:

Select the appropriate formula. Based on the principle of mass conservation:

$$Q_1 \times C_1 = Q_2 \times C_2$$

STEP 3:

Modify the equation algebraically for the desired term (unknown).

$$Q_1 = \frac{C_2 \times Q_2}{C_1}$$

STEP 4:

Substitute known numerical values, as well as units.

$$Q_1 = \frac{C_2 \times Q_2}{C_1} = \frac{2.5 \text{ mg}}{L} \times \frac{L}{200 \text{ g}} \times \frac{45 \text{ L}}{s}$$

STEP 5:

Cancel units which appear in both the numerator and denominator and check the final units are that of the feed pump rate, that is, volume per unit time.

$$Q_1 = \frac{2.5 \text{ mg}}{\cancel{L}} \times \frac{\cancel{L}}{200 \text{ g}} \times \frac{45 \text{ L}}{s} = \frac{2.5 \text{ mg}}{200 \text{ g}} \times \frac{45 \text{ L}}{s}$$

STEP 6:

Since the final units are that of the pump rate, it is a good clue that we are heading in the right direction. If you do not end up with units of flow rate, go back and find the mistake before you proceed further. After it is confirmed that equation is dimension-ally correct, the next step is to express the final results in desirable units.

   Decide on the proper units for the result, L/d in this case, and make appropriate conversions.

STEP 7:

Use conversion factors to eliminate unwanted units. Place the conversion factors as vertical fractions such that unwanted units appear opposite each other and cancel out. If a given unit has an exponent, make sure to raise the conversion factor to the same exponent. Failure to do so will lead you to wrong units and results. Experience has shown that this type of mistake is very common, so you need to be extra careful.

$$Q_1 = \frac{2.5 \, \cancel{mg}}{200 \, \cancel{g}} \times \frac{\cancel{g}}{1000 \, \cancel{mg}} \times \frac{45 \, L}{\cancel{s}} = \frac{3600 \, \cancel{s}}{\cancel{h}} \times \frac{24 \, \cancel{h}}{d}$$

STEP 8:

Perform the calculation and express the result, retaining an appropriate number of significant figures. The final answer is usually underlined.

$$Q_1 = \frac{2.5 \, \cancel{mg}}{200 \, \cancel{g}} \times \frac{\cancel{g}}{1000 \, \cancel{mg}} \times \frac{45 \, L}{\cancel{s}} = \frac{3600 \, \cancel{s}}{\cancel{h}} \times \frac{24 \, \cancel{h}}{d} = 48.6 = \underline{49 \, L/d}$$

2 Area and Volume

In water and wastewater processing, various shapes of tanks and pipes are used. Calculations of area and volume are often needed to calculate capacity, surface area overflow rates, hydraulic and organic loading, detention times and so on. The two common shapes in water and wastewater treatment are rectangular and circular tanks.

Math constant pi

$$\pi = \frac{circumference}{diameter} = \frac{22}{7} = 3.14$$

2.1 AREA

Area refers to the surface, and hence the units of area are length units squared. Area formulas of some of common geometrical shapes are as follows.

2.1.1 RECTANGULAR TANKS

The area thus calculated will yield the surface area. To calculate the velocity of flow in a section, we need to know the area of the cross-section or water section perpendicular to the direction of flow. For a rectangular section of channel with depth of flow d, the area of cross-section is given by the following equation:

Cross-sectional, Surface area

$$A_X = L \times d \quad A_s = L \times W$$

Cross-sectional, Trapezoidal section

$$A_X = 0.5(b_1 + b_2) \times h$$

In Figure 2.1, rectangular and trapezoidal shapes are compared.

2.1.2 CIRCULAR SHAPES

Circular shapes are defined by one dimension, radius R or diameter D, since $D = 2R$. In Figure 2.2, circular and triangular shapes are compared.

Area of a circle   *Area of a triangle*

$$A = \pi R^2 = \frac{\pi D^2}{4} = 0.785 D^2 \qquad\qquad A = \frac{1}{2} \times b \times h$$

DOI: 10.1201/9781003468745-2

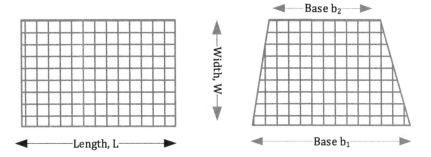

FIGURE 2.1   Area of a rectangle and a trapezoid.

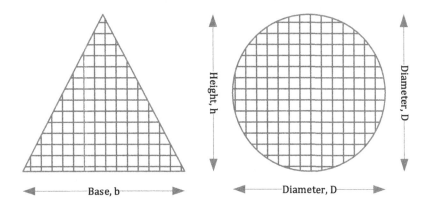

FIGURE 2.2   A circle and a triangle.

The mathematical constant π is equal to the ratio of circumference to diameter of a circle; hence the circumference or periphery of a circle is:

Circumference

$$\boxed{P = \pi D = 2\pi R}$$

A length unit of 1 hm = 100 m; hence 1 hm² (10 000 m²) represents an area of a square with each side of 100 m. This is more commonly called a hectare (abbreviated as ha), as shown in Figure 2.3. The units of area are length units squared, that is, cm² or m². When doing the conversions for the area units, make sure the conversion factor is raised to the power of 2.

Area unit SI

$$\boxed{1\,km^2 = 1\,km^2 \times \frac{(1000\,m)^2}{1\,km^2} = 1\,000\,000\,m^2 = 100\,ha}$$

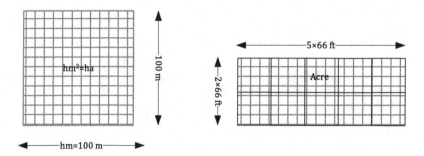

FIGURE 2.3   Units of area.

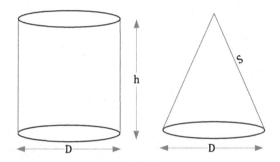

FIGURE 2.4   A cone and a cylinder.

Area unit of acre

$$1 \ acre = 10 \times chain^2 = 10 \times (66 \ ft)^2 = 43560 \ ft^2$$

2.1.3   CYLINDER

A circular cylinder total surface includes the top, bottom and lateral surfaces or the area of the wall. The top and bottom each are equal to the area of a circle. If you cut the cylinder wall and spread it out, it will form a rectangle with length equal to perimeter of the circle of diameter, *D*, and width equal to height *h*, or the depth of the cylinder (Figure 2.4).

Lateral surface of a cylinder

$$A = \pi D \times h = 2 \pi R \times h$$

2.1.4   CONE

The lateral area of a cone is half that of a rectangle with length equal to the perimeter of the circle and width equal to slant height *S* (Figure 2.4).

Lateral surface of a cone

$$A = \pi D \times S/2 = \pi R \times S$$

Slant height

$$s = \sqrt{(R^2 + h^2)}$$

A hectare is a unit of area and it is wrong to include the exponent 2.

2.1.5 SPHERE

The surface area of a sphere or ball is equal to pi times the diameter squared.

Surface of a sphere

$$A = \pi D^2 = 4\,\pi R^2$$

EXAMPLE PROBLEM 2.1

Determine the water surface area of a rectangular clarifier of length 55 m and width 20 m.

Given:

$L = 35$ m $\qquad W = 12$ m $\qquad A_s = ?$

Solution:

$$A = L \times W = 35\ m \times 12\ m = 420.0 = \underline{420\ m^2}$$

EXAMPLE PROBLEM 2.2

It is desired to have a rectangular settling basin of surface area of 330 ft². If the width is kept to 12 ft, what should be the length of the basin?

Given:

$L = ?$ $\qquad W = 12$ ft $\qquad A_s = 330$ ft

Solution:

$$L = \frac{A}{W} = \frac{330\ ft^2}{12\ ft} = 26.6 = \underline{27\ ft}$$

EXAMPLE PROBLEM 2.3

The surface area of a circular clarifier is known to be 1500 ft². What is the diameter?

Given:

$$D = ? \qquad A_s = 1500 \text{ ft}^2$$

Solution:

Diameter of the clarifier

$$D = \sqrt{\frac{4A_s}{\pi}} = \sqrt{\frac{4}{\pi} \times 5000 \, ft^2} = 79.78 = \underline{80 \, ft}$$

EXAMPLE PROBLEM 2.4

Find the lateral surface area of a cylindrical chemical tank of diameter 5.0 m and height of 3.5 m.

Given:

$$D = 5.5 \text{ m} \qquad h = 3.5 \text{ m} \qquad A_s = ?$$

Solution:

Lateral surface of the cylindrical tank

$$A = \pi D \times h = \pi \times 5.5 \, m \times 3.5 \, m = 60.47 = \underline{60 \, m^2}$$

EXAMPLE PROBLEM 2.5

The bottom of a 21-ft-diameter tank is conical with a height of 8.5 ft. To order paint, your supervisor want to know the lateral area of the cone.

Given:

$$D = 21 \text{ ft} \qquad R = 21/2 = 10.5 \text{ ft} \qquad h = 8.5 \text{ ft} \qquad A_s = ?$$

Solution:

Slant of the cone

$$S = \sqrt{(R^2 + h^2)} = \sqrt{(10.5 \, ft)^2 + (8.5 \, ft)^2} = 13.5 = \underline{14 \, ft}$$

Surface area of the cone

$$A = \pi R \times S = \pi \times 10.5 \, ft \times 13.5 \, ft = 445.62 = \underline{450 \, ft^2}$$

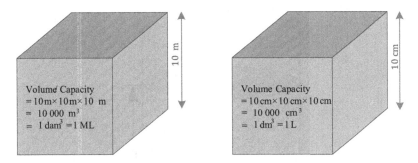

FIGURE 2.5 Units of volume, litre and mega litre.

2.2 VOLUME

The volume of a tank, container or pipe refers to its capacity to hold liquids. Note that area refers to two dimensions, whereas volume refers to three dimensions. It is for this reason the units of volume are those of length units raised to the power of 3, for example, cm^3 and m^3. Litre is a unit of volume and is equal to a volume of a cube with each side 10 cm, as illustrated in Figure 2.5.

Cubic metre is a standard unit of volume.

$$1\, m^3 = 1\, m^3 \times \frac{(100\ cm)^3}{m^3} \times \frac{mL}{1\ cm^3} \times \frac{L}{1000\ mL} = 1000\, L = 1\, kL$$

Though litre is not a standard SI unit, it is very commonly used. The equivalent SI unit of volume is dm^3. When expressing large quantities of water, units like ML and MG are used.

$$1\, dam^3 = 1\, dam^3 \times \frac{(10\ m)^3}{dam^3} \times \frac{1000\ L}{1\ m^3} = 1000000\, L = 1\, ML$$

That is to say a cubic container measuring 10 m on each side will have a full capacity of 1 million litres. Plant capacities are usually expressed in ML/d.

EXAMPLE PROBLEM 2.6

Convert a volume of 10 acre ft into million gallons.

Given:

V = 10 acre.ft   V = MG

Solution:

$$V = 10\ acre.ft \times \frac{43560\ ft^2}{acre} \times ft \times \frac{7.48\ gal}{ft^3} = \frac{\pi}{4} \times (11\ ft)^2$$

$$= 3258000\ gal = 3.3\ MG$$

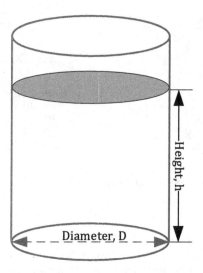

$$m^3 = 1000\,L = kL \qquad dam^3 = 1000\,m^3 = ML$$
$$hm^2 = 10000\,m^2 = ha \qquad km^2 = 10000\,000\,m^2 = 100\,ha$$

FIGURE 2.6   Volume of a cylinder.

2.2.1 CYLINDER

The volume of a cylinder is found by the multiplication of bottom area and height or depth.

Volume or capacity of a cylinder

$$V = A \times h = \frac{\pi D^2}{4} \times h = \pi R^2 \times h$$

Figure 2.6 illustrates the volume occupied by a cylindrical shape.

EXAMPLE PROBLEM 2.7

Determine the area of the bottom and capacity of a cylindrical chemical storage tank with a diameter of 11 ft and height of 13 ft.

Given:

$D = 11\ \text{ft} \qquad H = 13\ \text{ft} \qquad A_s = ? \qquad V = ?$

Solution:

Area of the bottom of the tank

$$A = \frac{\pi D^2}{4} = \frac{\pi}{4} \times (11 \ ft)^2 = 95.03 = \underline{95 \ ft^2}$$

Retain the value in the calculator's memory, as we are going to need this when doing the volume calculations. This way not only do we save the time of punching in the numbers again, it also preserves all the accuracy and avoid any chances of making a mistake while punching in the numbers again.

Capacity of the cylindrical tank

$$V = A_s \times H = 9.5033 \ ft^2 \times 13 \ ft = 1235 = \underline{1200 \ ft^3}$$

Volume in gallons.

$$V = 1235 \ ft^3 \times \frac{7.48 \ gal}{ft^3} = 9237 = \underline{920 \ gal}$$

EXAMPLE PROBLEM 2.8

A section of 200-mm-diameter pipeline is to be filled with chlorinated water for disinfection. If 200 m of pipeline is to be disinfected, how many litres of water will be required?

Given:

$$D = 200 \ mm = 0.2 \ m \qquad L = 200 \ m \qquad V = ?$$

Solution:

Volume of water required to fill the pipe

$$V = \frac{\pi D^2}{4} \times L = \frac{\pi}{4} \times (0.2 \ m)^2 \times 200 \ m \times \frac{1000 \ L}{m^3} = 6283 = \underline{6300 \ L}$$

2.2.2 VOLUME AND AREA

For symmetrical shapes like cylinders, volume or capacity can be worked out by multiplying top or bottom surface with the third dimension, height or length.

Volume and Area

$$\boxed{V = A_b \times h \quad or \quad A_x \times L}$$

A section of the channel can be other than circular or rectangular, for example, trapezoidal or parabolic. Based on the geometry, the area can be calculated using the formula for that shape.

EXAMPLE PROBLEM 2.9

A 4.5-m-deep rectangular aeration tank is to be sized to hold 9.0 ML of mixed liquor. If it is planned to keep the length to width ratio of 3 to 1, determine the dimensions of the tank.

Given:

$$V = 9.0 \text{ ML} = 9000 \text{ m}^3 \qquad d = 4.5 \text{ m}$$

Solution:

Surface area

$$A_s = \frac{V}{d} = \frac{9.0 \text{ ML}}{4.5 \text{ m}} \times \frac{1000 \text{ m}^3}{ML} = 2000 \text{ m}^2$$

$$W = x, \quad L = 3x; \quad A_s = 3x^2$$

Width and length

$$x = W = \sqrt{\frac{A_s}{3}} = \sqrt{\frac{2000 \text{ m}^2}{3}} = 25.82 = \underline{26 \text{ m}}$$

$$L = 3x = 3 \times 25.82 = 77.46 = \underline{77 \text{ m}}$$

EXAMPLE PROBLEM 2.10

A trench 4.0 ft wide, 8.0 ft deep and 75 ft long is to be backfilled. How much fill is required?

Given:

$$L = 75 \text{ ft} \qquad W = 8.0 \text{ ft} \qquad d = 4.0 \text{ ft} \qquad V = ?$$

Solution:

Volume of fill required

$$V = L \times W \times d = 75 \text{ ft} \times 8.0 \text{ ft} \times 4.0 \text{ ft} = 2400.0 = \underline{2400 \text{ ft}^3}$$

EXAMPLE PROBLEM 2.11

How many cubic meters of paving material will be required to pave over a trench 780 m long and 1.0 m wide using a layer of 10 cm deep patch?

Given:

$L = 780 \quad W = 1.0\ m \quad d = 10\ cm \quad V = ?$

Solution:

Volume of paving material required

$$V = L \times W \times d = 780\ m \times 1.0\ m \times 10\ cm \times \frac{m}{100\ cm} = 78.0 = \underline{78\ m^3}$$

EXAMPLE PROBLEM 2.12

A trench 1.3 m wide, 2.5 m deep and 21 m long is to be backfilled. How many cubic metres of fill are required?

Given:

$L = 21\ m \quad W = 2.5\ m \quad d = 1.3\ m \quad V = ?$

Solution:

Volume of fill required

$$V = L \times W \times d = 21\ m \times 2.5\ m \times 1.3\ m = 68.25 = \underline{68\ m^3}$$

EXAMPLE PROBLEM 2.13

How many cubic ft of sand will be required to prepare a surface for paving over a trench half a mile long and 5.0 ft wide using a layer 4.0 in deep?

Given:

$L = 0.50\ mile \quad W = 5.0\ ft \quad d = 4.0\ in \quad V = ?$

Solution:

Volume of sand needed

$$V = L \times W \times d = 0.50\ mile \times \frac{5280\ ft}{mile} \times 5.0\ ft \times 4.0\ in \times \frac{ft}{12\ in}$$

$$= 4.40 \times 10^3 = \underline{4400\ ft^3}$$

EXAMPLE PROBLEM 2.14

The interior of a 90 m of 300-mm-diameter pipe is uniformly coated with 20 mm of grease. How many kL of wastewater can be held by this pipe when completely full?

Given:

$D = 300$ mm   Effective $D = 300 - 2 \times 20 = 260$ mm $L = 90$ m $V = ?$

Solution:

Capacity of pipe to hold wastewater

$$V = \frac{\pi}{4} \times D^2 \times L = \frac{\pi}{4} \times (0.26\ m)^2 \times 90\ m = 4.77 = \underline{4.8\ m^3}$$

EXAMPLE PROBLEM 2.15

The interior of a 1200 ft of 12-in-diameter water main is uniformly coated with a scale of half an inch. How many gallons of water capacity are lost due to scaling?

Given:

$D_1 = 12$ in $= 1.0$ ft $D_2 = 12$ in $- 2 \times 0.5$ in $= 11$ in $L = 1200$ ft $\Delta V = ?$

Solution:

Capacity of pipe without scaling

$$V_1 = \frac{\pi D_1^2}{4} \times L = \frac{\pi}{4} \times (1.0\ ft)^2 \times 1200\ ft \times \frac{7.48\ gal}{ft^3} = 7049.7 = \underline{7050\ gal}$$

Capacity of pipe with scaling

$$V_2 = \frac{\pi D_2^2}{4} \times L = \frac{\pi}{4} \times \left(\frac{11}{12}\ ft\right)^2 \times 1200\ ft \times \frac{7.48\ gal}{ft^3} = 5923.7 = \underline{5920\ gal}$$

Capacity lost due to scaling

$$\Delta V = V_1 - V_2 = 7049.7 - 5923.7 = 1126 = \underline{1130\ gal}$$

EXAMPLE PROBLEM 2.16

What is the water holding capacity of a 100 m of 200-mm line?

Given:

$D = 200$ mm $= 0.20$ m $L = 100$ m $V = ?$

Solution:

Capacity of pipe

$$V = \frac{\pi D^2}{4} \times L = 0.785(0.20\ m)^2 \times 100\ m = 3.14 = \underline{3.1\ m^3}$$

EXAMPLE PROBLEM 2.17

What is the full capacity of a 500-m-long sewer main with a diameter of 250 mm?

Given:

$D = 250$ mm $= 0.25$ m $L = 500$ m $V = ?$

Solution:

Capacity of the sewer pipe to hold wastewater

$$V = \frac{\pi D^2}{4} \times L = 0.785(0.25 \ m)^2 \times 500 \ m = 24.5 = \underline{25 \ m^3}$$

EXAMPLE PROBLEM 2.18

What is the storage capacity of a 10-ft-diameter circular tank in gal/ft depth?

Given:

$D = 10$ ft $V/d = A = ?$

Solution

Capacity of the tank per unit depth

$$\frac{V}{d} = A_s = \frac{\pi}{4} \times D^2 = \frac{\pi}{4} \times (10 \ ft)^2 \times \frac{7.48 \ gal}{ft^3} = 5.87.4 = \underline{590 \ gal/ft}$$

EXAMPLE PROBLEM 2.19

During a 15 min time span, the water level in a 1.1-m-diameter tank rose by 80 cm. What is the rate of flow into the tank?

Given:

$D = 1.1$ m $d = 80$ cm $= 0.80$ m $t = 15$ min $Q = ?$

Solution:

Rate of flow

$$Q = A \times v = A \times \frac{d}{t} = 0.785 \ D^2 \times \frac{d}{t} = \frac{\pi}{4} \times (1.1 \ m)^2 \times \frac{0.80 \ m}{15 \ min} \times \frac{60 \ min}{h}$$

$$= 3.04 = \underline{3.0 \ m^3/h}$$

2.2.3 CONE

The volume of a cone is equal to one-third of the volume of a cylinder of the same dimensions.

Volume of a cone

$$V = \frac{\pi D^2}{12} \times h = \frac{\pi D^2}{12} \times h$$

2.2.4 SPHERE

A sphere is defined by only one dimension, that is, radius or diameter. The volume of a sphere can be found using the following expression.

Volume of a sphere

$$V = \frac{\pi D^3}{6} = \frac{4\pi R^3}{3}$$

EXAMPLE PROBLEM 2.20

What is the volume of a conical tank in cubic feet that is 3.5 m in diameter and 3.2 m in height?

Given:

$$D = 3.5 \text{ m} \quad h = 3.2 \text{ m} \quad V = ?$$

Solution:

Volume of conical tank

$$V = \frac{\pi}{12} \times D^2 \times h = \frac{\pi}{12} \times (3.5\,m)^2 \times 3.2\,m = 10.26 = \underline{10\,m^3}$$

EXAMPLE PROBLEM 2.21

A dry chemical feed tank is conical at the bottom and cylindrical at the top. If the diameter of the cylinder is 22 ft with a depth of 36 ft and the cone depth is 11 ft, what is the volume of the tank?

Given:

$$D = 22 \text{ ft} \quad d = 36 \text{ ft} \quad d(\text{cone}) = 11 \text{ ft} \quad V = ?$$

Solution:

Volume of cylindrical portion

$$V = \frac{\pi}{4} \times D^2 \times d = \frac{\pi}{4} \times (22\,ft)^2 \times 36\,ft \times \frac{7.48\,gal}{ft^3} = 102362\,gal$$

Volume of conical portion

$$V = \frac{\pi}{12} \times D^2 \times d = \frac{\pi}{4} \times (22\,ft)^2 \times 11\,ft \times \frac{7.48\,gal}{ft^3} = 31277\,gal$$

Total Volume of the tank

$$V = 102362\,gal + 31277\,gal = 133639 = \underline{130\,000\,gal}$$

2.2.5 VOLUME AND DEPTH

Hydrologic variables like rain and evaporation are usually expressed as depth. To express it as volume of water, it is multiplied by the area. If the volume of water is known, then depth is found from volume over area. It is further illustrated by the following example problems.

EXAMPLE PROBLEM 2.22

Water from a rainstorm was collected in a bucket with a top diameter of 14 inches. The collected rainwater was measured to be 3.5 gal. Find the depth of the rainfall.

Given:

$$D = 14\ in = 1.0\ ft \qquad V = 3.5\ gal \qquad d = ?$$

Solution:

Depth of rainfall

$$d = \frac{V}{A} = \frac{4V}{\pi D^2} = \frac{4}{\pi} \times \frac{3.5\,gal.ft^3}{7.48\,gal} \times \frac{1}{(14\,in)^2} \times \left(\frac{12\,in}{ft}\right)^3 = 5.25 = \underline{5.3\,in}$$

EXAMPLE PROBLEM 2.23

A watershed that feeds a water supply has a drainage area of 320 miles². On average, the area receives 35 in of annual rainfall, 15% of which becomes runoff. What volume of water is discharged per year? Also find the average discharge rate in MGD.

Given:

$$A = 320\ miles^2 \qquad d = 35\ in/y \qquad V = ? \qquad Q = ?$$

Solution:

Volume of rainfall

$$V = A \times d = 320\ mile^2 \times 35\ in \times \frac{ft}{12\ in} \times \left(\frac{5280\ ft}{mile}\right)^2 = 2.601 \times 10^6 = \underline{2.6 \times 10^{10}\ ft^3/y}$$

Volume of runoff

$$V = 15\% \text{ of rain} = \frac{15\%}{100\%} \times \frac{2.601 \times 10^6 \, ft^3}{y} \times \frac{7.48 \, gal}{ft^3} = \underline{2.92 \times 10^{10} \, gal/y}$$

Average annual discharge rate

$$Q = \frac{2.919 \times 10^{10} \, gal}{y} \times \frac{y}{365 \, d} = 79.98 \times 10^6 = \underline{80 \, MGD}$$

EXAMPLE PROBLEM 2.24

An inland lake spread over an area of 65 km² is a source of water supply for a town. The annual average precipitation for the area is 755 mm/a. Water lost due to evaporation and seepage is about 350 mm/y. What is the average rate at which water can be withdrawn without causing a net depletion of storage?

Given:

$$A = 65 \text{ km}^2 \quad d_{in} = 755 \text{ mm/a} \quad d_{out} = 350 \text{ mm/a} \quad Q = ?$$

Solution:

Net depth of precipitation

$$d = d_{in} - d_{out} = 755 \, mm/a - 350 \, mm/a = 405 = 410 \, mm/a$$

Average annual rate of withdrawal

$$Q = A \times d = 65 \, km^2 \times \frac{405 \, mm}{a} \times \frac{(1000 \, m)^2}{km^2} \times \frac{m}{1000 \, mm} \times \frac{a}{365 \, d} \times \frac{ML}{1000 \, m^3}$$

$$= 72.12 = \underline{72 \, ML/d}$$

EXAMPLE PROBLEM 2.25

A volume of 110 m³ of digested sludge is spread on a field measuring 25 m × 12 m. What is the depth of sludge?

Given:

$$V = 11 \, 0 \, m^3 \quad A = 25 \text{ m} \times 15 \text{ m} \quad d = ?$$

Solution:

Depth of sludge

$$d = \frac{V}{A} = \frac{110 \, m^3}{25 \, m \times 12 \, m} \times \frac{100 \, cm}{m} = 36.66 = \underline{37 \, cm}$$

PRACTICE PROBLEMS

PRACTICE PROBLEM 2.1

The surface area of a rectangular basin is estimated to be 100 m². If the length is twice the width, what is the length of the basin? (14 m)

PRACTICE PROBLEM 2.2

A rectangular settling basin is designed such that length is thrice the width. If the width of the tank is kept at 15 ft, what is the surface area of the tank? (680 ft²)

PRACTICE PROBLEM 2.3

You are planning to paint the walls of a cylindrical tank with a diameter of 5.5 m and height of 3.5 m. If one 4-L paint can covers a surface of 45 m², how many cans would you need? (2)

PRACTICE PROBLEM 2.4

You want to know the bottom area of a cylindrical tank. By running a tape around the tank, you read the perimeter to be 79 ft. What is the area of the bottom of the chemical tank? (500 ft²)

PRACTICE PROBLEM 2.5

The bottom of a 6.0-m-diameter tank is conical, with a height of 2.0 m. To order paint, your supervisor wants to know the lateral area of the cone. (34 m²)

PRACTICE PROBLEM 2.6

How many acre-ft will make 1 MG of water? (3.1 acre-ft)

PRACTICE PROBLEM 2.7

Determine the volume of water that a circular clarifier with a diameter of 52 ft and side water depth of 11 ft can hold. (23 000 ft³)

PRACTICE PROBLEM 2.8

A new 30-cm-diameter water main is to be tested before it is put into service. What volume of water will be needed to fill a 600-m-long water main? (42 m³)

PRACTICE PROBLEM 2.9

Size a 2-m-deep square-shaped contact chamber to provide a capacity of 50 m³. (5.0 m)

PRACTICE PROBLEM 2.10

A trench 5.0 ft wide, 10 ft deep and 1000 ft long is to be backfilled. How many cubic yards of fill are required? (1900 yd³)

PRACTICE PROBLEM 2.11

An excavation 1.1 m wide, 2.0 m wide and 15 m long is to be filled with sand. How many cubic metres of sand are required? (33 m³)

PRACTICE PROBLEM 2.12

Estimate cubic metres of AC paving material required to lay an 80-mm patch over an area of 1.0 m × 1.3 m. (100 m³)

PRACTICE PROBLEM 2.13

How many cubic ft of sand will be required to prepare a surface for paving over a trench half a mile long and 5.0 ft wide using a layer of 4.0 in deep?

PRACTICE PROBLEM 2.14

The interior of 120 m of a 300-mm-diameter sewer pipe is uniformly coated with 15 mm of grease. Find the capacity of the sewer pipe when half full. (3.4 m³)

PRACTICE PROBLEM 2.15

The interior of 1500 ft of a 16-in-diameter water main is uniformly coated with a tubercle of quarter of an inch. How many gallons of water capacity are lost due to tuberculation? (960 gal)

PRACTICE PROBLEM 2.16

A wet well 6.0 m in diameter is filled to a depth of 3.0 m. How many kL of water are held? (85 kL)

PRACTICE PROBLEM 2.17

How much wastewater would 300 m of 200-mm-diameter sewer pipe hold? (9.4 kL)

PRACTICE PROBLEM 2.18

What diameter tank will store 500 gal/ft depth of storage? (9.2 ft)

PRACTICE PROBLEM 2.19

In 5 min time, the water level in a 1.8-m-diameter tank rose by 40 cm. What is the rate of flow into the tank? (3.4 L/s)

PRACTICE PROBLEM 2.20

What volume a conical tank 3.2 m in diameter can hold when filled to a level of 2.8 m? (7.5 m³)

PRACTICE PROBLEM 2.21

A soda ash tank is conical at the bottom and cylindrical at the top. If the diameter of the cylinder is 18 ft with a depth of 32 ft and the cone depth is 15 ft, what is the volume of the tank? (9400 ft^3)

PRACTICE PROBLEM 2.22

Water from a rainstorm was collected in a bucket that has a circular opening of 20 in. What should be the volume of collected rain water to indicate a storm rainfall depth of 4.0 in? (5.4 gal)

PRACTICE PROBLEM 2.23

A watershed that feeds a water supply has a drainage area of 230 miles^2. On average, the area receives 25 in of annual rainfall, 12% of which becomes runoff. What volume of water is discharged per year? Also find the average discharge rate in ft^3/s. (50 ft^3/s)

PRACTICE PROBLEM 2.24

A water filtration plant takes its supply from an inland lake spread over an area of 55 km^2. Annual average demand is estimated to be 65 ML/d. What should be the minimum net depth of annual precipitation to meet this demand without causing a net depletion of storage? (430 mm)

PRACTICE PROBLEM 2.25

Compute the depth of water when 100 kL of a chemical is stored in a 10-m-diameter tank. (1.3 m)

3 Ratio and Proportions

3.1 RATIOS

Ratios are a quick and easy way to solve simple problems when a particular relationship of two variables is known and one of those variables is changed to a known value. This relationship in the form of a ratio allows us to solve problems when, on changing one variable, a change in the related variable is required. Suppose you are given the mass of a chemical required to make a certain volume of solution. Using the rule of ratios, it is easy to work out the amounts of chemical required to make up different quantities of solutions of the same strengths. This type of problem can be easily solved using simple algebra. The following problems are examples of ratio problems that water and wastewater personnel may find useful in their work.

The division of two numbers is actually a **ratio**. For example, a ratio of two to three can also be expressed as 2/3 or 2:3. As an example, 2.0 g of chemical in 3.0 L of water to make a solution, or two spoons of sugar for every 3 cups of milk for a recipe.

A fraction is the ratio of component divided by the whole. A fraction can be expressed as decimal or a percentage. A decimal fraction is when the whole is 1, and a percentage is when the whole is 100.

$$Fraction = \frac{numerator}{denominator} = \frac{component}{whole}$$

Let us say if the fraction is 2 out of 5

$$Fraction = \frac{numerator}{denominator} = \frac{2}{5} = \frac{0.40}{1} = \frac{40}{100} = 40\%$$

To convert a decimal fraction to percentage fraction, multiply by 100 and vice versa. Hence 0.50 is the same thing as 50% or ½. A quantity expressed as 25% is the same thing as 0.25 or ¼. Remember, 100% is the same thing as unity.

EXAMPLE PROBLEM 3.1

Your annual budget for purchase of chemicals is $98,000. You have already spent 37,780 in the first 6 months. What percent of your budget did you spend?

Solution:

$$Fraction = \frac{\$37780}{\$98000} \times 100\% = 38.5 = \underline{39\%}$$

DOI: 10.1201/9781003468745-3

EXAMPLE PROBLEM 3.2

A pipeline project requires the installation of 3642 ft of 6-in pipe. If your crew can install 120 ft in a day, how many days are required to complete the project?

Solution:

$$= 3642\,ft \times \frac{d}{120\,ft} = 30.3 = \underline{30\,d}$$

3.2 PROPORTIONS

Proportions, on the other hand, are the equating of ratios. For example, ½ is the same thing as 2/4 or 5/10. Proportions are written as follows:

$$2:4::5:10 \quad or \quad 2/4 = 5/10$$

The end terms 2 and 10 in this case are known as **extremes**, and the middle terms 4 and 5 are called **means**. In proportions, the product of extremes is always equal to the product of means

$$2 \times 10 = 4 \times 5 = 20$$

This property helps us to solve one of the unknowns. If you cross-multiply, you will get the same thing as before regarding the product of extremes and means. Ratios and proportions are used to solve problems where both values are known for one set of conditions but one value of the second set is unknown.

3.3 DIRECT PROPORTIONS

Proportions are of two types, *direct* and *inverse*. In **direct proportion**, an increase in one causes a proportional increase in the other and vice versa. However, in **inverse proportions**, an increase in one value results in a proportional decrease in the other value. For example, for a set pumping rate, the volume of water V is directly proportional to the time t for which water is pumped. If subscripts 1 and 2 represent two sets of conditions, this relationship can be algebraically expressed as follows.

Direct proportions

$$\boxed{\frac{V_2}{V_1} = \frac{t_2}{t_1} \quad or \quad V_2 \times t_1 = t_2 \times V_1}$$

The same could be said about the speed of pump n and the discharge rate Q. If the speed of the pump is increased by 20%, the discharge rate would increase by 20%.

Pump speed and discharge

$$\frac{Q_2}{Q_1} = \frac{N_2}{N_1} \quad or \quad Q_2 \times N_1 = N_2 \times Q_1$$

Key Points:

Here are some important points regarding direct proportions.

- Mathematically, ratios and proportions have the same value.
- In the case of direct proportions, an increase in one causes an increase in other, for example, time to fill a tank: more capacity means more time.
- When written in the stacked form, the subscripts of the numerator and denominator are the same on both sides of the equation.
- When written in linear or horizontal form, the subscripts of the two variables in question on each side of the equation are opposite each other, as follows.

Direct Proportions

$$\frac{V_2}{V_1} = \frac{t_2}{t_1} \quad or \quad V_2 \times t_1 = t_2 \times V_1$$

EXAMPLE PROBLEM 3.3

If 5.0 lb of a chemical are mixed with 100 gallons of water to make a solution of a desired strength, how many lb of chemical will be required to make 2000 gallons of solution of the same concentration?

Solution:

It is a case of direct proportions

$$m_2 = m_1 \times \frac{V_2}{V_1} = 5.0 \ lb \times \frac{2000 \ gal}{100 \ gal} = 100.0 = \underline{100 \ lb}$$

EXAMPLE PROBLEM 3.4

A raw water flow of 31 ML/d is pre-chlorinated with 71 kg of chlorine gas. If the flow is changed to 27 ML/d, what should be the adjustment to the chlorinator?

Solution:

New setting on the chlorinator

$$m_2 = m_1 \times \frac{Q_2}{Q_1} = 71 \ kg \times \frac{27 \ MLD}{31 \ MLD} = 61.8 = \underline{62 \ kg}$$

3.4 INDIRECT PROPORTIONS

In indirect proportions, an increase in the independent variable causes a decrease in the dependent variable, also called an inverse relationship, for example, the flow velocity v and flow area A relationship. Due to the fact that flow velocity increases as the area decreases, this is a case of inverse proportions. Mathematically this relationship can be expressed as follows:

Indirect proportions

$$\frac{v_2}{v_1} = \frac{A_1}{A_2} \quad or \quad v_2 \times A_2 = v_1 \times A_1$$

Another example of an inverse relationship is the dilution formula. Adding more water or increasing the volume of solution, the concentration or strength of the solution drops. In other words, the solution gets diluted. This formula or equation is also known as the dilution formula.

Volume vs concentration

$$V_1 \times C_1 = V_2 \times C_2$$

Feeding rates

$$Q_1 \times C_1 = Q_2 \times C_2$$

Key Points:

Here are some key points you should keep in mind when solving problems applying the method of inverse proportions.

- Mathematically, ratios and proportions have the same value.
- In indirect proportions, subscripts in numerators and denominators on both sides of the equation are opposite.
- When inverse proportions are written in linear or horizontal form, subscripts of the two variables on each side of the equation are the same or say the product of the extremes must be equal to the product of the means.

EXAMPLE PROBLEM 3.5

A pump running at 85% efficiency fills the tank in 47 min. How long it will take if the pump is only 56% efficient?

Solution:

This is a case of indirect proportions, as with lower efficiency, the pump rate will be lowered, thus taking more time to fill the tank.

New time

$$t_2 = t_1 \times \frac{E_1}{E_2} = 47\,min \times \frac{85\%}{56\%} = 71.3 = \underline{71\,min}$$

EXAMPLE PROBLEM 3.6

If 5 g of chemical are mixed with 10 L of water to make a desired solution, how many g of chemical would be mixed with 200 L of water?

Solution:

It is a case of direct proportions

$$m_2 = m_1 \times \frac{V_2}{V_1} = 5.0\,g \times \frac{200\,L}{10\,L} = 100.0 = \underline{100\ g}$$

EXAMPLE PROBLEM 3.7

If 40 kg of a chemical costs \$50, how many kg will \$400 buy?

Solution:

New amount of chemical

$$m_2 = m_1 \times \frac{Cost_2}{Cost_1} = 40\,kg \times \frac{\$400}{\$50} = 320.0 = \underline{320\,kg}$$

EXAMPLE PROBLEM 3.8

A pump running at 80 rpm discharges 100 gpm. If the speed is reduced to 50 rpm, what will the pump feed rate be?

Sol:

Reduced discharge rate

$$Q_2 = Q_1 \times \frac{N_2}{N_1} = 100\ gpm \times \frac{50\ rpm}{80\ rpm} = 62.5 = \underline{63\ gpm}$$

EXAMPLE PROBLEM 3.9

If a pump fills a tank in 13h @ 20 L/min, how long will it take to fill @50 L/min?

Sol:

New time

$$t_2 = t_1 \times \frac{Q_1}{Q_2} = 13h \times \frac{20\,L/min}{50\,L/min} = 5.20 = \underline{5.2\,h}$$

EXAMPLE PROBLEM 3.10

If a pipe flow area is reduced by a factor of 4, how many times is the flow velocity increased?

Sol:

Flow velocity and flow area

$$v_2 = v_1 \times \frac{A_1}{A_2} = v_1 \times \frac{1}{1/4} = 4 \times v_1$$

EXAMPLE PROBLEM 3.11

A primary clarifier provides a detention time of 2.5 h when the flow rate is 3.0 MGD. During the peak flow period, a flow of 4.5 MGD is observed. Find the reduced detention time during the peak flow hours.

Sol:

This is a case of inverse proportions, since hydraulic detention time is inversely proportional to flow rate.

New hydraulic time

$$t_2 = t_1 \times \frac{Q_1}{Q_2} = 2.5 h \times \frac{3.0 \, MGD}{4.5 \, MGD} = 1.66 = \underline{1.7 \, h}$$

EXAMPLE PROBLEM 3.12

A chemical pump discharges 128 mL of alum at a speed setting of 48% and a stroke setting of 30%. If the alum pump's speed is increased to 62% and the stroke setting remains the same, what would be the mL output from the pump? Assume the pump has a linear output.

Sol:

New feed rate

$$Q_2 = Q_1 \times \frac{S_2}{S_1} = 120 \, mL \times \frac{62\%}{48\%} = 155.0 = \underline{155 \, mL}$$

EXAMPLE PROBLEM 3.13

At a wastewater treatment plant, sludge with SS concentration of 3.2% is thickened to 5.1% solids. What is the reduction in the volume of sludge?

Sol:

Reduction in volume of sludge

$$\frac{V_2}{V_1} = \frac{SS_1}{SS_2} = \frac{3.2\%}{5.1\%} = 0.675 = 68\% \quad or \quad \underline{32\% \, reduction}$$

EXAMPLE PROBLEM 3.14

Liquid polymer is supplied to a water filtration plant as a 12% solution. How many litres of liquid polymer should be mixed in a tank with water to produce 500 L of 0.50% polymer solution?

Sol:

Volume of polymer

$$V_1 = V_2 \times \frac{C_2}{C_1} = 500 \, L \times \frac{0.5\%}{12\%} = 20.8 = \underline{21 \, L}$$

EXAMPLE PROBLEM 3.15

At a sewage treatment plant, sludge with SS concentration of 2.2% is thickened to 4.1% solids. What is the reduction in the volume of sludge?

Sol:

Reduction in volume of sludge

$$\frac{V_2}{V_1} = \frac{SS_1}{SS_2} = \frac{2.2\%}{4.1\%} = 0.536 = 54\% \quad or \quad \underline{46\% \, reduction}$$

PRACTICE PROBLEMS

PRACTICE PROBLEM 3.1

The cost of a chemical was $1500/container. This year it has gone up by $120. What is the percent increase in the cost of the chemical? (8%)

PRACTICE PROBLEM 3.2

The chlorine feed rate at your plant is 35 lb/d. Find the number days, a one ton (2000 lb) chlorine cylinder will last? (57 d)

PRACTICE PROBLEM 3.3

Based on your calculations, you need 2.0 lb of chlorine to disinfect 90 ft of the new water main. If you need to disinfect 1420 ft of pipe, how many pounds of chlorine will you need? (32 lb)

PRACTICE PROBLEM 3.4

If 0.5 lb of a chemical are required to make 55 gallons of a solution, how many lb of chemical will be required to make 1000 gallons of a solution of the same concentration? (9.1 lb)

PRACTICE PROBLEM 3.5

A pump running at 85% efficiency fills the tank in 47 min. After a year, the same pump takes 78 min to fill the tank. What is the new efficiency of the pump? (51%)

PRACTICE PROBLEM 3.6

If the cost of electricity is $0.15/kW.h, what is the cost of running a pumping unit with a power demand of 45 kW running on average 20 h daily? ($135/d)

PRACTICE PROBLEM 3.7

You need to provide a minimum contact time for chlorine to adequately disinfect the water. For a given well pump rate, contact time in the pipeline is directly proportional to the volume of water before water is taken off for use. If the pipe size is increased from 350 mm to 400 mm, find how many times contact time will be increased. Hint: flow area is proportional to the square of the pipe diameter. (1.3×)

PRACTICE PROBLEM 3.8

The strength of a chemical solution is known to be 150 mg/L. If the solution is diluted by a factor of 2, what is the concentration of the diluted solution? (75 mg/L)

PRACTICE PROBLEM 3.9

Three men take 8 h to finish a job. If you would like to finish the job in 2 h, how many people you would hire? (12 people)

PRACTICE PROBLEM 3.10

A secondary clarifier is designed to provide a detention time of 2.0 h for the average daily flow of 2.8 MGD. During the early morning hours, flow is reduced to 1.2 MGD. Find the average detention time during the early morning hours. (4.7 h)

PRACTICE PROBLEM 3.11

The dosage rate of chlorine at a water treatment plant is 650 lb/d for a flow of 22 ft^3/s. If the flow is adjusted to 18 ft^3/s, what would the theoretical chlorine dosage be in pounds per day if everything else remains the same? (530 lb/d)

PRACTICE PROBLEM 3.12

Rework Example Problem 3.12 assuming speed is kept constant and the stroke length is increased from 30% to 45%. (180 mL)

PRACTICE PROBLEM 3.13

After pressure filtration, solids concentration increases from 8.0% solids to 22% solids. What is the percent reduction in the volume of sludge? (64%)

PRACTICE PROBLEM 3.14

Liquid polymer is supplied to a water filtration plant as a 10% solution. How many litres of liquid polymer should be mixed in a tank with water to produce 500 L of 0.1% polymer solution? (5.0 L)

PRACTICE PROBLEM 3.15

At a wastewater treatment plant, sludge with SS concentration of 2.5% is thickened to 4.8% solids. What is the percent reduction in the volume of sludge? (48%)

4 Logarithms

4.1 DEFINITION

Logarithms are a way of expressing a number as an exponent of another number, called a **base** value. A logarithm can be defined as the power of an exponent to which base must be raised to become equal to that number.

Definition of logarithm

$$x = b^y, \qquad log_b\ x = y$$

Commonly used bases are 10 and the mathematical constant e. When the base is e, the logarithm is called a *natural logarithm*, written as log_e or ln. If the base is not indicated, then it is assumed to be base 10, which is usually written as log or log_{10}.

- Since the log is the exponent of the base, the log of a number is a smaller number. For this reason, the log of 1 is zero, since any base raised to the power of zero equals one. Following the same logic, log of 10 is 1 and the log of 100 is 2:

$$1 = 10^0, \qquad log_{10}\ 1 = 0$$

- The natural logarithm of a number is 2.302 times larger than the log to the base 10.
- The natural log of the constant e, ln(e) = 1, where e = 2.718.
- The antilog of a number is the base raised to the number; for example, the antilog of 1 is 10. Basically, the antilog is the exponential form. Antilog is also called inverse log.
- The log of zero is meaningless

Antilog or log inverse

$$Alog\ y = 10^y; \quad Alog\ 3 = 10^3 = 1000$$

Addition with logs

$$log\ A + log\ B = log(A \times B)$$

Subtraction with logs

$$log\ A - log\ B = log\left(\frac{A}{B}\right)$$

DOI: 10.1201/9781003468745-4

Logarithms of exponents

$$\boxed{log\ x^n = n\ log\ x}$$

To fully understand some terms in water and wastewater, you need knowledge of logarithms. This may include pH, log removals and exponential and power law functions.

EXAMPLE PROBLEM 4.1

The log ten of the number is 0.325. What is the number?

Given:

$Log\ x = 0.325 \qquad x = ?$

Sol:

$x = 10^{0.325} = 2.113 = \underline{2.11}$

EXAMPLE PROBLEM 4.2

The natural log (ln) of the number is 0.325. What is the number?

Given:

$ln\ x = 0.325 \qquad x = ?$

Sol:

$x = e^{0.325} = 1.384 = \underline{1.38}$

EXAMPLE PROBLEM 4.3

The value of the expression $ln(x)^{2.5}$ is known to be 0.456. What is the value of x?

Given:

$ln(x)^{2.5} = 0.456 \qquad x = ?$

Sol:

$$2.5\ ln\ x = 0.456 \quad or \quad ln\ x = \frac{0.456}{2.5} = 0.182$$

$x = e^{0.182} = 1.200 = \underline{1.20}$

EXAMPLE PROBLEM 4.4

What is the log of 10 to the base 2?

Sol:

By the definition of logarithms

$$log_2\ x = y \quad or \quad x = 2^y \quad or \quad y = log\ x/log2$$

$$y = log\ x/log2 = \frac{log10}{log2} = \underline{2.32}$$

4.2 pH

In the math sense, the letter p stands for the negative logarithm. Hence pH is defined as negative logarithm of the hydrogen ion concentration in given water.

Definition of pH

$$pH = -log[H^+] = log\left[\frac{1}{[H^+]}\right]$$

For example, if the hydrogen ion concentration in a water sample is 10^{-6} mol/L, the pH of the sample is:

$$pH = -log[H^+] = log\left[\frac{1}{10^{-6}}\right] = 6$$

Similarly, if pH is known, the concentration of hydrogen ion can be found by taking the antilog of the pH value. To remind you, antilog is the exponential form, that is, the base raised to the power

$$[H^+] = 10^{-pH} = 10^{-6}\ mol/L.$$

If you are observing the pH of a water stream and want to calculate the average value, it will not be a simple average, as many people think. For example, if the two values are 6 and 7, then the average pH is not 6.5, since pH value are based on logarithm. In the math sense, the average pH will be the negative log of the average hydrogen ion concentration, as shown.

$$pH = -log\left(5.5 \times 10^{-7}\right) = 6.259 = 6.3$$

4.3 pOH

Similar to pH, pOH is defined as the negative logarithm of the hydroxyl ion concentration. When water gets ionized, it possesses both hydrogen ions and hydroxyl ions. In neutral water, the concentration of both hydrogen and hydroxyl ions equals 10^{-7} mol/L. Due to this fact, the following holds true.

$$pH + pOH = 14$$

If the pH of water is 6, then the pOH must be 8. When the hydrogen ion concentration exceeds 10^{-7} mol/L, water is said to be acidic. Similarly, if the hydrogen ion concentration falls below 10^{-7} mol/L, water is said be basic because the balance has turned in favour of hydroxyl ions.

EXAMPLE PROBLEM 4.5

If the pH of a wastewater sample is 7.2, what is the pOH of wastewater?

Given:

pH = 7.2   pOH = ?

Sol:

$pOH = 14 - pH = 14 - 7.2 = \underline{6.8}$

EXAMPLE PROBLEM 4.6

For the data of Example Problem 4.5, find the hydrogen ion and hydroxyl ion concentration.

Given:

pH = 7.2 $OH^-, H^+ = ?$

Sol:

Hydroxyl ion concentration
$pOH = 14 - pH = 14 - 7.2 = 6.8$

$[H^+] = 10^{-pH} = \underline{10^{-7.2}\ mol/L}$

$[OH-] = 10^{-pOH} = \underline{10^{-6.8}\ mol/L}$

EXAMPLE PROBLEM 4.7

Two aqueous solutions have pH values of 6.5 and 8.5. If equal volumes of these solutions are mixed, what will be the pH of the mixture solution?

Given:

$pH_1 = 6.5$ $pH_2 = 8.5$ $pH_{mix} = ?$

Sol:

Caution: It is not the simple average, since pH is logarithms of H^+ concentrations. Thus, before taking the average, we need to express it as an H^+ concentration.

pH of the mixture

$$pH_1 = 6.5, \quad so \ H^+_1 = 10^{-6.5} \ mol/L$$

$$pH_2 = 8.5, \quad so \ H^+_2 = 10^{-8.5} \ mol/L$$

$$H^+_{mix} = (10^{-6.5} + 10^{-8.5})/2 = 1.60 \times 10^{-7} \ mol/L$$

$$pH_{mix} = -log(1.60 \times 10^{-7}) = \underline{6.8}$$

4.4 LOG REMOVALS

In the majority of cases, removal efficiencies of plant processes are defined in the form of percent removals. However, when the reduction in a given parameter is from a very large number to a low number, as in case of microorganisms, this may not be the appropriate way. Log removals are more commonly used in such cases. What is log removal? If a contaminant is removed such that effluent concentration is one-tenth that of the influent concentration, it will be called 1-log removal, since the log of 10 is 1. The equivalent percent removal would be 90%. Following the same logic, 99% removal would be equivalent to 2-log removal. Domestic cleaning chemicals like Lysol claim to have germ removal of 99.9%. In terms of log removal, it is 3-log removal.

Log Removal, $LR = log\left(C_i/C_e\right)$

Depending on the source of water supply, most water regulations state log removal requirements for virus, bacteria and crypto-type organisms. For groundwater supplies, it is usually 2-log virus removal, while it is 4-log removal in surface water supplies.

EXAMPLE PROBLEM 4.8

A disinfection process reduces the virus content by a factor of 3500 times. Express it as log removal.

Given:

$$C_i/C_e = 3500 \qquad Log \ removal = ?$$

Sol:

Log removal

$$LR = log\left(\frac{C_i}{C_e}\right) = log(3500) = 3.54 = \underline{3.5}$$

EXAMPLE PROBLEM 4.9

Based on operating data of a chlorination process in a water treatment plant, 3.25-log removal of virus is achieved. What is that expressed as percent removal?

Given:

$LR = 3.25$ Percent removal $= ?$

Sol:

$$\frac{C_i}{C_e} = Alog(LR) = Alog(3.25) = 2.238 \times 10^3 = 2238$$

Percent removal

$$PR = \left(1 - \frac{C_e}{C_i}\right) \times 100\% = \left(1 - \frac{1}{2238}\right) \times 100\% = \underline{99.96\%}$$

EXAMPLE PROBLEM 4.10

A chlorine dioxide disinfection process ensures 99.7% removal of crypto. Find the log removal.

Given:

$LR = ?$ $PR = 99.7\%$

Sol:

Log removal

$$LR = -Log\left(1 - \frac{PR}{100}\right) = -Log\left(1 - \frac{99.7\%}{100\%}\right) = 2.52 = \underline{\underline{2.5}}$$

4.5   LOG SCALE

As compared to **linear scale**, where the interval is uniform, in *log scale*, the interval is in multiples or in order of the magnitude of the physical quantity. In log scale, the interval is according to the percentage increase.

For example, on a linear scale, intervals between 5 and 10 would be the same as between 25 and 30. However on a log scale, intervals between 25 and 30 would be much smaller since the percentage increase is 20% as compared to 100% in the first case. In Figure 4.1, you see semi-log graph paper, such that the x-axis is a log scale and the y-axis is a linear scale.

As is evident, the interval shrinks as the quantity value increases. Also note that on a log scale, scale interval varies with magnitude, but multiples or ratios remain constant. When the multiple or ratio is 10, as in case of 1 to 10, 10–100 and 0.2 to 2,

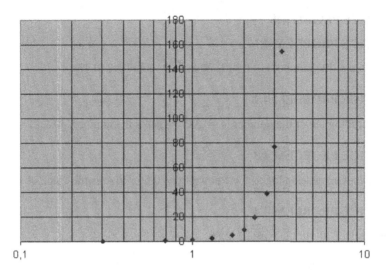

FIGURE 4.1   Linear scale and log scale.

the interval called a **log cycle** remains the same. Notice that the interval 0.5 to 1 is the same as the interval 5 to 10 and 50 to 100 and so on.

Presentation of data on a **logarithmic scale** can be helpful when the data covers a large range of values. The use of the logarithms of the values rather than the actual values reduces a wide range to a more manageable size, for example, particle size distribution and reduction in number of microorganisms as water passes through various processes. If you have to plot numbers varying from single digits to billions on a linear scale, imagine how large a graph sheet you would need. An exponential function like BOD remaining versus time will plot as a curve on an arithmetic graph paper but yield a straight line when plotted on semi-log graph paper.

PRACTICE PROBLEMS

PRACTICE PROBLEM 4.1

The log ten of the number is −0.325. What is the number? (0.47)

PRACTICE PROBLEM 4.2

The natural log ten of the number is −0.325. What is the number? (0.722)

PRACTICE PROBLEM 4.3

The value of expression $\log(x)^{1.5}$ is known to be 0.56. What is the value of x? (2.36)

PRACTICE PROBLEM 4.4

What is log of 100 to the base 5? (2.86)

PRACTICE PROBLEM 4.5

After chlorination, the pH of the water is read to be 6.95. Find the pOH. (7.05)

PRACTICE PROBLEM 4.6

For the data of practice problem 4.4, find the hydrogen ion and hydroxyl ion concentration. (10−6.95, 10−7.05 mol/L)

PRACTICE PROBLEM 4.7

Two aqueous solutions have pH value of 5.5 and 7.5. If equal volumes of these solutions are mixed, what will be the pH of the mixture solution? (5.8)

PRACTICE PROBLEM 4.8

A disinfection process reduces the virus content by a factor of 2500×. What is that expressed as log removal? (3.4)

PRACTICE PROBLEM 4.9

To meet the regulatory requirements, 2-log removal is required. The plant operation manager has decided to operate the plant to achieve 2.5-log removal, just to be on the safer side. What will that be as percent removal? (99.7%)

PRACTICE PROBLEM 4.10

What is 88.8% removal expressed as log removal? (0.95)

5 Density

In layman's terms, density indicates lightness or heaviness of an object, but in technical terms, it is the quantity of the matter per unit volume. Since quantity can be expressed as mass or weight, density can be described as mass density or weight density.

5.1 MASS DENSITY

When we say density, it usually means mass density. In engineering literature, mass density is commonly represented by the Greek letter rho $\rho = m/V$. Commonly used units are kg/m³, g/cm³ or g/mL. The density of solutions with low concentrations can be safely assumed to be equal to that of water, as a large fraction of the solution is water.

Density of water

$$\rho_w = \frac{1.0\,g}{mL} = \frac{1.0\,kg}{L} = \frac{1000\,kg}{m^3}$$

You can estimate the density of a substance by weighing a known volume of it. For example, to estimate the density of sludge being pumped, weigh a known volume of it.

5.2 WEIGHT DENSITY

Weight density is commonly represented by the Greek letter gamma, γ. It is the weight per unit volume of a substance. This term is also called specific weight. The weight density of water is 9.81 kN/m³. Unlike US customary units, weight density has different units compared to that of mass density. In the English system of units, it is confusing when the same unit of lb is used for both mass as well as weight.

5.2.1 WEIGHT DENSITY AND PRESSURE

Since weight is force due to gravity and pressure is force per unit area, these two terms are related. This will be discussed in more detail in the unit on hydraulics.

5.3 SPECIFIC GRAVITY

Although there may be many numbers that express the density of the same substance (depending on the unit used), there is only one relative density associated with each substance. This number is called specific gravity. The density of water is established as the reference (standard), and all other densities are then compared to that of water. The specific gravity is the ratio of the density of a substance to the density of water at 4°C.

DOI: 10.1201/9781003468745-5

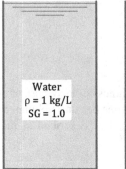
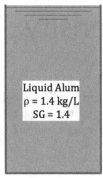

FIGURE 5.1   Specific gravity and density.

Specific gravity

$$SG = \frac{\rho}{\rho_w} = \frac{\gamma}{\gamma_w}$$

Any substance with a density greater than that of water or relatively heavy will have a SG greater than 1.0. Based on the same logic, liquids lighter than water will have a SG less than 1.0. The specific gravity of mercury is 13.6; this to say a volume of mercury will weigh 13.6 as much as the same volume of water. The density concept is further illustrated in Figure 5.1.

For gases the SG is defined with respect to the density of air at 20°C and normal atmospheric pressure.

Density of air

$$\rho_{air} = 1.2\,kg/m^3 = 1.2\,g/L$$

EXAMPLE PROBLEM 5.1

The SG of liquid alum is 1.35. How much would 2.5 gal of liquid alum weigh?

Given:

$V = 2.5$ gal $SG = 1.35$ $m = ?$

Solution:

Mass of liquid alum

$$m = V \times \rho = V \times SG \times \rho_w = 2.5\,gal \times 1.35 \times \frac{8.34\,lb}{gal} = 28.1 = 28\,lb$$

EXAMPLE PROBLEM 5.2

After subtracting the mass of the empty cylinder, 1500 mL of a given sample of sludge weighed 1.52 kg. What is the SG of the sludge?

Given:

$$V = 1500 \text{ mL} \quad m = 1.52 \text{ kg} \quad SG = ?$$

Solution:

Specific gravity of the sludge

$$SG = \frac{\rho}{\rho_w} = \frac{m}{V \times \rho_w} = \frac{1.52 \text{ kg}}{1500 \text{ mL}} \times \frac{L}{1.0 \text{ kg}} \times \frac{1000 \text{ mL}}{L} = 1.031 = \underline{1.03}$$

This indicates the density of sludges is very much the same as that of water. Only for dewatered sludges can the SG be significantly more than 1.0.

EXAMPLE PROBLEM 5.3

An empty bottle weighs 200.4 g. The bottle filled with water weighs 305.2 g. The bottle filled with a given solution weighs 322.6 g. Calculate the SG of the solution.

Given:

$$m = (322.6 - 200.4) \text{ g} \quad m_w = (305.2 - 200.4) \text{ g}$$

Solution:

Specific gravity of the solution

$$SG = \frac{\rho}{\rho_w} = \frac{m}{m_w} \frac{(3.22.6 - 200.4)g}{(305.2 - 200.4)g} = 1.1660 = \underline{1.166}$$

5.4   WEIGHT DENSITY AND MASS DENSITY

Weight is the force exerted due to gravity. In mathematical terms:

Weight and mass

$$\boxed{w = m \times g \quad where \ g = 9.81 \ N/kg = 32.2 \ lb/slug}$$

For example, the weight of a body with a mass of 1 kg is roughly 10 N, as shown:

$$w = m \times g = 1.0 \, kg \times \frac{9.81 \, N}{kg} = 9.81 = 9.8 \, N \cong 10 \, N (approximately)$$

where N (newton) is a unit of weight or force and is defined as the force required to move a mass of 1 kg at an acceleration of 1 m/s². The **mass** of a body relates to the matter content, and it does not dependent on the pull of gravity. The **weight** of a body, on the other hand, is the force due to the pull of gravity. Hence, the weight depends on acceleration due to gravity. A weight of the same mass will be much less on the moon, as the gravitational pull on the moon is about 1/6 of that of the earth.

Weight density/specific weight

$$\gamma = \rho \times g = \frac{w}{V} = \frac{9.81\ kN}{m^3} = \frac{9.81\ kPa}{m} = \frac{0.433\ psi}{ft}$$

5.4.1 WEIGHT DENSITY OF WATER

The weight density of water refers to the unit weight of water. It is also called **specific weight**, and it is commonly represented by the symbol gamma, γ. Like weight and mass, weight density is equal to mass density multiplied by acceleration due to gravity.

$$\gamma = \rho_w g = \frac{kg}{L} \times \frac{9.81\ N}{kg} = 9.81\ N/L = 9.81\ kN/m^3$$
$$= 9.81 \cong 10\ kPa/m = 0.433\ psi/ft$$

5.4.2 SPECIFIC GRAVITY AND WEIGHT DENSITY

As discussed, **specific gravity** being the ratio of the density of a substance to that of water, it tells us how many times a given substance is denser compared to water. This relationship allows us to calculate the density knowing the SG of a given substance.

Weight density

$$\gamma = SG \times \gamma_w$$

As discussed before, weight density and hydrostatic pressure are related terms.

Pressure and height

$$p = \gamma \times h$$

where h is the height of the liquid column.

Weight density as hydrostatic pressure

$$\gamma = \frac{p}{h} = 9.81\ kN/m^2/m = 9.81\ kPa/m \cong 10\ kPa/m = 0.433\ psi/ft$$

Key points:

- SG is also called relative density.
- Weight density is sometimes referred to as specific weight
- When the word density is used alone, it usually means mass density.
- Density expressed as ratio is the relative density.

5.5   MASS AND WEIGHT FLOW RATE

Mass flow rate

$$M = \rho \times Q$$

Weight flow rate

$$W = \gamma \times Q$$

Knowing the SG of a given liquid, the density can be found by multiplying SG by the density of water. The density value is used to express volume flow rate (Q) as mass flow rate (M) or weight flow rate (W).

EXAMPLE PROBLEM 5.4

Given that the SG of a liquid alum solution is 1.35, calculate its mass density and weight density.

Given:

$$SG = 1.35 \qquad \rho = ? \qquad \gamma = ?$$

Solution:

Mass density and weight density of liquid alum

$$\rho = SG \times \rho_w = 1.35 \times 1.00 \ kg/L = 1.350 = \underline{1.35 \ kg/L}$$

$$\gamma = SG \times \gamma_w = 1.35 \times 9.81 \ kN/m^3 = 13.24 = \underline{13.2 \ kN/m^3}$$

EXAMPLE PROBLEM 5.5

The commercial liquid alum delivered to the plant has SG = 1.3. What is the density of the liquid alum expressed as lb/gal?

Given:
$$SG = 1.35 \qquad \rho = ? \qquad \gamma = ?$$

Solution:

Weight density of liquid alum

$$\gamma = SG \times \gamma_w = 1.3 \times 8.34 \ lb/gal = 10.84 = \underline{11 \ lb/gal}$$

EXAMPLE PROBLEM 5.6

The liquid alum (SG = 1.35) feed rate is 200 mL/min. Calculate the dosage rate in kg/d.

Given:

$Q = 200$ mL/min $SG = 1.35$ $M = ?$

Solution:

Mass density of liquid alum

$$\rho = SG \times \rho_w = 1.35 \times 1.00 \ kg/L = 1.350 = \underline{1.35 \ kg/L}$$

Dosage rate of alum

$$M = \rho Q = \frac{1.35 \ kg}{L} \times \frac{200 \ mL}{min} \times \frac{L}{1000 \ mL} \times \frac{1440 \ min}{d}$$

$$= 388.8 = \underline{390 \ kg/d}$$

EXAMPLE PROBLEM 5.7

What is the pressure at the bottom of a liquid alum (SG = 1.35) tank filled to a height of 2.2 m?

Given:

$SG = 1.35$ $p = ?$

Solution:

Weight density of liquid alum

$$\gamma = SG \times \gamma_w = 1.35 \times 9.81 \ kPa/m = 13.24 = \underline{13.2 \ kPa/m}$$

Pressure at the bottom of the tank

$$p = \gamma h = \frac{13.24 \ kPa}{m} \times 2.2 \ m = 29.1 = \underline{29 \ kPa}$$

EXAMPLE PROBLEM 5.8

The gauge attached to the bottom of a chemical tank reads a pressure of 25 psi. If the chemical's specific gravity is 1.5, what is the height of the chemical in the tank?

Given:

$SG = 1.5$ $p = 15$ psi $h = ?$

Solution:

Weight density of the chemical

$$\gamma = SG \times \gamma_w = 1.5 \times 0.433 \, psi/ft = 0.6495 = 0.650 \, psi/ft$$

Height of the liquid

$$h = \frac{p}{\gamma} = 25 \, psi \times \frac{ft}{0.6495 \, psi} = 23.09 = \underline{23 \, ft}$$

EXAMPLE PROBLEM 5.9

A coagulant of SG = 1.2 is pumped into water at a rate of 3.5 gal/h. What is the dosage rate in lb/d?

Given:

$$Q = 3.5 \, gal/h \quad SG = 1.2 \quad M = ?$$

Solution:

Mass density of the coagulant

$$\rho = SG \times \rho_w = 1.20 \times 8.34 \, lb/gal = 10.00 = \underline{10.0 \, lb/gal}$$

Dosage rate of the coagulant

$$M = \rho Q = \frac{10 \, lb}{gal} \times \frac{3.5 \, gal}{h} \times \frac{24 \, h}{d} = 840.6 = \underline{840 \, lb/d}$$

EXAMPLE PROBLEM 5.10

A coagulant of SG = 1.1 is pumped into water at a rate of 250 mL/min. What is dosage rate?

Given:

$$Q = 250 \, mL/min \quad SG = 1.1 \quad M = ?$$

Solution:

Mass density of liquid alum

$$\rho = SG \times \rho_w = 1.1 \times 1.0 \, kg/L = 1.10 = \underline{1.1 \, kg/L}$$

Dosage rate

$$M = \rho Q = \frac{1.1 \, kg}{L} \times \frac{250 \, mL}{min} \times \frac{L}{1000 \, mL} \times \frac{1440 \, min}{d}$$

$$= 396 = \underline{400 \, kg/d}$$

PRACTICE PROBLEMS

PRACTICE PROBLEM 5.1

What volume of ferric chloride (SG = 1.39) would weigh 0.55 kg? (400 mL)

PRACTICE PROBLEM 5.2

The commercial liquid alum delivered to the plant has SG = 1.3. What is the density of the liquid alum expressed as lb/gal? (10.8 lb/gal)

PRACTICE PROBLEM 5.3

A plant produces on average 50 m^3/d of sludge. The sludge is directly spread on land. How many metric tons of sludge will need to be transported every week? (350 t/week)

PRACTICE PROBLEM 5.4

For the data in the previous example problem, calculate the volume of the bottle. (105 mL)

PRACTICE PROBLEM 5.5

Sea water is a bit heavier than water, with a SG of 1.03. Calculate its mass density and weight density. (1.03 kg/L, 10.1 N/L)

PRACTICE PROBLEM 5.6

For the data of Example Problem 5.6, calculate the weight flow rate. (3.8 kN/d)

PRACTICE PROBLEM 5.7

In a water tower, overflow is set at 118 ft. What will be the pressure at the ground level? (51 psi)

PRACTICE PROBLEM 5.8

The gauge attached to the bottom of an elevated tank reads a pressure of 55 psi. What is the level of the water above the gauge? (130 ft)

PRACTICE PROBLEM 5.9

A coagulant of SG = 1.1 is pumped into water to achieve a dosage rate of 25 lb/h. What should the feed pump rate be? (2.7 gal/h)

PRACTICE PROBLEM 5.10

A coagulant of SG = 1.3 is pumped into water at a rate of 110 mL/min. What is the dosage rate? (210 kg/d)

6 Flow Velocity and Flow Rate

Volume flow rate is the volume of water flowing in a unit of time. Flow rate or simply flow is generally denoted by the letter Q. The commonly used units are m^3/s and ML/d, gpm, MGD. Flow rate is related to flow velocity, as shown in Figure 6.1.

Flow rate and flow velocity

$$Q = \frac{V}{t} = v \times A \quad or \quad v = \frac{Q}{t}$$

6.1 AVERAGE FLOW RATE

Flow rates through channels and pipelines are normally measured by some type of flow measuring device like a venturi meter, a weir or a flume. The flow rate at a particular moment in time as measured by one of these devices is called *instantaneous flow rate*. Instantaneous flow rates in a treatment system, especially wastewater plants, vary considerably during the course of the day. Calculating an average flow rate is a way to determine a typical flow rate for a given time frame such as average daily flow or average weekly flow.

There are two ways to calculate the average flow rate. In the first method, several flow rate values are used to determine an average value. In the second method, a total flow (volume) is used. The total flow volume over a specified period is called *totalized flow*. Water meters in homes are a totalizer, so that the difference in the readings between two dates gives you the volume of water used over that time period.

Average flow rate

$$Q_{Avg} = \frac{\Sigma Q_i}{N} = \frac{Q_1 + Q_2 + \ldots\ldots + Q_N}{N}$$

$N = number\ of\ flow\ rate\ values\ i = ith\ value\ of\ flow\ rate$

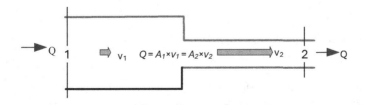

FIGURE 6.1 Flow velocity and flow area.

DOI: 10.1201/9781003468745-6

EXAMPLE PROBLEM 6.1

The daily flow in ML/d recorded for a given week starting Monday is 8.6, 7.6, 7.2, 7.8, 8.4, 8.6 and 7.5, respectively. Calculate the average daily flow for this week.

Solution:

Average flow rate

$$Q_{Avg} = \frac{\sum Q_i}{N} = \frac{8.6+7.6+7.2+7.8+8.4+8.6+7.5}{7} = 7.957 = \underline{7.96\,ML/d}$$

Average is expressed with one decimal more precision than the original data.

Any type of unit can be selected for volume and time. The previous value can be expressed as 330 m³/h. This does not mean that every hour of the day the flow volume is 330 m³.

EXAMPLE PROBLEM 6.2

The totalizer reading for the month of June is 552 ML (difference in readings at the beginning and end of the month). Calculate the average daily flow for the month.

Given:

V = 552 ML   t = 1-month   Q = ?

Solution:

Average flow for the month

$$Q = \frac{V}{t} = \frac{552\,ML}{mo} \times \frac{mo}{30\,d} = 18.4 = \underline{18\,ML/d}$$

6.2   INSTANTANEOUS FLOW

Instantaneous flow rate refers to the flow at a given time and can be measured directly by various flow measuring devices. Or if the average flow velocity at a given section is known, then instantaneous flow rate can be calculated by using the following equation:

Volume flow rate

$$\boxed{Q = v \times A \quad or \quad v = Q/A}$$

v = *average flow velocity at a given section*

A = *Flow area*

The area in this expression refers to flow area perpendicular to the direction of flow. The continuity equation relates the flow rate in each pipe or channel to its flow area and flow velocity. There are three variables, so, knowing any two, the third one can be found. For example, the size of a pipe to carry a given flow without exceeding a certain velocity can be determined. Dividing Q by v calculates the flow area. Depending on the shape of the conduit, area can be expressed in terms of the dimensions of the pipe or channel.

Required flow area and diameter

$$A = \frac{\pi D^2}{4} = \frac{Q}{v} \quad or \quad D = \sqrt{\frac{1.27Q}{v}}$$

The equation $Q = A \times v$ also allows us to calculate flow rate by measuring the depth of flow and flow velocity in a given channel. Velocity of flow can be estimated by use of a float or dye placed in the water. Then, by timing the distance travelled, the velocity can be determined. In addition to estimating flow, the continuity equation can be used to estimate the changes in velocity as the flow area changes (Figure 6.1).

Flow velocity and flow area

$$A_1 \times v_1 = A_2 \times v_2 \quad or \quad \frac{v_2}{v_1} = \frac{A_1}{A_2}$$

Which shows flow velocity varies inversely with the flow area. In circular pipes, area is proportional to the square of the diameter. Hence, change in pipe size will cause a bigger change in flow velocity.

Flow velocity changes

$$\frac{v_2}{v_1} = \frac{A_1}{A_2} = \left(\frac{D_1}{D_2}\right)^2$$

Flow velocity in a pipe is inversely proportional to the square of the diameter. If the diameter of the pipe reduces to one half, velocity increases four times.

$$\frac{v_2}{v_1} = \frac{A_1}{A_2} = \left(\frac{D_1}{D_2}\right)^2 = 2^2 = 4$$

6.3 VOLUMETRIC FLOW MEASUREMENT

When emptying or filling a tank, the flow rate into or out of the tank can be found by observing the rise rate or fall rate in the tank. In fact, rise or fall rate represents the flow velocity in the vertical direction.

EXAMPLE PROBLEM 6.3

A rectangular channel 1 m wide has water flowing to a depth of 0.5 m. If the average flow velocity through the channel is 0.4 m/s, what is the flow rate in L/s?

Given:

$W = 1\,m \qquad d = 0.5\,m \qquad v = 0.4\,m/s \qquad Q = ?$

Solution:

Flow rate

$$Q = v \times A = \frac{0.40\,m}{s} \times 1.0\,m \times 0.50\,m \times \frac{1000\,L}{m^3} = 200.0 = \underline{200\,L/s}$$

EXAMPLE PROBLEM 6.4

Calculate the flow carrying capacity of a 200-mm-diameter pipe if the flow velocity is not to exceed 2.0 m/s.

Given:

$D = 200\,m = 0.20\,m \qquad v = 2.0\ m/s \qquad Q = ?$

Solution:

Flow carrying capacity

$$Q = \frac{\pi D^2}{4} \times v = 0.785 \times (0.20\,m)^2 \times \frac{2.0}{s} \times \frac{1000\,L}{m^3} = 62.1 = \underline{62\ L/s}$$

EXAMPLE PROBLEM 6.5

Determine the flow carrying capacity of a 12-in water main when flowing at the rate 2.0 ft/s.

Given:

$D = 12\,in = 1\,ft \qquad v = 2.0\ ft/s \qquad Q = ?$

Solution:

Flow carrying capacity

$$Q = \frac{\pi D^2}{4} \times v = \frac{\pi}{4} \times (1.0\,ft\,)^2 \times \frac{2.0\,ft}{s} \times \frac{7.48\,gal}{ft^3} \times \frac{60\,s}{min} = 704.9 = \underline{700\,gpm}$$

EXAMPLE PROBLEM 6.6

What size of pipe is needed to carry a flow of 650 gpm without exceeding the flow velocity of 2.5 ft/s?

Given:

$Q = 650\,gpm \qquad v = 2.5\,ft/s \qquad D = ?$

Solution:

Diameter of the pipe

$$D = \sqrt{\frac{1.27Q}{v}} = \sqrt{\frac{1.27 \times 620\,gal}{min} \times \frac{ft^3}{7.48\,gal} \times \frac{s}{2.5\,ft} \times \frac{min}{60\,s} \times \frac{144\,in^2}{ft^2}}$$

$$= 10.05 = \underline{10\ in}$$

EXAMPLE PROBLEM 6.7

What size of pipe is needed to carry a water flow of 50 L/s without exceeding the flow velocity of 2.0 m/s?

Given:

$Q = 50\ L/s$ $v = 2.0$ m/s $D = ?$

Solution:

Diameter of the pipe

$$D = \sqrt{\frac{1.27Q}{v}} = \sqrt{\frac{1.27 \times 50\,L}{s} \times \frac{m^3}{1000\,L} \times \frac{s}{2.0\,m}} = 0.1785\ m = \underline{200\ mm}$$

EXAMPLE PROBLEM 6.8

A weighted float travels 100 m in a water channel in 6 min and 42 seconds. The depth of flow in the channel is measured to be 28 cm, and the width of the channel is 0.5 m. Calculate the flow rate in L/s.

Given:

$s = 100$ m $t = 6$ min 42 s $d = 28$ cm $W = 0.50$ m $Q = ?$

Solution:

Float velocity

$$v = \frac{s}{t} = \frac{100\ m}{(6 \times 60 + 42)\,s} = 0.248 = 0.25\ m/s$$

Flow rate

$$Q = Av = 0.50\ m \times 0.28\ m \times \frac{0.248\ m}{s} \times \frac{1000\ L}{m^3} = 34.8 = \underline{35\ L/s}$$

EXAMPLE PROBLEM 6.9

A tank is 4.0 m by 4.0 m. With the discharge value closed, the influent to the tank causes the water level to rise by 1.0 m in 2 min and 22 s. Calculate the flow into the tank in L/s.

Given:

A = 4.0 m × 4.0 m   s = 1.0 m   t = 2 min 22 s   Q = ?

Solution:

Flow velocity

$$v = \frac{s}{t} = \frac{1.0\ m}{(2 \times 60 + 22)s} = 7.02 \times 10^{-3}\ m/s$$

Flow rate

$$Q = Av = 4.0\,m \times 4.0\,m \times \frac{7.02 \times 10^{-3}\ m}{s} \times \frac{1000\ L}{m^3}$$

$$= 112.3 = \underline{110\ L/s}$$

EXAMPLE PROBLEM 6.10

A wet well measures 15 ft × 15 ft. With no pump running, the wastewater influent to the well causes the water level to rise by 3.0 ft in 2 min and 10 s. Calculate the wastewater flow rate.

Given:

A = 15 ft × 15 ft   t = 2 min 10 s = 130 s   Q = ?

Solution:

Flow velocity

$$v = \frac{s}{t} = \frac{3.0\ ft}{130\,s} = 2.307 \times 10^{-2}\ ft/s$$

Flow rate

$$Q = Av = 15\ ft \times 15\ ft \times \frac{2.307 \times 10^{-2}\ ft}{s} \times \frac{7.48\ gal}{ft^3} \times \frac{60\,s}{min}$$

$$= 2330 = \underline{2300\ gpm}$$

EXAMPLE PROBLEM 6.11

During a 15-min time span, the water level in a 1.1-m-diameter tank rose by 80 cm. What is the rate of flow into the tank?

Given:

D = 1.1 m   d = 80 cm = 0.80 m   t = 15 min   Q = ?

Solution:

Flow rate

$$Q = A \times v = A \times \frac{d}{t} = \frac{\pi}{4} D^2 \times \frac{d}{t}$$

$$= \frac{\pi}{4} \times (1.1\ m)^2 \times \frac{0.80\ m}{15\ min} \times \frac{60\ min}{h} = 3.04 = \underline{3.0\ m^3/h}$$

EXAMPLE PROBLEM 6.12

A 300-mm-diameter water main is connected to a 250-mm-water main with a reducer. During peak flow, water demand is 82 L/s. Find the flow velocities in the two segments of the water main.

Given:

$D_1 = 300$ mm $D_2 = 250$ mm $Q = 82$ L/s

Solution:

Flow velocity in the 300-mm pipe

$$v_1 = \frac{Q}{A_1} = \frac{4Q}{\pi D_1^2} = \frac{4}{\pi} \times \frac{82\,L}{s} \times \frac{1}{(0.30\,m)^2} \times \frac{m^3}{1000\,L} = 1.16 = \underline{1.2\,m/s}$$

Flow velocity in the 250-mm pipe

$$v_2 = \frac{Q}{A_1} = \frac{4Q}{\pi D_1^2} = \frac{4}{\pi} \times \frac{82\,L}{s} \times \frac{1}{(0.25\,m)^2} \times \frac{m^3}{1000\,L} = 1.67 = \underline{1.7\,m/s}$$

This can be also solved by the ratio proportions method.

$$V_2 = V_1 \times \left(\frac{D_1}{D_2}\right)^2 = \frac{1.16\,m}{s} \times \left(\frac{300\ mm}{250\ mm}\right)^2 = 1.67 = \underline{1.7\,m/s}$$

EXAMPLE PROBLEM 6.13

To find the pumping rate, inflow to a 10 ft × 12 ft wet well is blocked. It took 10 min and 15 s to lower the water level by 5.0 ft. Calculate the pumping rate.

Given:

$A = 10$ ft × 12 ft = 120 ft² $v = 5.0$ ft in 10 min 15 s(615 s) $Q = ?$

Solution:

Pumping rate

$$Q=A\times v=120\,ft^2\times\frac{5.0\,ft}{615\,s}\times\frac{7.48\,gal}{ft^3}\times\frac{60\,s}{min}=437.8=\underline{440\,gpm}$$

EXAMPLE PROBLEM 6.14

A 150-m-long section of a 300-mm water main is made ready for a leakage test. During filling, it is recommended not to exceed the flow velocity of 0.3 m/s. What should be the rate of inflow?

Given:

$D = 300\,mm = 0.30\,m \qquad L = 150\,m \qquad v = 0.30\,m/s \qquad Q = ?$

Solution:

Inflow rate

$$Q=Av=\frac{\pi(0.30\,m)^2}{4}\times\frac{0.30\,m}{s}\times\frac{1000\,L}{m^3}=20.1=\underline{20\,L/s}$$

Time to fill the pipe

$$t=\frac{L}{v}=\frac{150\,m.s}{0.3\,m}\times\frac{min}{60\,s}=8.33=\underline{8.3\,min}$$

EXAMPLE PROBLEM 6.15

A 12-in sewer main carries a flow of 350 gpm during peak flow conditions. What flow velocity is achieved?

Given:

$D = 12\,in = 1.0\,ft \qquad Q = 350\,gpm \qquad v = ?$

Solution:

Flow velocity

$$v=\frac{Q}{A}=\frac{8Q}{\pi D^2}=\frac{8}{\pi}\times\frac{350\,gal}{min}\times\frac{1}{1.0\,ft^2}\times\frac{ft^3}{7.48\,gal}=1.98=\underline{2.0\,ft/s}$$

PRACTICE PROBLEMS

PRACTICE PROBLEM 6.1

The hourly flows for an 8-h shift are 1.2, 2.5, 1.7, 2.1, 2.5, 1.0, 0.5 and 3.1 L/s. Calculate the average hourly flow rate. (1.95 L/s)

PRACTICE PROBLEM 6.2

The total flow for a given day at a water filtration plant is 4.6 ML. What is the average flow for that day expressed as L/s? (53 L/s)

PRACTICE PROBLEM 6.3

A rectangular channel is 1.5 m wide and is flowing with a water depth of 0.45 m. Determine the flow rate in m^3/h if the average velocity is 0.5 m/s. (1200 m^3/h)

PRACTICE PROBLEM 6.4

Determine the flow carrying capacity of an 8-in water main when flowing at the rate 2.5 ft/s. (390 gpm)

PRACTICE PROBLEM 6.5

What size water main is required to carry 750 gpm without exceeding the flow velocity of 3.0 ft/s? (10 in)

PRACTICE PROBLEM 6.6

Find the flow carrying capacity of a 400-mm-diameter pipe at a flow velocity of 2.0 m/s. (250 L/s)

PRACTICE PROBLEM 6.7

What minimum diameter pipe is required to carry a flow of 200 L/s without exceeding the flow velocity of 2.0 m/s? (375 mm)

PRACTICE PROBLEM 6.8

A ping-pong ball is used to estimate the velocity of flow in a 300-mm sewer. It takes 122 s for the ball to travel between two manholes 90 m apart. Assuming the average flow velocity is 85% of the float velocity, what is the flow velocity? (0.63 m/s)

PRACTICE PROBLEM 6.9

A pump discharges into a 75-cm-diameter cylindrical tank. If the rise rate is observed to be 50 cm in 33 s, what is the pump discharge rate in L/min? (400 L/min)

PRACTICE PROBLEM 6.10

A wet well measures 12 ft × 12 ft. With no pump running, the wastewater influent to the well causes the water level to rise at the rate of 1.0 ft/min. Calculate the wastewater flow rate. (1100 gpm)

PRACTICE PROBLEM 6.11

A 12-in water main is connected to a 10-in water main with a reducer. During peak flow, water demand is 1300 gpm. Find the flow velocities in the two segments of the water main. (3.7 ft/s, 5.3 ft/s)

PRACTICE PROBLEM 6.12

A 200-mm water main is flushed to achieve flow velocity exceeding 0.80 m/s. Find the minimum discharge rate at the end hydrant. (25 L/s)

PRACTICE PROBLEM 6.13

When a pump is on, the water level in a 10 ft × 12 ft wet well drops by 1.0 ft in 2 min 35 s. If the inflow to the wet well is 220 gpm, find the pumping rate. (570 gpm)

PRACTICE PROBLEM 6.14

A 120-m-long section of a 250-mm water main is made ready for a leakage test. During filling it is recommended not to exceed the flow velocity of 0.3 m/s. What should be the rate of inflow? (15 L/s)

PRACTICE PROBLEM 6.15

A 16-in sewer main carries a peak flow of 750 gpm flowing 3/4 full. What flow velocity is achieved? (2.4 ft/s)

7 Hydraulic Loading

Hydraulic loading is a way of expressing the volume flow rate of water to which a unit process device is subjected. The unit device may be, for example, a sedimentation tank, an aeration tank, a flash mixer or a filter. It is an important design and operating parameter, as it will affect the operating efficiency of a given process. Excessive hydraulic loading tends to wash out the tanks, thus reducing the removal efficiency of a given process. Calculations that reflect various types of water loading include hydraulic loading rate, weir overflow rate, surface overflow rate, hydraulic detention time and filtration velocity.

7.1 OVERFLOW RATE

Overflow rate (v_o) is a way to express surface loading and is defined as the flow rate per unit surface area of the unit device. In the case of sedimentation tanks or clarifiers, it represents the rise or average velocity with which water is rising upward before it flows over the weirs. The settling velocity of the particles has to be more than the overflow rate to settle at the bottom.

Overflow rate

$$v_o = \frac{Q}{A_s}$$

Units are the same as that of velocity: $m^3/m^2{\cdot}d$ or simply m/d.

7.2 FILTRATION RATE/VELOCITY

Filtration velocity is flow per unit surface area of the filter. Filtration velocity will be less than the actual flow velocity of water through the filter media because a significant portion of the filter area is occupied by the filter media.

Filtration rate

$$v_F = \frac{Q}{A_F}$$

7.3 HYDRAULIC LOADING RATE

Hydraulic loading rate is a term used to indicate flow per unit surface area, as used to define loading on biological filters, RBCs, and stabilization ponds. Recirculated flows must be included as part of the total flow. This is important when calculating

DOI: 10.1201/9781003468745-7

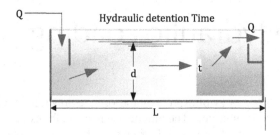

FIGURE 7.1 Detention time.

the hydraulic loading rate on trickling filters. Commonly used units are L/s.m^2 or m^3/m^2·h.

7.4   HYDRAULIC DETENTION TIME

Detention time is another way to express the hydraulic loading on unit devices including sedimentation tanks and flocculators in water treatment and grit tanks and aeration tanks in wastewater treatment. It refers to the length of time a given parcel of water remains in the tank, which is essentially the same as the hours or minutes required for filling a tank. Another way of saying the same thing would be the time between the entry and exit of water from a given tank. This is a useful concept in the sense that it indicates average time for which water will be subjected to a given treatment process (Figure 7.1).

High flow rates for the same size tanks will result in lower detention times and thus yield lower treatment efficiencies. The equation for calculating detention time is very simple. As detention time refers to the length of time it takes to fill up a tank at a specified flow rate, it can be calculated by dividing the volume of the tank by the water flow rate.

Hydraulic detention time

$$HDT \quad or \quad t_d = \frac{V}{Q} = \frac{SWD}{v_o}$$

Where SWD = side water depth and v_o = overflow rate

Volume divided by volume flow rate will yield units of time. Commonly used units are hours, except for grit tanks and chlorine contact chambers, when it is usually expressed in minutes. Hydraulic detention time (HDT) in the case of aeration tanks is commonly known as **aeration period**.

7.5   WEIR OVERFLOW RATE

Weir overflow rate is defined as the flow per unit length of the weir over which water overflows. This term is used to express weir loading in the case of clarifiers. Typical units are m^3/m·d or L/s·m or gpd/ft. For circular clarifiers, the length of the weir is

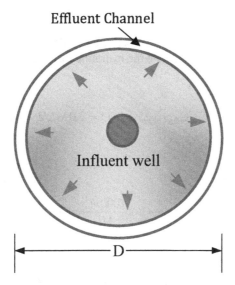

FIGURE 7.2 Weir loading.

equal to the circumference of the circle. In terms of diameter D, length of weir, $L_w = \pi D$. Weir loading in clarifiers is illustrated in Figure 7.2.

Weir loading rate

$$WL = \frac{Q}{L_W} = \frac{Q}{\pi D}$$

EXAMPLE PROBLEM 7.1

A rectangular clarifier has a weir length of 175 ft. What is the weir overflow rate if the flow is 0.95 MGD?

Given:

$$L = 175\,ft \qquad Q = 0.95\,MG/d \qquad WL = ?$$

Solution:

Weir loading

$$WL = \frac{Q}{L_W} = \frac{0.95 \times 10^6\,gal}{d} \times \frac{1}{175\,ft} = 5428 = \underline{5400\,gal/\,ft.d}$$

$$= \frac{5428\,gal}{ft.d} \times \frac{ft^3}{7.48\,gal} = 725 = \underline{730\,ft^3/\,ft.d}$$

EXAMPLE PROBLEM 7.2

A trickling filter 25 m in diameter treats a primary effluent flow of 8.0 ML/d. If recirculated flow is 25% of this, calculate the hydraulic loading on the filter.

Given:

$$Q = 8.0\,\text{ML/d} \quad D = 25\,\text{m} \quad Q_{RS} = 25\% \text{ of } Q = 2.0\,\text{ML/d} \quad Q_{RS} + Q = 10\,\text{ML/d}$$

Solution:

Hydraulic loading rate

$$HLR = \frac{(Q + Q_{RS})}{A_S} = \frac{10\,ML}{d} \times \frac{4}{\pi(25m)^2} \times \frac{1000\,m^3}{ML} = 20.38 = \underline{20\,m^3/m^2.d}$$

EXAMPLE PROBLEM 7.3

A pond receives a flow of 9500 m³/d. If the surface area of the pond is 6.5 ha, what is the hydraulic loading rate in cm/d?

Given:

$$Q = 9500\,\text{m}^3/\text{d}, \quad A_s = 6.5\,\text{ha} \quad HLR = ?$$

Solution:

Hydraulic loading rate

$$HLR = \frac{Q}{A_s} = \frac{9500\,m^3}{d} \times \frac{1}{6.5\,ha} \times \frac{ha}{10000\,m^2} \times \frac{100\,cm}{m} = 14.6 = \underline{15\,cm/d}$$

EXAMPLE PROBLEM 7.4

A circular clarifier receives a flow of 12 ML/d. If the diameter of the weir circle is 25 m, what is the weir overflow rate in m³/m·d?

Given:

$$Q = 12\,\text{ML/d} = 12000\,\text{m}^3/\text{d} \quad D = 25\,\text{m}$$

Solution:

Weir loading

$$WL = \frac{Q}{\pi D} = \frac{12000\,m^3}{d} \times \frac{1}{\pi(25m)} = 152.8 = \underline{150\,m^3/m^2.d}$$

EXAMPLE PROBLEM 7.5

An aeration tank is 17 m by 7 m and 5 m deep. Calculate the aeration period if the flow to the tank is 26 L/s.

Given:

$$V_A = 17 \text{ m} \times 7 \text{ m} \times 5.0 \text{ m} \qquad Q = 26 \text{ L/s} \qquad AP = ?$$

Solution:

Aeration period

$$AP = \frac{V_A}{Q} = 17\,m \times 7.0\,m \times 5.0\,m \times \frac{s}{26\,L} \times \frac{h}{3600\,s} \times \frac{1000\,L}{m^3} = 6.36 = \underline{6.4\,h}$$

EXAMPLE PROBLEM 7.6

A rectangular basin is 100 ft × 50 ft × 20 ft. How long it will take for water at the rate of 3500 gpm to flow through the basin?

Given:

$$V = 100 \text{ ft} \times 50 \text{ ft} \times 20 \text{ ft} \qquad Q = 3500 \text{ gpm} \qquad t_d = ?$$

Solution:

Detention time

$$t_d = \frac{V}{Q} = 100\,ft \times 50\,ft \times 20\,ft \times \frac{min}{3500\,gal} \times \frac{h}{60\,min} \times \frac{7.48\,gal}{ft^3} = 3.56 = \underline{3.6\,h}$$

EXAMPLE PROBLEM 7.7

A chlorine contact chamber is 8.0 m long, 3.0 m wide and 2.0 m deep. Calculate the contact time in minutes for a peak flow of 26 L/s.

Given:

$$V_A = 8.0 \text{ m} \times 3.0 \text{ m} \times 2.0 \text{ m} \qquad Q = 26 \text{ L/s} \qquad t_d = ?$$

Solution:

Detention time

$$t_d = \frac{V}{Q} = 8.0\,m \times 3.0\,m \times 2.0\,m \times \frac{s}{26\,L} \times \frac{min}{60\,s} \times \frac{1000\,L}{m^3} = 30.76 = \underline{31\,min}$$

EXAMPLE PROBLEM 7.8

What is the minimum flow capacity of a chlorine contact tank required to provide a minimum contact time of 30 min to chlorinate a flow of 2.5 MGD?

Given:

$V = ?$ $Q = 2.5\,\text{MGD}$ $t_d = 30\,\text{min}$

Solution:

Capacity of chlorine contact tank

$$V = Q \times t_d = \frac{2.5\,MG}{d} \times 30\,min \times \frac{d}{1440\,min} = 52083 = \underline{52\,000\,gal}$$

EXAMPLE PROBLEM 7.9

A 15-m-diameter clarifier is 3.0 m deep. What is the detention time in hours for an average daily flow of 8.0 ML/d? If the peak flow rate is 2.5 times the average flow rate, what will be the detention time achieved during this period?

Given:

$D = 15\,\text{m}\ d = 3.0\,\text{m}$ $Q = 8.0\,\text{ML/d}$ $HDT = ?$

Solution:

Detention time

$$t_d = \frac{V}{Q} = \frac{\pi(15\,m)^2 \times 3.0\,m}{4} \times \frac{d}{8.0\,ML} \times \frac{24\,h}{d} \times \frac{ML}{1000\,m^3} = 1.58 = \underline{1.6\,h}$$

Since detention time is inversely proportional to flow rate, an increase in flow rate will reduce effective detention time.

$$t_{d(peak)} = t_{d(avg)} \times \frac{Q_{avg}}{Q_{peak}} = \frac{1.58\,h}{2.5} = 0.632 = \underline{0.63\,h}$$

7.6   DETENTION TIME AND OVERFLOW RATE

When expressing hydraulic loading, generally one of the terms HDT or v_0 is used, not both. The two parameters are directly related by the depth of water in the tank, as shown:

Detention time and overflow rate

$$\boxed{HDT = \frac{V}{Q} = \frac{SWD}{v_o}}$$

This shows that specifying overflow rate and depth of tank, detention time can be worked out.

EXAMPLE PROBLEM 7.10

A rectangular clarifier measures 100 ft × 50 ft and is operated to maintain a side water depth of 10 ft. What is the surface loading (overflow rate) for an average daily flow of 4.5 MGD?

Given:

$$A_s = 100 \text{ ft} \times 50 \text{ ft} \qquad Q = 4.5 \text{ MGD} \qquad v_o = ?$$

Solution:

Overflow rate

$$v_o = \frac{Q}{A_s} = \frac{4.5 \times 10^6 \, gal}{d} \times \frac{1}{100 \, ft \times 50 \, ft} = 900.0 = \underline{900 \, gpd/ft^2}$$

EXAMPLE PROBLEM 7.11

A 50-ft-diameter clarifier is 10 ft deep. What are the overflow rate and detention time in hours for an average daily flow of 2.0 MGD?

Given:

$$D = 50 \text{ ft} \qquad Q = 2.0 \text{ MGD} \qquad v_o = ? \qquad t_d = ?$$

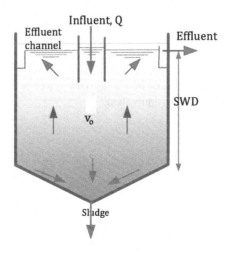

FIGURE 7.3 Detention time vs overflow rate.

Solution:

Overflow rate and detention time

$$v_0 = \frac{Q}{A_s} = \frac{4Q}{\pi D^2} = \frac{2.0 \times 10^6 \ gal}{d} \times \frac{4}{\pi \times (50 \ ft)^2} = 1018.5 = \underline{1000 \ gpd/ ft^2}$$

Detention time

$$t_d = \frac{SWD}{v_o} = 10 \ ft \times \frac{d.ft^2}{1018.5 \ gal} = \frac{24 \ h}{d} \times \frac{7.48 \ gal}{ft^3} = 1018.5 = 1.76 = \underline{1.8 \ h}$$

EXAMPLE PROBLEM 7.12

A primary clarifier is 26 m in diameter with a side water depth of 2.1 m. For a wastewater flow of 13 ML/d, calculate the overflow rate.

Given:

$$D = 26 \ m \quad d = 2.1 \ m \qquad Q = 13 \ ML/d \qquad v_0 = ?$$

Solution:

Overflow rate

$$v_o = \frac{Q}{A_s} = \frac{13 \ ML}{d} \times \frac{4}{\pi (26 m)^2} \times \frac{1000 \ m^3}{ML} = 24.5 = \underline{25 \ m^3/m^2.d}$$

7.7 HYDRAULIC POPULATION EQUIVALENT

The hydraulic population equivalent (HPE) is an indirect way of expressing industrial flows in terms of number of people that would produce the same volume of wastewater. A typical per-capita flow in North America is 450 L/c.d (120 gal/c.d).

EXAMPLE PROBLEM 7.13

Wastewater discharge from a dairy into a sanitary sewer is on average 1500 m³/d. Express this flow as a population equivalent.

Given:

$$Q = 1500 \ m^3/d \qquad HPE = ?$$

Solution:

Hydraulic population equivalent

$$HPE = \frac{1500 \ m^3}{d} \times \frac{p.d}{450 \ L} \times \frac{1000 \ L}{m^3} = 3333 = \underline{3300 \ p}$$

EXAMPLE PROBLEM 7.14

At a given pumping station, the daily average pumping rate is 200 L/s. What is the wastewater production in L/c.d? Assume this station serves a community of 27,000 people.

Given:

$Q = 200$ L/s   Population $= 27000$ Q/capita $= ?$

Solution:

Per capita flow contribution

$$\frac{Flow}{Pop.} = \frac{200\,L}{s} \times \frac{3600\,s}{h} \times \frac{24\,h}{d} \times \frac{1}{27000\,p} = 640.0 = \underline{640\,L/c.d}$$

EXAMPLE PROBLEM 7.15

A wastewater plant is to be designed for a community of 4200 people. If the average annual per capita wastewater flow is 420 L/c·d, what is the design daily average flow to the plant?

Given:

$Q = 420$ L/c·d   Population $= 4200$ $Q_{design} = ?$

Solution:

Design flow

$$Q_{design} = \frac{420\,L}{p.d} \times 4200\,p \times \frac{ML}{10^6\,L} = 1.764 = \underline{1.8\,ML/d}$$

EXAMPLE PROBLEM 7.16

If per capita flow is assumed to be 110 gal/c·d, what is the daily average flow from a community of 5300 people?

Given:

$Q = 110$ gal/c·d   Population $= 5300$ $Q_{design} = ?$

Solution:

Design flow

$$Q_{design} = \frac{110\,gal}{p.d} \times 5300\,p \times \frac{MG}{10^6\,gal} = 0.583 = \underline{0.58\,MGD}$$

PRACTICE PROBLEMS

PRACTICE PROBLEM 7.1

For the data of Example Problem 7.1, work out the weir loading for the peak flow hour assuming peak flow is 1.8× the average flow. (9800 gal/ft·d)

PRACTICE PROBLEM 7.2

In Example Problem 7.2, calculate hydraulic loading if the recirculation rate is 30% and the trickling filter is 20 m in diameter. (33 m³/m²·d)

PRACTICE PROBLEM 7.3

In an RBC treatment system, the average daily flow is 12 ML/d. The total disk surface area according to the manufacturer's specifications is 80 000 m². Calculate the hydraulic loading on the RBC. (0.15 m³/m²·d)

PRACTICE PROBLEM 7.4

The 25-m-diameter clarifier in the example problem is replaced by two 15-m-diameter clarifiers. Calculate the weir loading. (130 m³/m·d)

PRACTICE PROBLEM 7.5

An aeration tank is to be sized to provide an aeration period of 8.0 h when treating a flow of 3.0 ML/d. What is the minimum capacity required? (1.0 ML)

PRACTICE PROBLEM 7.6

A rectangular basin is 100 ft × 50 ft × 20 ft. What is the maximum flow (MGD) that can be treated to provide a detention time of 3.0 h? (6.0 MGD)

PRACTICE PROBLEM 7.7

A chlorine contact chamber is 10 m long, 3.0 m wide and 2.1 m deep. Calculate the contact time in minutes for a peak flow of 2.5 ML/d. (36 min)

PRACTICE PROBLEM 7.8

What minimum length chlorine contact tank of 10 ft × 7.0 ft sections is required to provide a minimum contact time of 30 min to chlorinate a flow of 1.5 MGD? (60 ft)

PRACTICE PROBLEM 7.9

An alum solution is fed from a 5.4-m-diameter storage tank at the rate of 2.5 L/min to treat lake water. How long will it take to drop the liquid level in the tank by 3.0 m before it is refilled? (19 d)

PRACTICE PROBLEM 7.10

A rectangular clarifier measures 80 ft × 40 ft and is operated to maintain a side water depth of 10 ft. What is the surface loading (overflow rate) for an average daily flow of 3.0 MGD? (940 gpd/ft²)

PRACTICE PROBLEM 7.11

For Example Problem 7.11, find the detention time. (2.1 h)

PRACTICE PROBLEM 7.12

For the data of Example Problem 7.9, what must be the side water depth if a detention time of 2.5 h is desired, maintaining the same overflow rate? (14 ft)

PRACTICE PROBLEM 7.13

A brewery's daily flow is 2.5 ML/d. Find the hydraulic population equivalent. (5600 p)

PRACTICE PROBLEM 7.14

A city of 22,000 people contributes an average of 8.2 ML/d of wastewater. How much wastewater is produced per capita? (370 L/p·d)

PRACTICE PROBLEM 7.15

A plant is designed for a daily average flow of 4.8 ML/d. Based on a per capita wastewater production of 400 L/c·d, what is the population it can serve? (12 000 p)

PRACTICE PROBLEM 7.16

If per capita flow is assumed to be 100 gal/c·d, what is the daily average flow from a community of 6500 people? (0.65 MGD)

8 Concentration and Solutions

8.1 MASS CONCENTRATION

The mass concentration of a solution is defined as the mass of the solute (chemical constituent) in a given volume of solution or sample.

Mass concentration

$$C_{m/v} = \frac{m}{V} \quad or \quad m = C \times V$$

$C = concentration\ of\ the\ chemical$
$m = mass\ of\ the\ chemical$
$V = volume\ of\ the\ solution$

8.2 UNITS OF EXPRESSION

Units of concentration will be units of mass divided by units of volume. Commonly used units are g/m³, kg/m³ and mg/L.

Unit equivalences

$$\frac{mg}{L} = \frac{g}{m^3} = \frac{kg}{1000\,m^3} = \frac{kg}{ML} = ppm$$

EXAMPLE PROBLEM 8.1

5.0 lb kg of caustic (NaOH) is added to a circular tank with a diameter of 7.5 ft and a water level of 5.0 ft. What is the concentration of caustic achieved in the water?

Given:

$m = 5.0\ \text{Ib}$ \quad $D = 7.5\ \text{ft}$ \quad $H = 5.0\ \text{ft}$ \quad $C = ?$

Solution:

Volume of water in the tank

$$V = \frac{\pi D^2}{4} \times H = \frac{\pi}{4} \times (7.5\,ft)^2 \times 5.0\,ft = 220.89 = \underline{220\,ft^3}$$

DOI: 10.1201/9781003468745-8

Concentration of caustic solution

$$C = \frac{m}{V} = \frac{5.0\ lb}{220.89\ ft^3} \times \frac{ft^3}{62.4\ lb} \times \frac{10^6}{10^6} = 36.7\ lb/million\ lb = \underline{360\ ppm}$$

EXAMPLE PROBLEM 8.2

To make 2.5 L of a 0.10% solution, how many grams of the dry chemical should be weighed out?

Given:

V = 2.5 L   C = 0.10%   m = ?

Solution:

Mass concentration of the desired solution

$$C = \frac{0.10\%}{100\%} \times \frac{1.0\ g}{mL} \times \frac{1000\ mL}{L} = 1.00 = \underline{1.0\ g/L}$$

Mass of dry chemical

$$m = C \times V = \frac{1.0\ g}{L} \times 2.5\ L = 2.00 = \underline{2.5\ g}$$

8.3   PERCENTAGE CONCENTRATION

For high concentrations, as in the case of liquid chemicals and sludges, the concentration is usually expressed as mass of the chemical in a given mass of the solution.

Mass to mass concentration

$$\boxed{C_{m/m} = \frac{m}{m} = \frac{C_{m/V}}{\rho} \quad or \quad C = C_{m/m} \times \rho}$$

Note the difference, over here, concentration is mass of the chemical to mass of the solution. As discussed earlier, the property relating the mass and volume of a given liquid is called density. Knowing density ρ, concentration as mass to mass can be expressed as mass to volume concentration. Let us consider a solution with a concentration of 10 g/L and express it as mass to mass concentration. Assume the density of the solution is the same as that of water which is true as a small portion that is 10 g in a litre is that of the chemical and rest 990 g is all water.

Density of the dilute solution

$$\boxed{\rho = \rho_W = 1.0\ g/mL = 1.0\ kg/L}$$

The concentration of 10 g/L is equal to 1.0% when expressed as mass to mass.

$$C = \frac{10\,g}{L} \times \frac{L}{kg} \times \frac{kg}{1000\,g} \times 100\% = 1.0\%\left(pph\right)$$

Concentration of 1%

$$\boxed{1\% = \frac{10\,g}{L} = \frac{10\,kg}{m^3} = 1\,pph\ (parts\ per\ hundred)}$$

A 1% solution is when one mass unit of a chemical is dissolved in 100 mass units of solution or sample (pph). Parts per million (ppm) or billion (ppb) will be more appropriate when speaking about very dilute solutions or water samples with very low concentrations.

EXAMPLE PROBLEM 8.3

If 100 g of dry hypochlorite (60% available chlorine) is dissolved in 20 L of water, what is the strength of the solution in terms of available chlorine?

Given:

m = 100 g   Availability = 60%   V = 20. L   C = ?

Solution:

Strength of the solution

$$C = \frac{100\,g}{20\,L} \times \frac{60\%\,chlorine}{100\%\,hypo} \times \frac{L}{1000\,g} \times 100\% = 0.500 = \underline{0.50\%}$$

EXAMPLE PROBLEM 8.4

A hypochlorite solution has 12% available chlorine. If 10 lb of available chlorine are needed to disinfect water well, how many gallons of solution required?

Given:

m = 10 lb   Availability = 12%   V = ?

Solution:

Volume of the solution required

$$V = \frac{m}{C} = \frac{10\,lb}{0.12} \times \frac{gal}{8.34\,lb} = 9.99 = \underline{10\,gal}$$

EXAMPLE PROBLEM 8.5

24.5 g of dry polymer is weighed. It is required to prepare a 0.20% solution of this polymer. What volume of solution will you make?

Given:

$m = 24.5$ g $C = 0.20\%$ $V = ?$

Solution:

Volume of the solution to be made

$$V = \frac{m}{C} = \frac{24.5g}{0.002} \times \frac{L}{1\ kg} \times \frac{kg}{1000\ g} = 12.3 = \underline{12\ L}$$

8.4   DENSITY CONSIDERATION

Liquid chemical solutions can be lighter or heavier than water; it is incorrect to assume that the density of a liquid as is the same as that of water. Knowing the specific gravity of the liquid, the density can be found. Concentration expressed as mass to mass can be converted to mass per unit volume by multiplying it by the density of the solution.

Expressing concentration

$$\boxed{C_{m/V} = C_{m/m} \times \rho = C_{m/m} \times SG \times \rho_W}$$

EXAMPLE PROBLEM 8.6

What is the concentration in mg/L of 48.5% liquid alum with a specific gravity of 1.35? What volume in mL of liquid alum is required to make 2.5 L of 0.1% alum solution?

Given:

$C = 48.5\%$ $SG = 1.35$ (1.35 g/mL) $V = ?$

Solution:

The SG of liquid alum is 1.35, meaning it is 1.35 denser than water.

Mass concentration of alum solution

$$C = \frac{48.5\%}{100\%} \times \frac{1.35\,g}{mL} \times \frac{1000\ mL}{L} = 654.8 = \underline{655\ g/L}$$

Volume of the liquid alum required

$$V = \frac{m}{C} = \frac{1.0 \, g}{L} \times 2.5 \, L \times \frac{L}{654.8 \, g} \times \frac{1000 \, mL}{L} = 1568.0 = \underline{1570 \, mL}$$

EXAMPLE PROBLEM 8.7

What is the concentration in mg/L of 48% liquid alum with a specific gravity of 1.3?
What volume in mL of liquid alum is required to make 5.0 L of 0.1% alum solution?

Given:

$C = 48\%$ $SG = 1.3 \, (1.3 \, g/mL)$ $V = ?$

Solution:

Mass concentration of alum solution

$$C = \frac{48\%}{100\%} \times \frac{1.3 \, g}{mL} \times \frac{1000 \, mL}{L} = 624.0 = 624 \, g/L$$

Volume of liquid alum required

$$V = \frac{m}{C} = \frac{1.0 \, g}{L} \times 5.0 \, L \times \frac{L}{624 \, g} \times \frac{1000 \, mL}{L} = 208.3 = \underline{210 \, mL}$$

EXAMPLE PROBLEM 8.8

Determine the strength of a polymer solution as percent if 0.25 lb of dry polymer is
mixed with 15 gallons of water.

Given:
$m = 0.25 \, lb$ $V = 15 \, gal$

Solution:

Mass to mass concentration

$$c_{m/m} = \frac{m}{m_{sol}} \frac{0.25 \, lb}{(0.25 + 15 \, gal \times 8.34 \, lb/gal)} \times 100\% = 0.199 = \underline{0.20\%}$$

EXAMPLE PROBLEM 8.9

What is the concentration in ppm of a 2.0% sodium hypochlorite solution with a
specific gravity of 1.05?

Given:
$C = 2\%$ $SG = 1.05$ $C = ?$

Solution:

Mass to mass concentration

$$C_{m/m} = \frac{2\%}{100\%} \times \frac{1.5 \times 8.34 \, lb}{gal} \times \frac{gal}{8.34 \, lb} \times \frac{10^6}{Million \, l \, lb} = \frac{21000 \, lb}{million \, lb}$$

$$= \underline{21000 \, ppm = 2.1\%}$$

EXAMPLE PROBLEM 8.10

A volume of half a gallon of a solution weighs 4.8 lb. What is the specific gravity of the solution?

Given:

m = 4.8 lb   V = 0.50 gal   SG = ?

Solution:

Specific gravity of the solution

$$SG = \frac{\gamma}{\gamma_W} = \frac{4.8 \, lb}{0.50 \, gal} \times \frac{gal}{8.34 \, lb} = 1.15 = \underline{1.2}$$

EXAMPLE PROBLEM 8.11

A liquid chemical tank 5.5 ft in diameter is filled with a liquid polymer with a SG of 1.2. A pressure gauge attached to the bottom of the tank reads 5.5 psi. What is the level of the liquid in the tank?

Given:

D = 5.5 ft   p = 5.5 psi   SG = 1.2   h = ?

Solution:

Height of polymer solution in the tank

$$h = \frac{p}{\gamma} = \frac{p}{SG \times \gamma_w} = \frac{5.5 \, psi.ft}{1.2 \times 0.433 \, psi} = 11.2 = \underline{11 ft}$$

EXAMPLE PROBLEM 8.12

A day tank is 85 cm in diameter. A chemical weighing 0.45 kg is added, and the tank is filled with water to a level of 1.0 m. The solution is continuously stirred. What is the strength of the solution? If it is desired to make 0.1% solution, how much chemical should be added?

Given:

H = 1.2 m   D = 85 cm = 0.85 m   C = ?

Solution:

Strength of the solution

$$C = \frac{m}{V} = \frac{4m}{\pi D^2 H} = \frac{4}{\pi} \times \frac{0.45\ kg}{(0.85\,m)^2 \times 1.0\ m} \times \frac{1000\,g}{kg} = 793 = \underline{790\ g/m^3}$$

Mass of chemical to be weighed

$$m = C \times V = \frac{1.0\ g}{L} \times \frac{\pi}{4} \times (0.85\ m)^2 \times 1.0\,m \times \frac{1000\,L}{m^3} = 567.4 = \underline{570\,g}$$

This can be also solved by doing ratio and proportions.

$$m_2 = m_1 \times \frac{C_2}{C_1} = 450\,g \times \frac{1000\,mg/L}{793\,mg/L} = 567.4 = \underline{570\,g}$$

PRACTICE PROBLEMS

PRACTICE PROBLEM 8.1

For the data of Example Problem 8.1, if you are required to dose the water at the rate of 100 ppm, how many pounds of caustic are required? (1.4 lb)

PRACTICE PROBLEM 8.2

If you were to make 2.5 L of a 1.5% solution, how many grams of the chemical would you weigh out? (37.5 g)

PRACTICE PROBLEM 8.3

What quantity of dry hypochlorite (65% available chlorine) needs to be weighed to prepare 25 L of 1.0% available chlorine solution? (385 g)

PRACTICE PROBLEM 8.4

A hypochlorite solution has 5% available chlorine. If 10 lb of available chlorine are needed to disinfect water well, how many gallons of solution are required? (24 gal)

PRACTICE PROBLEM 8.5

How many litres of hypochlorite are required to make 200 L of 1.0% chlorine solution? (14 L)

PRACTICE PROBLEM 8.6

What volume of 13.7% concentrated ferric chloride solution (SG = 1.39) is required to prepare 1.0 L of 1% solution of ferric chloride? (52.5 mL)

PRACTICE PROBLEM 8.7

The label on the container indicates that the specific gravity of a commercial solution of sodium hypochlorite is 1.1. How many mL of this solution would weigh 0.80 kg? What volume of the chemical is required to make 200 L of 1.0% chlorine solution? (730 mL)

PRACTICE PROBLEM 8.8

If you mix 0.20 lb of granulated 65% calcium hypochlorite in a 55-gallon drum containing 50 gallons of water, what will be the concentration of the solution? (310 ppm)

PRACTICE PROBLEM 8.9

What is the strength in g/L of a 3.5% sodium hypochlorite solution with a specific gravity of 1.1? (39 g/L)

PRACTICE PROBLEM 8.10

A volume of 550 mL of a solution is weighed to be 650 g. What is the specific gravity of the solution? (1.2)

PRACTICE PROBLEM 8.11

A liquid chemical tank 18 ft in diameter is filled with a liquid chemical with a SG of 1.34. A pressure gauge attached to the bottom of the tank reads 7.5 psi. What is the level of the liquid in the tank? (13 ft)

PRACTICE PROBLEM 8.12

A day tank is 95 cm in diameter. A chemical weighing 470 g is added, and the tank is filled with water to a level of 1.2 m. The solution is continuously stirred. What is the strength of the solution? If it is desired to make a 0.2% solution, how much chemical should be added? (550 mg/L, 1.7 kg)

9 Feeding of Chemicals

9.1 LIQUID CHEMICALS

When a liquid chemical is used to make up a solution of a desired strength, dilution is required. The key point is that the mass of a chemical remains the same; only the volume of the solution is increased by adding more water, as seen in Figure 9.1.

As the mass remains the same, in terms of concentration, the concentration of the diluted solution is inversely proportional to the volume of water added.

Dilution formula

$$V_1 \times C_1 = V_2 \times C_2$$

Dilution factor

$$\frac{V_2}{V_1} = \frac{C_1}{C_2}$$

where V_2 represents the diluted solution, V_2 / V_1 is the dilution factor, that is, the number of times the concentrated solution has to be diluted. As shown in the figure, the volume of the solution is increased by a factor of two, resulting in two times dilution.

Key Points:

- For accurate feeding, liquid polymers are diluted to make a feed solution of strength matching the feed pump rate and the dosage.
- When the strength of the solution is less than 10%, the density of the solution can be safely assumed to be equal to that of water, that is, 1.0 g/L.

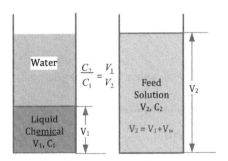

FIGURE 9.1 Diluting solutions.

DOI: 10.1201/9781003468745-9

EXAMPLE PROBLEM 9.1

If a polymer pump is delivering 12 gpd, what is the feed rate in mL/min?

Solution:

Feed pump rate

$$Q = \frac{12\,gal}{d} \times \frac{3.78\,L}{gal} \times \frac{d}{1440\,min} \times \frac{1000\,mL}{L} = 31.5 = \underline{32\,mL/min}$$

EXAMPLE PROBLEM 9.2

Express a feed pump rate of 65 mL/min as gal/h.

Solution:

Feed pump rate

$$Q = \frac{65\,mL}{min} \times \frac{L}{1000\,mL} \times \frac{60\,min}{h} \times \frac{gal}{3.78\,L} = 1.03 = \underline{1.0\,gal/h}$$

EXAMPLE PROBLEM 9.3

What volume of 13.7% concentrated ferric chloride solution is required to prepare 10.0 L of 1% solution?

Given:

Solution	Conc., %	SG	Density, kg/L	Volume, L
Liquid chemical	13.7	1.39	1.39	?
Desired solution	1.0	1.0	1.0	10.0

Solution:

Concentrated ferric chloride solution

$$C_1 = \frac{13.7\%}{100\%} \times \frac{1.39\,kg}{L} \times \frac{1000}{kg} = 190.4 = 190\,g/L$$

Since the desired solution is 99% water, 1.0% = 10 g/L.

Volume of concentrated solution required

$$V_1 = \frac{C_2}{C_1} \times V_2 = \frac{10\,g/L}{190.4\,g/L} \times 100\,L \times \frac{1000\,mL}{L} = 525 = \underline{520\,mL}$$

EXAMPLE PROBLEM 9.4

Calculate the volume of liquid alum (48.5%, SG = 1.35) required to make 5.0 gal of 0.1% solution.

Given:

Solution	Conc., %	SG	Volume, gal
Liquid alum	48.5	1.35	?
Alum solution	0.10	1.0	5.0

Solution:

Mass concentration of alum solution

$$C_1 = \frac{48.5\%}{100\%} \times \frac{1.35 \times 8.34\, lb}{gal} = 5.46 = 5.6\ lb/gal$$

Volume of concentrated solution required

$$V_1 = \frac{C_2}{C_1} \times V_2 = \frac{0.1\%}{100\%} \times \frac{8.34\ lb}{gal} \times \frac{gal}{5.46\ lb} \times 5\ gal \times \frac{3780\ mL}{gal}$$

$$= 28.86 = \underline{29\ mL}$$

EXAMPLE PROBLEM 9.5

A 10% liquid polymer will be used in making a 0.4% solution. How many gallons of liquid polymer should be added to water to make up 55 gal of solution? The liquid polymer has a specific gravity of 1.1.

Given:

Parameter	Liquid = 1	Solution = 2
Concentration	10%	0.40% (4 g/L)
Volume	?	55 gal
SG	1.1	10%

Solution:

Mass concentration of polymer solution

$$C_1 = \frac{10\%}{100\%} \times \frac{1.1\,kg}{L} \times \frac{1000\,g}{kg} = \underline{110\,g/L}$$

Volume of concentrated solution required

$$V_1 = \frac{C_2}{C_1} \times V_2 = \frac{4.0\ g}{L} \times \frac{gal}{110\ g} \times 55\ gal = 2.00 = \underline{2.0\ gal}$$

9.2 CHEMICAL FEED RATE

In chemical treatment, it is required to dose the water or wastewater at a certain rate. The chemical feed rate can be expressed as mass (kg/d) or volume (L/d). The amount of chemical required called dosage rate to add desired concentration of the chemical in a given flow of water.

Dosage rate   *Unknown concentration*

$$\boxed{M = Q_1 \times C_1 = Q_2 \times C_2} \qquad \boxed{C_2 = C_1 \times \frac{Q_1}{Q_2} = \frac{M}{Q_2}}$$

M = *mass feed or dosage rate*
Q_2 = *water flow rate*
C_2 = *dosage of the chemical applied*

Dosage rate can be thought to be the dry chemical feed rate of the pure chemical. Knowing the amount of chemical required, the desired pump rate of the chemical solution can be worked out in the same way.

Feed pump rate

$$\boxed{Q_1 = Q_2 \times \frac{C_2}{C_1}}$$

EXAMPLE PROBLEM 9.6

A well produces 16 L/s of water, and the chlorine dosage is setting is 6.0 kg/d. What is the chlorine dosage in mg/L?

Given:

M = 6.0 kg/d   Q = 16 L/s   C = ?

Solution:

Pumping rate of well water

$$Q = \frac{16}{s} \times \frac{3600\ s}{h} \times \frac{24\ h}{d} \times \frac{ML}{1000\,000\ L} = 1.38 = 1.4ML/d$$

Chlorine dosage

$$C = \frac{M}{Q} = \frac{6.0 \; kg}{d} \times \frac{d}{1.36 \; ML} = 4.41 = 4.4 \; kg \, / \, ML = \underline{4.4 \; mg/L}$$

EXAMPLE PROBLEM 9.7

A chlorinator is set to feed chlorine at the rate of 45 lb/d for a duration of 20 min. If chlorine gas is applied to a 50-ft-diameter water tank filled to a depth of 20 ft, what is the chlorine dosage?

Given:

$M = 45 \; lb \, / \, d$ for $t = 20 \; min$ $D = 50 \; ft$ $H = 20 \; ft$ $C = ?$

Solution:

Total mass of chlorine applied

$$m = M \times t = \frac{45 \; lb}{d} \times 20 \; min \times \frac{d}{1440 \; min} = 0.625 = 0.63 \; lb$$

Chlorine dosage applied

$$C = \frac{m}{V} = 0.625 \, lb \times \frac{4}{\pi} \times \frac{1}{(50 \; ft)^2 \times 20 \; ft} \times \frac{ft^3}{7.48 \; gal} \times \frac{8.34 lb/MG}{mg/L}$$

$$= 0.255 = \underline{0.26 \; mg/L}$$

EXAMPLE PROBLEM 9.8

The desired dose of a polymer is 8.0 mg/L. Commercially available polymer is only 60% active. If a flow of 16 ML/d is to be treated, how many kg of the commercial polymer are required per day?

Given:

$C = 8.0 \; mg/L = 8.0 \; kg/ML$ Purity $= 60\%$ $Q = 16 \; ML/d$ $M = ?$

Solution:

Dosage rate of polymer

$$M = \frac{16 \, ML}{d} \times \frac{8.0 \; kg}{ML} \times \frac{100\% \; commercial}{60\% \; pure} = 213 = \underline{210 \; kg/L}$$

EXAMPLE PROBLEM 9.9

A well pump delivers 250 gpm during its normal operation. What should be the chlorine feed rate in lb/d if the desired dosage is 2.0 mg/L?

Given:

$C = 2.0$ mg/L $Q = 250$ gpm $M = ?$

Solution:

Chlorine feed rate

$$M = \frac{250 \, gal}{min} \times \frac{2.0 \, kg}{L} \times \frac{8.34 \, lb / MG}{mg / L} \times \frac{10^6 \, gal}{MG} \times \frac{1440 \, min}{d}$$

$$= 6.00 = \underline{6.0 \, lb/d}$$

EXAMPLE PROBLEM 9.10

It is required to dose a water supply at the rate of 2.0 mg/L chlorine by applying hypochlorite solution with 20% available chlorine. For a water supply rate of 90 L/s, find the required feeder setting in L/d.

Given:

$Q_2 = 90$ L/s $C_2 = 2.0$ mg/L $Q_1 = ?$ $C_1 = 20\% = 200$ g/L

Solution:

Feed pump setting

$$Q_1 = \frac{C_2}{C_1} \times Q_2 = \frac{2.0 \, mg}{L} \times \frac{L}{200 \, g} \times \frac{90 \, L}{s} \times \frac{g}{1000 \, mg} \times \frac{3600 \, s}{h} \times \frac{24 \, h}{d}$$

$$= 77.76 = \underline{78 \, L/d (78 \, L/24h)}$$

EXAMPLE PROBLEM 9.11

The flow to be treated is 8.0 ML/d, and the optimum dosage is 4.0 mg/L. The strength of the chemical solution is 45% with a SG of 1.37. Determine feed pump rate and dosage rate.

Given:

Parameter	Solution = 1	Water = 2
Conc., C	45%	4.0 mg/L = 4.0 kg/L
Q	?	8.0 ML/d
SG	1.37 (1.37 kg/L)	1.0

Solution:

Feed pump rate

$$Q_1 = \frac{C_2}{C_1} \times Q_2 = \frac{4.0 \, kg}{ML} \times \frac{100\%}{45\%} \times \frac{L}{1.37 \, kg} \times \frac{8.0 \, ML}{d} = 51.9 = \underline{52 \, L/d}$$

Dosage rate

$$M = \frac{8.0\,ML}{d} \times \frac{4.0\,kg}{ML} = 32.0 = \underline{32\ kg/d}$$

EXAMPLE PROBLEM 9.12

The well chlorination pump is set to feed at the rate of 120 L/d. The strength of the sodium hypochlorite solution is 12.5%, with a SG of 1.21. If the well pumping rate is 16 ML/d, what is the dosage of chlorine applied?

Given:

Parameter	Solution = 1	Well Water = 2
Conc., C	12.5%	?
Q	120 L/d	16 ML/d
SG	1.21 (1.21kg/L)	1.0

Solution:

Chlorine dosage applied

$$C_2 = C_1 \times \frac{Q_1}{Q_2} = \frac{12.5\%}{100\%} \times \frac{1.21\,kg}{L} \times \frac{120\,L}{d} \times \frac{d}{16\,ML} = 1.13\ kg/ML = \underline{1.1\ mg/L}$$

EXAMPLE PROBLEM 9.13

2.5 MGD of water is to be dosed with alum at the rate of 20 ppm. What should the setting on the dry alum feeder be?

Given:

$$C = 20\ ppm \qquad Q = 2.5\ MG/d \qquad M = ?$$

Solution:

Feeder setting

$$M = Q \times C = \frac{2.5 \times 10^6\,gal}{d} \times \frac{20}{10^6} \times \frac{8.34\,lb}{gal} = 417 = \underline{420\ lb/d}$$

EXAMPLE PROBLEM 9.14

A polymer feed pump delivers a flow of 800 L/d containing a 5.0% polymer solution with a SG of 1.05. Estimate the kg of polymer fed per day.

Given:

$$Q = 800\ L/d \qquad C = 5.0\% \qquad SG = 1.05 \qquad M = ?$$

Solution:

Mass concentration of polymer solution

$$C = C_{m/m} \times SG \times \rho_w = \frac{5.0\%}{100\%} \times 1.05 \times \frac{1\,kg}{L} \times \frac{1000}{kg} = 52.5 = 53\ g/L$$

Dosage rate of polymer

$$M = Q \times C = \frac{800\,L}{d} \times \frac{0.053\,kg}{L} = 42.0 = \underline{42\,kg/d}$$

9.3 CHEMICAL SOLUTIONS

9.3.1 DRY CHEMICALS

Polymers are frequently used as coagulant aids. Commonly a dry polymer is used to prepare the stock solution. The solution concentration of the solution depends on the many factors, including the type of the polymer and its molecular mass, dosage rate and maximum output of the chemical feed pump. Anionic and non-ionic dry polymers used as coagulant aids are often prepared as very dilute (weak) solutions in the range of 0.1–1%. To prepare a batch of solution, work out the unknown quantity using the following formula:

Percent concentration

$$\boxed{C\% = \frac{m}{(m + m_w)} \times 100\%}$$

M = mass of the pure chemical
m_w = mass of water added

Mass of water

$$\boxed{m_w = \frac{m}{C\%\,/\,100\%} - m}$$

Mass of chemical

$$\boxed{m = \frac{C\%}{(100\% - C\%)} \times m_w}$$

9.3.2 LIQUID CHEMICALS

When starting from liquid solutions, the concentration of the liquid is known, as indicated by the supplier. Often the unknown is the volume of liquid solution required to make a dilute solution of desired concentration.

Dilution formula

$$\boxed{V_1 \times C_1 = V_s \times C_s}$$

EXAMPLE PROBLEM 9.15

Determine the strength of a polymer solution when 0.2 lb of dry polymer are mixed with 4.0 gal of water.

Solution:

Mass of water

$$m_w = 4.0 \, gal \times \frac{8.34 \, lb}{gal} = 33.36 = 33 \, lb$$

Strength of polymer solution

$$C\% = \frac{m}{(m + m_w)} \times 100\% = \frac{0.20 \, lb}{(0.2 \, lb + 33.36 \, lb)} \times 100\% = 0.595 = \underline{0.60\%}$$

EXAMPLE PROBLEM 9.16

How many litres of water should be mixed with 1.0 kg of dry polymer to produce a 0.50% polymer solution?

Solution:

Mass of water

$$m_w = \frac{m \times 100\%}{C\%} - m = \frac{1.0 \, kg \times 100\%}{0.50\%} \times 1.0 \, kg = 199 = \underline{200 \, kg}$$

Volume of water

$$V_w = 199 \, kg \times \frac{L}{1.0 \, kg} = 199.0 \, L = \underline{200 \, L}$$

EXAMPLE PROBLEM 9.17

How many kg of dry polymer must be added to 2000 L of water a produce a 0.50% polymer solution?

Given:

$$V_W = 2000 \, L = 2000 \, kg \quad C = 0.50\% \quad m = ?$$

Solution:

Mass of dry polymer required

$$m = \frac{C\%}{(100\% - C\%)} - m_w = \frac{0.50\%}{(100\% - 0.50\%)} \times 2000 \ kg = 10.05 = \underline{10.1 \ kg}$$

EXAMPLE PROBLEM 9.18

You are going to perform a jar test, and you have a 0.10% alum solution prepared. You have decided to dose the jars, each containing 1.0 L of water, in multiples of 10 mg/L: 10, 20, 30 and so on. How many mL of stock solution (1 g/L) should be added to each jar to achieve a dosage of 10 mg/L?

Given:

Parameter	Solution = 1	Water = 2
Concentration	1.0 g/L	10 mg/L
Volume	?	1.0 L

Solution:

Volume of stock solution to be added to jar

$$V_1 = V_2 \times \frac{C_2}{C_1} = 1.0 \, L \times \frac{10 \, mg}{L} \times \frac{L}{1.0 \, g} \times \frac{g}{1000 \, mg} \times \frac{1000 \, mL}{L} = 10.0 = \underline{10 \, mL}$$

EXAMPLE PROBLEM 9.19

How many mL of liquid alum 48% in strength with a SG of 1.34 are required to make up 1.0 gal of the 1.0% solution required for jar testing?

Given:

Parameter	Solution	
	Chemical = 1	Stock = 2
Concentration, %	48	1.0 (10 g/L)
SG	1.34	1.0
Volume	?	1.0 gal

Solution:

Mass concentration of liquid alum

$$C = C_{m/m} \times SG \times \rho_w = \frac{48\%}{100\%} \times 1.34 \times \frac{1000 \, g}{L} = 643.2 = 640 \, g/L$$

Volume of liquid alum needed

$$V_1 = V_2 \times \frac{C_2}{C_1} = 1.0\,gal \times \frac{10\,g}{L} \times \frac{L}{643.2\,g} \times \frac{3.78\,L}{gal} \times \frac{1000\,mL}{L} = 58.76 = \underline{59\,mL}$$

EXAMPLE PROBLEM 9.20

Liquid polymer is supplied to a water filtration plant as a 10% solution. How many litres of liquid polymer should be mixed in a tank with water to produce 1000 L of 0.50% polymer solution?

Given:

Parameter	Solution	
	Polymer = 1	Feed = 2
Concentration	10%,100g/L	0.50%, 5.0g/L
Volume	?	1.0 L

Solution:

Volume of concentrated solution required

$$V_1 = V_2 \times \frac{C_2}{C_1} = 1000\,L \times \frac{5.0\,g}{L} \times \frac{L}{100\,g} = 50.0 = \underline{50\,L}$$

EXAMPLE PROBLEM 9.21

A sodium hypochlorite (NaOCl) solution is going to be prepared in a 55-gal drum. If 5.0 gal of a 12% solution is added to the drum, how much water should be added to make a sodium hypochlorite solution of 2.0%?

Given:

Parameter	Solution	
	Chemical = 1	Solution = 2
Concentration, %	12%	2.0%
Volume	?	1.0 gal

Solution:

Volume of solution

$$V_2 = V_1 \times \frac{C_1}{C_2} = 5.0\,gal \times \frac{12\%}{2\%} = 30.0 = 30\,gal$$

Volume of water

$$V_w = V_2 - V_1 = 30 - 5 = \underline{25\,gal}$$

EXAMPLE PROBLEM 9.22

A water treatment plant has a flow of 25 MGD. A chemical solution storage tank has a capacity of 5000 gallons. If it is desired to use 90% of the maximum capacity on a single day, what would be the strength of the solution to provide a dosage of 2.5 mg/L?

Given:

Parameter	Solution = 1	Treated Water = 2
Concentration	?	2.5 mg/L
Volume	90% of 5000 gal	25 MG/d

Solution:

Volume of solution

$$V_1 = \frac{90\%}{100\%} \times 5000 \ gal = 54500 \ gal$$

Strength of the solution

$$C_1 = C_2 \times \frac{V_2}{V_1} = \frac{2.5\,mg}{L} \times \frac{25\,MG}{d} \times \frac{d}{4500\,gal} \times \frac{g}{1000\,mg} = 13.8 \ g/L = \underline{1.4\%}$$

> *Weigh 5.0 g and add water to make up a volume of 5.0 L to make 0.1% solution.*

> *Choose strength of solution that corresponds to dosing volume of 1.0 mL for jar testing.*

9.4 CHEMICAL FEEDER SETTING

In chemical dosing, a measured amount of chemical is added to the water. The amount of chemical required depends on such factors as the type of chemical used, the reason for dosing and the flow rate being treated. After evaluation of the jar test results, we select the optimum dosage to be applied. Settings on a given chemical feeder depend on whether you are using a dry chemical feeder or a liquid chemical feeder. Dry chemical feeders are often set on the basis of kg of chemical fed per day.

9.4.1 LIQUID CHEMICAL

Liquid chemical feeders are set on the basis of volume of solution delivered over a given period, more often as mL/min or L/d, as shown in Figure 9.2. Based on the principle of mass conservation:

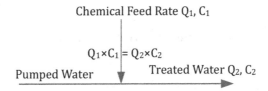

FIGURE 9.2 Chemical feeding.

Dosage rate

$$M = Q_1 \times C_1 = Q_2 \times C_2$$

Q_1 = *chemical feed rate*
Q_2 = *water flow rate*
C_1 = *solution strength*
C_2 = *dosage*

The SG of the liquid chemical must be considered when doing the conversion.

Conc. of the chemical

$$C = C_{m/m} \times SG \times \rho_w$$

EXAMPLE PROBLEM 9.23

8.0 gal of a solution are used up over a 12-hour period. Express the feed pump rate in gal/h.

Given:

$V = 8.0 \, gal \quad t = 12 \, h \quad Q = ?$

Solution:

Feed pump rate

$$Q = \frac{V}{t} = \frac{8.0 \, gal}{12 \, h} = 6.66 = \underline{6.7 \, gal/h}$$

EXAMPLE PROBLEM 9.24

Based on jar testing, the optimum alum dosage for turbidity removal is found to be 12 mg/L. What should the feed pump rate be of 48% (SG = 1.3) liquid alum? The daily water flow is 6500 m³/d.

Given:

Parameter	Liquid Alum = 1	Treated Water = 2
Pump or flow rate	?	6500 m³/d
Specific gravity	1.3	1.0
Strength/dosage	48%	12 g/m³

Solution:

Mass concentration of liquid alum

$$C = C_{m/m} \times SG \times \rho_w = \frac{48\%}{100\%} \times 1.3 \times \frac{1000\,g}{L} = 624 = 620\ g/L$$

Feed pump rate of liquid alum

$$Q_1 = Q_2 \times \frac{C_2}{C_1} = \frac{6500\ m^3}{d} \times \frac{12\,g}{m^3} \times \frac{L}{624\,g} = 125.0 = \underline{130\,L/d}$$

EXAMPLE PROBLEM 9.25

A water treatment plant has a flow of 25 MGD and is being chlorinated applying a dosage of 2.5 mg/L. If the feed pump rate is 3.0 gpm, determine the required concentration of the hypochlorite solution in percent.

Given:

Parameter	Solution Fed = 1	Treated Water= 2
Pump or flow rate	25 MGD	3.0 gpm
Strength/dosage	%?	2.5 mg/L

Solution:

Required strength of chlorine solution

$$C_1 = C_2 \times \frac{Q_2}{Q_1} = \frac{2.5\,mg}{L} \times \frac{25 \times 10^6\ gal}{d} \times \frac{min}{3.0\,gal} \times \frac{d}{1440\ min} \times \frac{g}{1000\,mg}$$

$$= \frac{14.5\ g}{L} \times \frac{L}{kg} \times \frac{kg}{1000\,g} \times 100\% = 1.45 = \underline{1.5\%}$$

EXAMPLE PROBLEM 9.26

A jar test indicates that the best alum dosage for treating Spanish river water is 10 mg/L. Determine the setting on the liquid alum feeder in gal/h for treating a flow of 2.0 MGD. The liquid alum delivered to the plant contains 0.64 kg of alum per litre of liquid solution.

Given:

Parameter	Solution Fed = 1	Treated Water = 2
Pump or flow rate	gal/h	2.0 MG/d
Strength/dosage	0.64 kg/L	10 mg/L

Solution:

Feed pump setting

$$Q_1 = Q_2 \times \frac{C_2}{C_1} = \frac{2.0 M\,gal}{d} \times \frac{10\,kg}{ML} \times \frac{L}{0.64\,kg} \times \frac{d}{24\,h} = 1.30 = \underline{1.3\,gal/h}$$

EXAMPLE PROBLEM 9.27

The maximum output (100% setting) of a feed pump is 250 mL/min. Water is treated at an average rate of 2.0 ML/d. The optimum dose to be applied is 10 mg/L. What strength solution should be prepared to feed the solution at the 50% pump setting?

Given:

Variable	Feed Solution = 1	Treated Water = 2
Pump or flow rate	125 mL/min (50%)	2.0 ML/d
Strength/dosage	?	10 mg/L

Solution:

Required strength of chlorine solution

$$C_1 = C_2 \times \frac{Q_2}{Q_1} = \frac{10\,g}{m^3} \times \frac{2.0\,ML}{d} \times \frac{min}{125\,mL} \times \frac{d}{1440\,min} \times \frac{1000\,m^3}{ML} \times \frac{mL}{g}$$
$$= 0.111 \times 100\% = 11.1 = \underline{11\%}$$

EXAMPLE PROBLEM 9.28

In a drawdown test, a volume of 162 mL of a 98% cationic polymer is being used in 4 min to treat the raw water. The specific gravity of the polymer is 1.37. If the plant flow rate at the time is 7300 gpm, what is the polymer dosage?

Given:

Parameter	Cationic Polymer = 1	Treated Water = 2
Feed/flow rate	162 mL in 4 min?	7300 gpm
Specific gravity	1.37	1.0
Strength/dosage	98%	?

Solution:

Chemical feed pump rate

$$Q_1 = \frac{V}{t} = \frac{162\ mL}{4\ min} = 40.5\ mL/min$$

Strength of the solution

$$C_1 = C_{m/m} \times SG = 0.98 \times 1.37\ kg/L = 1.34 = 1.3\ kg/L$$

Polymer dosage

$$C_2 = \frac{1.34\ kg}{L} \times \frac{40.5\ mL}{min} \times \frac{min}{7300\ gal} \times \frac{gal}{3.78\ L} \times \frac{L}{1000\ mL} \times \frac{10^6\ mg}{kg}$$

$$= 1.96 = \underline{2.0\ mg/L}$$

EXAMPLE PROBLEM 9.29

Determine the feed rate for alum in gal/h when daily flow is 26 MGD and the desired alum dosage is 15 mg/L. The liquid alum strength is 48% with a specific gravity of 1.26.

Given:

Parameter	Alum = 1	Treated Water = 2
Feed/flow rate	gal/h?	26 MGD
Specific gravity	1.26	1.0
Strength/dosage	48%	15 mg/L

Solution:

Strength of the liquid alum

$$C_1 = C_{m/m} \times SG = \frac{48\%}{100\%} \times \frac{1.26 \times 8.34\ lb}{gal} = 5.04 = \underline{5.0\ lb/gal}$$

Alum feed rate

$$Q_1 = Q_2 \times \frac{C_2}{C_1} = \frac{26\ MG}{d} \times \frac{15\ mg}{L} \times \frac{8.34\ lb/MG}{mg/L} \times \frac{gal}{5.04\ lb} \times \frac{d}{24\ h}$$

$$= 26.86 = \underline{27\ gal/h}$$

EXAMPLE PROBLEM 9.30

A 5-min drawdown test result showed that 85 mL of a polymer aid was being used to treat the raw water. The purity of the polymer is 98%, and the specific gravity of the polymer aid is 1.23. If the plant flow rate is 210 L/s, what is the polymer dosage in milligrams per litre?

Parameter	Polymer = 1	Treated Water = 2
Feed/flow rate	210 L/s	85 mL/5 min
Specific gravity	1.23	1.0
Strength/dosage	100%	?

Solution:

Mass concentration of liquid polymer

$$C_1 = C_{m/m} \times SG \times \rho_w \times \frac{98\%}{100\%} \times 1.26 \times \frac{1\ kg}{L} \times 1.205 = 1.2\ kg/L$$

Polmer dosage

$$C_2 = C_1 \times \frac{Q_1}{Q_2} = \frac{1.205\ kg}{L} \times \frac{85\ mL}{5\ min} \times \frac{s}{210\ L} \times \frac{L}{1000\ mL} \times \frac{min}{60\ s} \times \frac{10^6\ mg}{kg}$$

$$= 1.626 = \underline{1.6\ mg/L}$$

9.5   CHEMICAL FEEDER CALIBRATION

Calibration of a feeder is comparing the actual feed rate to the rate the feeder is set on. To calculate the actual feed rate for a dry chemical feeder, place an already weighed bucket under the feeder. Collect the chemical over a specified length of time, such as 20 min, and weigh the bucket again. The difference in the two readings will give you the mass of the chemical actually fed over the specified period. The actual feed rate can be determined as follows.

Dry feed rate	*Chemical dosage*	*Feed pump rate*
$M = m/t$	$C = M/Q$	$Q = V/t$

Knowing the dosage rate M and water flow rate Q, the actual dosage in mg/L can be determined. Calibration of a liquid chemical feeder is done in a similar way. During a known time, the volume of the chemical delivered is noted. Dividing this volume by the time period, the feed pump rate can be calculated. Indirectly, calibration can be performed by observing the drop in level of chemical storage tanks over a given period of time. This information will enable the plant personnel to forecast the expected chemical use. Comparing it with the chemical in the inventory, additional chemical supplies can be ordered.

EXAMPLE PROBLEM 9.31

A bucket weighing 135 g was placed under the feeder. After collecting the chemical for 30 min, the bucket weighed 1.2 kg. Calculate the actual feed rate in kg/d.

Solution:

Mass of the chemical collected

$m = 1.2\ kg - 0.135\ kg = 1.07 = \underline{1.1\ kg}$

Feed rate of chemical

$Feed = \dfrac{m}{t} = \dfrac{1.07\ kg}{30\ min} \times \dfrac{1440\ min}{d} = 51.3 = \underline{51\ kg/d}$

EXAMPLE PROBLEM 9.32

It is observed that the liquid chemical level in the 1-m-diameter feed tank lowered by 45 cm during an 8-hour period. Determine the actual feed pump rate.

Solution:

Drop in volume in the tank

$V = \dfrac{\pi(1.0\ m)^2 \times 45\ cm}{4} \times \dfrac{m}{100\ cm} \times \dfrac{1000\ L}{m^3} = 353 = 350\ L$

Actual feed [ump rate

$Q = \dfrac{V}{t} = \dfrac{353\ L}{8.0\ h} \times \dfrac{h}{60\ min} \times \dfrac{1000\ mL}{min} = 735 = \underline{740\ mL/min}$

EXAMPLE PROBLEM 9.33

The average use of a polymer solution in a water plant is 225 L/d. The chemical feed tank has a diameter of 1.8 m and contains solution to a depth of 1.4 m. How many days will a tank last?

Solution:

Volume of polymer solution in the tank

$V = \dfrac{\pi(1.2\ m)^2 \times 1.4\ m}{4} \times \dfrac{1000\ L}{m^3} = 1583.3 = 1600\ L$

Number of days to empty the tank

$t = \dfrac{V}{Q} = 1583.3\ L \times \dfrac{d}{22.5\ L} = 70.37 = \underline{70\ d}$

EXAMPLE PROBLEM 9.34

You are required to graduate a chemical feed tank in terms of volume in litres. The circumference of the tank is measured to be 793 cm. What height in m will represent 1000 L of chemical in the tank?

Solution:

Diameter of the tank

$$D = \frac{P}{\pi} = \frac{793 \ cm}{\pi} \times \frac{m}{100 \ cm} = 2.524 = 2.52 \ m$$

Height of liquid indicating 1 kL

$$H = \frac{V}{A} = \frac{4V}{\pi D^2} = \frac{4 \times 1.0 \ m^3}{\pi (2.524 \ m)^2} \times \frac{100 \ cm}{m} = 19.98 = \underline{20 \ cm}$$

EXAMPLE PROBLEM 9.35

A wastewater treatment plant uses alum that is 5.37 lb of dry alum per gallon of solution. If it has been determined that 15 mg/L of alum is optimum, what should the chemical feed pump be set on in gpd if the plant is treating 1.5 MGD?

Solution:

Variable	Feed Solution = 1	Treated Water = 2
Flow rate	?	1.5 MGD
Strength/dosage	5.37 lb/gal	15 mg/L

Feed pump rate

$$Q_1 = \frac{1.5 \ MG}{d} \times \frac{15 \ mg}{L} \times \frac{8.34 \ lb \ / \ MG}{mg \ / \ L} \times \frac{gal}{5.37 \ lb} \times \frac{d}{24 \ h} = 1.46 = \underline{1.5 \ gal/d(gpd)}$$

PRACTICE PROBLEMS

PRACTICE PROBLEM 9.1

Express the feed pump rate of 12 gal/d in mL/min. (32 mL/min)

PRACTICE PROBLEM 9.2

A feed pump delivers 55 mL/min. Express the feed pump rate in gal/d. (21 gal/d)

PRACTICE PROBLEM 9.3

How many litres of 8% liquid polymer should be mixed with water to produce 75 L of a 0.5% polymer solution? The SG of the polymer liquid is 1.22. (3.8 L)

PRACTICE PROBLEM 9.4

What volume of 13.7% ferric chloride solution (SG = 1.39) is required to prepare 1.0 L of 1% solution? (52.5 mL)

PRACTICE PROBLEM 9.5

How many litres of 15% liquid polymer should be mixed with water to produce 75 gal of 0.5% polymer solution? The SG of the liquid polymer is 1.12. (2.2 gal)

PRACTICE PROBLEM 9.6

850 kg of copper sulphate is applied to a water reservoir to control algae. If the capacity of the reservoir is 170 ML, what is the dosage of copper sulphate? (5.0 mg/L)

PRACTICE PROBLEM 9.7

A chlorinator is set to feed chlorine at the rate of 85 lb/d. If chlorine gas is applied for 25 min to a 55-ft-diameter water tank filled to a depth of 25 ft, what is the chlorine dosage? (0.40 mg/L)

PRACTICE PROBLEM 9.8

A total chlorine dosage of 12 mg/L is required to treat particular surface water. If the flow is 5.0 ML/d and the hypochlorite has 65% available chlorine, what feed rate in kg/d of hypochlorite will be required? (92 kg/d)

PRACTICE PROBLEM 9.9

Find the setting on a chlorinator in lb/d if the well pump is pumping at the rate of 350 gpm and the desired chlorine dosage is 2.2 mg/L. (9.2 lb/d)

PRACTICE PROBLEM 9.10

In Example Problem 9.10, due to the increased chlorine demand, the required dosage has to be elevated by 1.0 mg/L, how much hypochlorite solution will be required? (39 L)

PRACTICE PROBLEM 9.11

The optimum dose has been determined to be 12 mg/L. The flow to be treated is 1.5 ML/d. If the solution to be used is 40% with a specific gravity of 1.5, what should the solution feeder setting be in mL/min? (21 mL/min)

PRACTICE PROBLEM 9.12

For the data of the example problem, if the dosage applied is observed to be 1.8 mg/L, what is the well pump rate in L/s? (120 L/s)

PRACTICE PROBLEM 9.13

At a water treatment facility, 2.8 MGD of water is to be dosed with alum at the rate of 25 mg/L. What should the setting be on the dry alum feeder? (580 lb/d)

PRACTICE PROBLEM 9.14

A polymer feed pump delivers a flow of 1.1 kL/d containing a 5.5% polymer solution with a SG of 1.05. Find the dosage rate of polymer in kg/d. (64 kg/d)

PRACTICE PROBLEM 9.15

Determine the strength of a polymer solution when 0.25 lb of dry polymer are mixed with 5.0 gal of water. (0.60%)

PRACTICE PROBLEM 9.16

A mass of 11,150 g of dry polymer is weighed. How many litres of water should be mixed with this polymer to produce a 0.5% polymer solution? (230 L)

PRACTICE PROBLEM 9.17

How many kg of dry polymer must be added to 2000 L of water a produce a 0.50% polymer solution? (10 kg)

PRACTICE PROBLEM 9.18

You have prepared 2.0 L of 0.5% alum solution for jar testing. You have decided to dose the jars, each containing 2.0 L of water, in multiples of 20 mg/L, that is, 20, 40,

60, . . . How many mL of alum solution should you add to the jar to achieve a dosage of 20 mg/L? (8.0 mL)

PRACTICE PROBLEM 9.19

You are going to perform jar testing. How many mL of liquid alum 48% in strength with SG of 1.34 are required to make up 0.5 gal of 1.0% solution required for jar testing? (29 mL)

PRACTICE PROBLEM 9.20

Liquid polymer is supplied to a water filtration plant as a 12% solution. How many litres of liquid polymer should be mixed in a tank with water to produce 500 L of 0.50% polymer solution? (21 L)

PRACTICE PROBLEM 9.21

A sodium hypochlorite (NaOCl) solution is going to be prepared in a 55-gal drum. If 8.0 gal of a 10% solution is added to the drum, how much water should be added to make a sodium hypochlorite solution of 2.0%? (32 gal)

PRACTICE PROBLEM 9.22

A water treatment plant has a filter effluent flow of 5400 gpm and is being treated with 850 gpd of a hypochlorite solution. If the desired dose is 2.25 mg/L, determine the concentration of the hypochlorite solution. (2.1%)

PRACTICE PROBLEM 9.23

A water treatment plant has a flow of 25 MGD. A chemical solution storage tank has a capacity of 5000 gallons. If it is desired to use 90% of the maximum capacity on a single day, what would be the strength of the solution to provide a dosage of 2.5 mg/L? (1.4%)

PRACTICE PROBLEM 9.24

The optimum alum dosage for turbidity removal is found to be 20 mg/L. What should be the feed pump rate of 47% (SG = 1.3) liquid alum? The daily water flow is 7.5 ML/d. (250 L/24h)

PRACTICE PROBLEM 9.25

What should be the feed pump rate of liquid alum in gal/h when daily flow is 36 MGD and the desired alum dosage is 20 mg/L? The liquid alum strength is 48% with a specific gravity of 1.26. (50 gal/h)

PRACTICE PROBLEM 9.26

The optimum dosage of the coagulant alum is 15 mg/L. Determine the setting on the liquid alum feeder in gal/h for treating a flow of 4.4 MGD. The liquid alum delivered to the plant contains 640 g of alum per litre of liquid solution. (4.3 gal/h)

PRACTICE PROBLEM 9.27

The maximum output (100% setting) of a feed pump is 500 mL/min. Water is treated at an average rate of 3.5 ML/d. The optimum dose to be applied is 20 mg/L. What strength solution should be prepared to feed the solution at the 50% pump setting? (19%)

PRACTICE PROBLEM 9.28

In a drawdown test, a volume of 162 mL of a cationic polymer is being used in 4 min to treat the raw water. The specific gravity of the polymer is 1.37. If the plant flow rate at the time is 13,000 gpm, what is the polymer dosage? (1.1 mg/L)

PRACTICE PROBLEM 9.29

What should be the feed rate of liquid alum in gal/h to treat a daily flow of 15 MGD? The optimum dosage of alum is 25 mg/L. Commercial liquid supplied to the plant is 48%, with a specific gravity of 1.26. (26 gal/h)

PRACTICE PROBLEM 9.30

A 5-min drawdown test result shows that 121 mL of a cationic polymer is being used to treat raw water. The specific gravity of the polymer is 1.31. If the plant is treating 9200 gpm, what is the polymer dosage in mg/L? (0.91 mg/L)

PRACTICE PROBLEM 9.31

A bucket weighing 135 g was placed under the feeder. After collecting the chemical for 25 min, the bucket weighed 992 g. Calculate the actual feed rate in kg/d. (49 kg/d)

PRACTICE PROBLEM 9.32

It is observed that the liquid chemical level in the 1.1-m-diameter feed tank lowered by 35 cm during an 8-hour shift. What is the actual feed pump rate in L/h? (42 L/h)

PRACTICE PROBLEM 9.33

The average use of a stock solution at a water filtration plant is 176 L/d. The chemical feed tank is 2.1 m in diameter and contains solution to a depth of 1.5 m. How many days will a tank last? (30 d)

PRACTICE PROBLEM 9.34

You are required to graduate a chemical feed tank in terms of volume in litres. By running a tape around the tank, the circumference of the tank is read as 920 cm. What height in m will represent 1000 L of chemical in the tank? (15 cm)

PRACTICE PROBLEM 9.35

A wastewater treatment plant uses alum for precipitation of phosphorus. The alum supplied to plant is 5.32 lb of dry alum per gallon of solution. If it has been determined that 20 mg/L of alum is optimum, what should the chemical feed pump be set on in gpd if the plant is treating 2.5 MGD? (3.3 gpd)

10 Organic Loading

Organic loading is a key parameter in design and operation of special wastewater treatment processes, including clarification, aeration and bio disk units. Usually the loading is expressed as the mass flow rate of some organic parameter like BOD, SS per unit volume or surface area of the treatment unit or device.

10.1 MASS LOADING

Mass loading of BOD, COD or suspended solids (SS) is generally expressed as kg/d. By knowing the water flow rate and the concentration of the constituent in question, the mass rate (loading) can be calculated by multiplying the two, as shown:

Mass loading

$$M = Q \times C \quad M_{BOD} = Q \times BOD \quad M_{SS} = Q \times SS$$

Depending on the units of flow rate Q, conversions will be needed to get the loading in the desired units. However, to make the task of doing conversions easier, use the following:

Unit equivalences

$$\frac{mg}{L} = \frac{g}{m^3} = \frac{kg}{1000\,m^3} = \frac{kg}{ML} = \frac{8.34\,lb}{MG} = ppm$$

The units of concentration should be selected such that the unit of volume in the concentration units is the same as that in volume flow rate units. This is illustrated further in the following examples.

EXAMPLE PROBLEM 10.1

Calculate the BOD loading in kg/d on a receiving stream if the final effluent flow is 10 ML/d and the BOD of the plant effluent is 20 mg/L.

Given:

$$Q = 10\,ML/d \quad BOD = 20\,mg/L \quad M_{BOD} = ?$$

Solution:

Since the plant flow is specified in ML/d, it will be wise to write the concentration of 20 mg/L as 20 kg/ML.

BOD mass loading

$$M_{BOD} = Q \times BOD = \frac{10\,ML}{d} \times \frac{20\,kg}{ML} = 200.0 = \underline{200\,kg/d}$$

EXAMPLE PROBLEM 10.2

A sewage treatment plant (STP) discharges @ 6.5 MGD into the receiving stream with BOD content of 15 mg/L. Calculate the BOD loading.

Given:

$$Q = 6.5\,MGD \quad BOD = 15\,mg/L \quad M_{BOD} = ?$$

Solution:

BOD loading rate

$$M_{BOD} = Q \times BOD = \frac{6.5\,MG}{d} \times \frac{15\,mg}{L} \times \frac{8.34\,lb/MG}{mg/L} = 813 = \underline{810\,lb/d}$$

EXAMPLE PROBLEM 10.3

The capacity of an aeration tank is 1.6 MG. If the concentration of solids in the mixed liquor is 2000 mg/L, how many lb of solids are held in the tank?

Given:

$$v = 1.6\,MG \quad MLSS = 2000\,mg/L \quad m = ?$$

Solution:

Dry mass of mixed liquor solids

$$m_{MLSS} = V \times MLSS = 6.5\,MG \times \frac{2000\,mg}{L} \times \frac{8.34\,lb/MG}{mg/L} = 1.08 \times 10^5 = \underline{1.1 \times 10^5\,lb}$$

EXAMPLE PROBLEM 10.4

The flow to an aeration tank is 26,000 m³/d. If the BOD of the primary effluent is 110 g/m³, how many kg of BOD are applied to the aeration tank daily?

Given:

$Q = 26000 \, m^3/d = 26 \, ML/d \qquad BOD = 110 \, g/m^3 \qquad M = ?$

Solution:

BOD loading of the aeration tank

$$M_{BOD} = Q \times BOD = \frac{26 \, ML}{d} \times \frac{110 \, kg}{ML} = 2860 = \underline{2900 \, kg/d}$$

10.2 BOD LOADING RATE

Organic loading on a biological process is indicated in terms of mass of BOD entering the process. In activated sludge processes, the BOD loading rate (BODLR) is defined as grams of BOD per unit volume of the aeration tank.

Aeration tank loading

$$\boxed{BODLR = \frac{M_{BOD}}{V_A} = \frac{Q \times BOD}{V_A} = \frac{BOD}{AP}}$$

$V_A = Volume \, of \, aeration \, tank$

$AP = Aeration \, Period$

Remember that BOD concentration in this case refers to the wastewater flowing into an aeration tank or primary effluent. Organic loading for trickling filters is calculated the same way, except the volume referred to is that of the filter.

Trickling filter loading

$$\boxed{BODLR = \frac{M_{BOD}}{V_F} \times \frac{Q \times BOD}{V_F}}$$

$V_F = Volume \, of \, trickling \, media$

In the case of stabilization ponds and rotating biological contactors (RBCs), organic loading rate is defined per unit surface area. In case of RBCs, sometimes BOD refers to soluble BOD rather than total BOD.

Stabilization ponds/RBC loading

$$\boxed{BODLR = \frac{M_{BOD}}{A_S} = \frac{Q \times BOD}{A_S}}$$

$A_S = Surface \, area \, (Ponds, RBC)$

EXAMPLE PROBLEM 10.5

For an aeration volume of 2100 m³, calculate the BOD loading rate for an average daily flow of 6.0 ML/d and BOD after settling of 120 mg/L.

Given:

$$V_A = 2100\,m^3 \quad Q = 6.0\,ML/d \quad BOD = 120\,mg/L = 120\,kg/ML$$

Solution:

BOD loading rate

$$BODLR = \frac{Q \times BOD}{V_A} = \frac{6.0\,ML}{d} \times \frac{120\,kg}{ML} \times \frac{1}{2100\,m^3} = 342 = \underline{340\,g/m^3 \cdot d}$$

EXAMPLE PROBLEM 10.6

The flow to a 7.5-acre lagoon is 1.2 MGD. Calculate the BOD loading rate in lb/acre knowing that the influent BOD concentration is 130 mg/L.

Given:

$$Q = 7.5\,MGD \qquad A_S = 7.5\,acre \qquad BOD = 130\,mg/L$$

Solution:

$$BODLR = \frac{Q \times BOD}{A_S} = \frac{1.2\,MG}{d} \times \frac{130\,mg}{L} \times \frac{8.34\,lb/MG}{mg/L} \times \frac{1}{7.5\,acre}$$
$$= 173.4 = \underline{170\,lb/acre}$$

EXAMPLE PROBLEM 10.7

The flow to a 1.5-ha lagoon is 400 m³/d. Calculate BOD loading given that the influent BOD concentration is 130 mg/L.

Given:

$$Q = 400\,m^3/d \quad A_S = 1.5\,ha \quad BOD = 130\,mg/L$$

Solution:

BOD loading rate

$$BODLR = \frac{Q \times BOD}{A_S} = \frac{400\,m^3}{d} \times \frac{130\,g}{m^3} \times \frac{1}{1.5\,ha} \times \frac{ha}{10000\,m^2} = 3.50 = \underline{3.5\,g/m^2 \cdot d}$$

10.2.1 Solids Loading Rate

This parameter is generally specified for secondary clarifiers of an activated sludge process and gravity sludge thickeners. The solids loading rate (SLR) indicates the kg/d solids loaded to each square metre of clarifier surface. In the case of secondary clarifier, SS concentration refers to mixed liquor (MLSS), and the flow stream also includes return flow, as shown in Figure 10.1.

Solids loading rate

$$SLR = \frac{M_{SS}}{A_S} = \frac{Q \times SS}{A_S}$$

Solids loading (secondary clarifiers)

$$SLR = \frac{M_{SS}}{A_S} = \frac{(Q + Q_{RS}) \times SS}{A_s}$$

Q = Wastewater flow
Q_{RS} = Return sludge
A_s = Surface area

EXAMPLE PROBLEM 10.8

A secondary clarifier 25 m in diameter receives a primary effluent flow of 12 ML/d, and the return sludge flow is 20% of this. If the MLSS concentration is 3000 mg/L, what is the solid loading rate on the clarifier?

Given:

D = 25 m   Q = 12 ML/d   MLSS = 3000 mg/L   R = 20% = 0.20

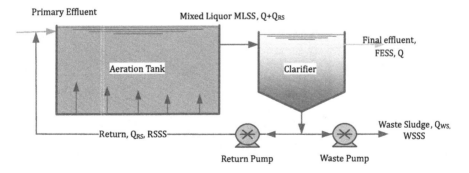

FIGURE 10.1   Solids loading on secondary clarifier.

Solution:

Solids loading rate

$$SLR = \frac{(Q + Q_{RS}) \times MLSS}{A_S} = \frac{1.2 \times 12\,ML}{d} \times \frac{3000\,kg}{ML} \times \frac{4}{\pi} \times \frac{1}{(25m)^2}$$

$$= 88.00 = \underline{88\ kg/m^2 \cdot d}$$

10.3 COMPOSITE CONCENTRATION

Municipal wastewater is a composite of domestic and industrial wastewater. In communities with significant amounts of industrial contributions, the BOD of the wastewater reaching the plant may be more than the typical value of 200 mg/L. To characterise the wastewater flow, composite concentration, should be determined.

Composite concentration

$$\bar{C} = \frac{Total\,mass}{Total\,Flow} = \frac{\Sigma Q_i \times C_i}{\Sigma Q_i}$$

Note the composite concentration is not the simple average of the concentrations of various flow compounds. As the various flow streams are not necessarily of the same volume, the concentration needs to be weighted with volume flow rate to arrive at the composite concentration (Figure 10.2).

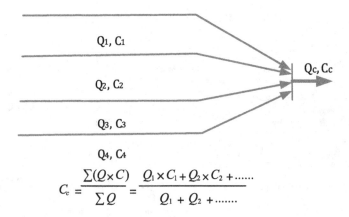

$$C_c = \frac{\Sigma(Q \times C)}{\Sigma Q} = \frac{Q_1 \times C_1 + Q_2 \times C_2 + \dots}{Q_1 + Q_2 + \dots}$$

FIGURE 10.2 Composite concentration.

EXAMPLE PROBLEM 10.9

The sanitary and industrial waste from a community consists of domestic wastewater flow of 3400 m³/d (7500 persons); potato processing waste of 120 m³/d with a BOD of 2100 mg/L; and dairy with a wastewater flow of 450 m³/d, with a BOD concentration of 1200 mg/L. Calculate the composite BOD concentration of the municipal wastewater.

Source	Q, m³/d	BOD, g/m³	M_{BOD}, kg/d (Q × BOD)
Domestic	3400	200	680
Potato	120	2100	252
Dairy	450	1200	540
Total	Σ = 3970	—	Σ = 1572

Solution:

Composite BOD

$$\overline{BOD} = \frac{\sum(Q_i \times BOD_i)}{\sum Q_i} = \frac{1572\ kg}{d} \times \frac{d}{3970\ m^3} \times \frac{1000\ g}{kg} = 396 = \underline{400\ g/m^3}$$

10.3.1 POPULATION EQUIVALENT

In general, industrial wastes are stronger than domestic waste. BOD population equivalent calculations equate these concentrated discharges with the number of people (equivalent) that would contribute the same amount of BOD. For a domestic wastewater system, each person typically contributes 450 L/d of wastewater containing BOD of 200 mg/L.

10.3.2 PER CAPITA MASS LOAD OF BOD

The BOD population equivalent of wastewater discharge can be found by knowing daily contribution of mass BOD by one person. Based on typical values of flow and BOD concentration, this comes out to be 90 g (0.20 lb), as shown:

$$\frac{450\ L}{person.d} \times \frac{200\ mg}{L} \times \frac{g}{1000\ mg} = \underline{90\ g/p.d}$$

$$\frac{120\ gal}{person.d} \times \frac{200\ mg}{L} \times \frac{8.34\ lb/MG}{mg/L} = 0.200 = \underline{0.2\ lb/p.d}$$

10.3.3   POPULATION EQUIVALENT

The BOD population equivalent is found from total mass of BOD divided by BOD contributed by one person in a day, that is, 90 g/p·d or 0.2 lb/p·d. Similarly, the hydraulic equivalent is based on a flow of 450 L/p·d or 120 gal/p·d. In other words, 1000 L of an industrial discharge contributes to hydraulic loading that is roughly equivalent to two persons in terms of sewage flow. When the BOD population equivalent is significantly larger than the hydraulic population equivalent, it is indicative of high-strength waste.

EXAMPLE PROBLEM 10.10

A 1.5 ML/d industrial flow has a BOD concentration of 1800 mg/L. What is the BOD population equivalent of this discharge?

Given:

$Q = 1.5$ ML/d   BOD $= 1800$ mg/L $= 1800$ kg/ML

Solution:

Population equivalent

$$PE = \frac{M_{BOD}}{M_{BOD}/p} = \frac{1.5\,ML}{d} \times \frac{1800\,kg}{ML} \times \frac{1000\,g}{kg} \times \frac{d.p}{90\,g} = \underline{30\ 000\ persons}$$

EXAMPLE PROBLEM 10.11

An industrial discharge of 0.65 MGD has BOD content of 1200 mg/L and a BOD concentration of 1800 mg/L. What is the BOD population equivalent of this discharge? Assume BOD equivalent of 0.20 lb/c·d.

Given:

$Q = 0.65$ MGD   BOD $= 2500$ mg/L

Solution:

BOD population equivalent

$$PE = \frac{M_{BOD}}{M_{BOD}/p} = \frac{0.65\,MG}{d} \times \frac{2500\,mg}{L} \times \frac{8.34\,lb/MG}{mg/L} \times \frac{d.p}{0.20\,lb} = 32520 = \underline{33000\,p}$$

10.4 MIXING SOLUTIONS

When two or more solutions of different strengths are mixed, the combined concentration in the mixture can be calculated as shown in the case of composite concentration before. It is assumed that no reaction takes place between the various streams mixed.

Concentration of the mixture

$$C_{mix} = \frac{Total\ mass}{Total\ Flow\ Volume} = \frac{\sum V_i \times C_i}{\sum V_i}$$

C_i, V_i = *Concentration and volume of the individual solution/liquid*

10.4.1 BLENDING OF SLUDGES

A sludge slurry is a concentrated solution of solids. The concept of solution mixing discussed in the preceding section can be applied when sludges are mixed together Remember for sludges with solids concentration less than 10%, the density can be safely assumed to be equal to that of water, that is, 1 kg/L = 1000 kg/m³. The concentration of the blended sludge (mixture) will be between the concentration of the mixing sludges and closer to the concentration of the sludge with the greater volume.

EXAMPLE PROBLEM 10.12

If 20 L of an 8% strength solution (SG = 1.4) is mixed with 150 L of a 0.5% strength solution (SG = 1.0), what is the strength of the solution mixture?

Given:

Parameter	Solution 1	Solution 2
Conc. C, %	8.0	0.50
Volume V, L	20	150
SG	1.4	1.0
Density, g/L	1400	1000

Solution:

Mass concentration of solution 1

$$C_1 = \frac{8.0\%}{100\%} \times \frac{1400\,g}{L} = 112.0 = 112\,g/L$$

Mass concentration of solution 2

$$C_2 = \frac{0.5\%}{100\%} \times \frac{1000\,g}{L} = 5.00 = 5.0\,g/L$$

Strength of solution mixture

$$C_{mix} = \frac{\sum V_i \times C_i}{\sum V_i} = \frac{(112 \times 20 + 5.0 \times 150)g}{(20 \times 150)L} = 13.5 = \underline{14\,g/L}$$

10.4.2 SOLUTION OF TARGET STRENGTH

Another type of problem may involve the calculation of the volume of one solution to be mixed with another solution of a given strength and volume to achieve a target strength of the mixture. The same formula applies; however, the V_1 becomes the unknown. The modified form of the equation is as follows:

Mix concentration

$$C_{mix} = \frac{V_1 \times C_1 + V_2 \times C_2}{V_1 + V_2}$$

$V_{mix} = V_1 + V_2 =$ *volume of the solution mixture*

Volume of solution 1

$$V_1 = V_2 \times \frac{(C_2 - C_{mix})}{(C_{mix} - C_1)}$$

EXAMPLE PROBLEM 10.13

What volume of a 7.0% solution (SG = 1.2) must be mixed with 50 L of a 2.0% solution to make up a 40 g/L solution? Data is presented in the following table.

Given:

Parameter	Solution 1	Solution 2	Mix
Conc. C,	7.0%	2.0%	40 g/L
Volume V, L	?	50	
SG	1.2	1.0	

Solution:

Mass concentration of 7.0% solution

$$C_1 = \frac{7.0\%}{100\%} \times \frac{1200\,g}{L} = 84.0 = \underline{84\,g/L}$$

Mass concentration of 2.0% solution

$$C_1 = \frac{2.0\%}{100\%} \times \frac{1000\,g}{L} = 20.0 = \underline{20\,g/L}$$

Volume of 2.0% solution required

$$V_1 = V_2 \times \frac{(C_2 - C_{mix})}{(C_{mix} - C_1)} = 50\,L \times \frac{(2.0 - 40)}{(40 - 84)} = 22.7 = \underline{23\,L}$$

EXAMPLE PROBLEM 10.14

A primary sludge of 20 m³ containing 5% solids is blended with a thickened secondary sludge of 15 m³ containing 3% solids. What is the solid content of the blended sludge?

Given:

Parameter	Primary = 1	Secondary = 2
SS, %	5.0	3.0
Volume V, L	20	15
SG	1.0	1.0
SS, kg/m³	50	30

Solution:

SS concentration of blended sludge

$$SS_{blend} = \frac{V_1 \times SS_1 + V_2 \times SS_2}{V_1 + V_2} = \frac{(20 \times 50 + 15 \times 30)}{(20 + 15)}$$
$$= 41.4 = 41\,kg/m^3 = \underline{4.1\%}$$

EXAMPLE PROBLEM 10.15

How much of primary sludge containing 5% solids should be mixed with 15 m³ of thickened sludge containing 3% solids to achieve 4.0% solid content of the blended sludge?

Given:

Parameter	Primary = 1	Thickened = 2	Blended = 3
SS, %	5.0	3.0	4.0%
Volume V, L	?	15	?
SG	1.0	1.0	1.0
SS, kg/m³	50	30	40

Solution:

Volume of primary sludge required

$$V_1 = V_2 \times \frac{(C_2 - C_{mix})}{(C_{mix} - C_1)} = 15 \ m^3 \times \frac{\left(30 \ kg/m^3 - 40 \ kg/m^3\right)}{(40 - 50)kg/m^3}$$

$$= 15 \ m^3 \times \frac{-10}{-10} = 15.0 = \underline{15 \ m^3}$$

PRACTICE PROBLEMS

PRACTICE PROBLEM 10.1

The SS of the wastewater entering the primary (plant influent) is 250 mg/L. If the plant flow is 16 ML/d, how many kg of SS enters the primary clarifier? (4000 kg/d or 4.0 t/d)

PRACTICE PROBLEM 10.2

A sewage treatment plant effluent contains 20 mg/L of BOD. If the daily average flow is 8.0 MGD, determine the BOD load into the receiving stream. (1300 lb/d)

PRACTICE PROBLEM 10.3

The capacity of an aeration tank is 1.3 MG. The plant is operated by maintaining a MLSS of 2200 mg/L. How many lb of solids are held in the tank? (2.4×10^4 lb)

PRACTICE PROBLEM 10.4

The daily flow to a trickling filter is 18 000 m^3/d. If the BOD concentration of the trickling filter's influent is 210 mg/L, how many kg of BOD enter the trickling filter daily? (3800 kg/d)

PRACTICE PROBLEM 10.5

A 30-m-diameter trickling filter with a media depth of 1.0 m receives a primary effluent flow of 6.5 ML/d with a BOD concentration of 110 mg/L. Calculate the organic loading on the trickling filter. (1000 g/m^3.d)

PRACTICE PROBLEM 10.6

The flow to a 6.5-acre lagoon is 1.0 MGD. Calculate the BOD loading rate in lb/acre knowing that the influent BOD concentration is 150 mg/L. (190 lb/acre)

PRACTICE PROBLEM 10.7

The flow to a 1.5-ha lagoon is 400 m^3/d. Calculate BOD loading given that the influent BOD concentration is 150 mg/L ($4.0 \ g/m^2 \cdot d$).

PRACTICE PROBLEM 10.8

An RBC receives a wastewater flow of 6.0 ML/d with a soluble BOD of 120 g/m^3 Total disc area is 6.0 hm^2; what is the BOD loading rate? (12 $g/m^2 \cdot d$)

PRACTICE PROBLEM 10.9

Total flow including the return to a 20-m-diameter secondary clarifier is 20 ML/d. Calculate the SLR for a MLSS concentration of 2000 mg/L. (130 $kg/m^2 \cdot d$)

PRACTICE PROBLEM 10.10

If a brewery with an average daily flow of 1 ML/d and BOD content of 2000 mg/L is allowed to discharge into the sanitary sewers of this community, what will be the BOD of the combined wastewater? (720 mg/L)

PRACTICE PROBLEM 10.11

A brewery discharges 0.35 MGD of wastewater into sanitary sewer containing BOD of 1500 mg/L. What is the BOD population equivalent of this discharge? Assume a BOD equivalent of 0.20 lb/c·d. (22,000 p)

PRACTICE PROBLEM 10.12

Calculate the BOD population equivalent of 400 m³/d of industrial flow with a BOD content of 2700 mg/L. (1200 persons)

PRACTICE PROBLEM 10.13

If 60 L of a 10% solution (SG = 1.2) is added to 200 L of 0.8% solution (SG = 1.05), what is the concentration of the solution mixture in g/L? (34 g/L = 3.4%)

PRACTICE PROBLEM 10.14

How many mL of a 20% solution (SG = 1.2) must be mixed with 10 L of a 1% solution to make a 2% solution? (455 mL)

PRACTICE PROBLEM 10.15

Primary and thickened secondary sludges are blended before being fed to the digester. The daily production of primary sludge is 22 m³ with 5.5% solids and 16 m³ of thickened secondary sludge with 4.0% solids. Calculate the solids concentration in the blended sludge. (4.9%)

11 Removal Efficiency

11.1 REMOVAL CALCULATIONS

To calculate the mass of BOD or SS removed daily, you will need to know the mg/L of BOD or SS removed and the plant flow. The removal concentration is the difference between the concentration of influent (entering) and effluent (exiting) of a given treatment system or process, as shown in Figure 11.1.

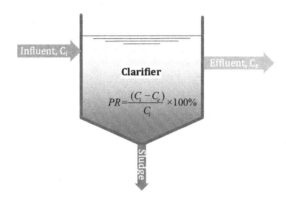

FIGURE 11.1 Percent removal.

Percent removal

$$PR = \frac{C_r}{C_i} \times 100\% = \frac{(C_i - C_e)}{C_i} \times 100\%$$

$C = Concentration$
Subs i, = influent, e = effluent, r = removed

Knowing the concentration removed, we can work out the mass of solids removed. The mass of solids thus removed is withdrawn in the form of liquid slurry or wet sludge.

EXAMPLE PROBLEM 11.1

Based on composite samples, SS in the raw wastewater (primary influent) and settled wastewater (primary effluent) are found to be 250 mg/L and 130 mg/L, respectively How many kg of SS are removed for a flow of 28 ML/d?

Given:

$Q = 28$ ML/d $\quad SS_i = 250$ mg/L $\quad SS_e = 130$ mg/L

DOI: 10.1201/9781003468745-11

Solution:

Suspended solids removed

$$SS_r = SS_i - SS_e = 250 \ mg/L - 130 \ mg/L = 120.0 = 120 \ mg/L$$

Mass rate of suspended solids removed

$$M_{SS(r)} = Q \times SS_r = \frac{28 \ ML}{d} \times \frac{130 \ kg}{ML} = 3360 = \underline{3400 \ kg/d}$$

11.2   VOLUME OF SLUDGE

Solids removed are withdrawn as sludge from the bottom of the clarifier. The volume of sludge produced can be estimated by knowing the mass of solids removed and concentration of solids (SS_{sl}) in the sludge withdrawn. See the illustration in Figure 11.2.

Volume of sludge

$$\boxed{Q_{Sl} = \frac{Q \times SS_r}{SS_{sl}} = \frac{Q \times (SS_i - SS_e)}{SS_{sl}}}$$

EXAMPLE PROBLEM 11.2

The flow to a primary clarifier is 800 m³/d. If the solid concentration in the influent and effluent is 190 mg/L and 90 mg/L, respectively, how many kg of solids are removed by the primary treatment?

Given:

$$Q = 6800 \ m^3/d \qquad SS_i = 190 \ mg/L \qquad SS_e = 90 \ mg/L$$

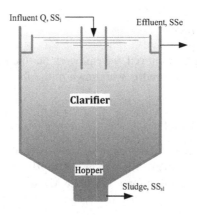

Influent Q, SS_i   Effluent, SSe

Clarifier

Hopper   Sludge, SS_sl

FIGURE 11.2   Volume of sludge produced.

Solution:

Suspended solids removed
$$SS_r = SS_i - SS_e = 190 - 90 = 100.0 = 100 \ mg/L$$

Mass rate of suspended solids removed
$$M_{SS(r)} = Q \times SS_r = \frac{6800 \ m^3}{d} \times \frac{100 \ g}{m^3} \times \frac{kg}{1000 \ g} = 680.0 = \underline{680 \ kg/d}$$

EXAMPLE PROBLEM 11.3

The flow to a primary clarifier is 0.80 MGD. If the BOD concentration in the influent and effluent is 190 mg/L and 120 mg/L, respectively, how many lb of BOD are removed by the primary treatment?

Given:

$Q = 0.80$ MG/d $\qquad$ $BOD_i = 190$ mg/L $\qquad$ $BOD_e = 120$ mg/L

Solution:

BOD removed
$$BOD_r = BOD_i - BOD_e = 190 - 120 = 70 \ mg/L$$

Mass rate of BOD removed
$$M_{BOD(r)} = Q \times BOD_r = \frac{0.80 \ MG}{d} \times \frac{70 \ mg}{L} \times \frac{8.34 \ lb/MG}{mg/L} = 467 = \underline{470 \ lb/d}$$

EXAMPLE PROBLEM 11.4

The flow to a primary clarifier is 20 ML/d. If the SS concentrations in the influent and effluent are 170 and 110 mg/L, respectively, calculate the percent removal of solids.

Given:

$Q = 20$ ML/d $\qquad$ $SS_i = 170$ mg/L $\qquad$ $SS_e = 110$ mg/L $\qquad$ PR = ?

Solution:

Percent removal
$$PR = \frac{(SS_i - SS_e)}{SS_i} \times 100\% = \frac{(170 - 110)}{110} \times 100\% = 54.5 = \underline{55\%}$$

EXAMPLE PROBLEM 11.5

The flow to a primary clarifier is 5.0 MGD. On average 50% of SS are removed when the raw wastewater contains 190 mg/L of SS. Estimate the volume of raw sludge produced, assuming raw sludge contains 3.0% of dry solids.

Given:

$Q = 5.0$ MG/d SS_i (influent) $= 190$ mg/L   Removal $= 50\%$ SS_{sl} (sludge) $= 3.0\%$

Solution:

Daily production of sludge

$$Q_{Sl} = \frac{Q \times SS_r}{SS_{sl}} = \frac{5.0 \, MG}{d} \times \frac{190 \, mg}{L} \times \frac{50\%}{3.0\%} \times \frac{8.34 \, lb/MG}{mg/L} \times \frac{gal}{8.34 \, lb}$$

$$= 25833 = \underline{25\,000 \, gal/d}$$

EXAMPLE PROBLEM 11.6

The flow to a primary clarifier is 5000 m³/d with solid content of 160 mg/L. If solids removal is 50%, find the volume of raw sludge produced daily based on an SS concentration in the raw sludge of 4.0%.

Given:

$Q = 5000$ m³/d $= 5.0$ ML/d SS_i (influent) $= 160$ mg/L $= 160$ kg/ML
SS_{sl} (sludge) $= 4.0\% = 40$ kg/m³

Solution:

Volume of sludge produced daily

$$Q_{Sl} = \frac{Q \times SS_r}{SS_{sl}} = \frac{5.0 \, ML}{d} \times \frac{50\%}{100\%} \times \frac{160 \, kg}{ML} \times \frac{m^3}{40 \, kg} = 10.0 = \underline{10 \, m^3/d}$$

EXAMPLE PROBLEM 11.7

The flow to a primary clarifier has a BOD concentration of 250 mg/L. The effluent from the plant contains BOD of 15 mg/L. What BOD removal is achieved?

Given:

$BOD_i = 250$ mg/L $BOD_e = 15$ mg/L   PR = ?

Solution:

Primary removal

$$PR = \frac{(BOD_i - BOD_e)}{BOD_i} \times 100\% = \frac{(250 - 15)}{250} \times 100\% = 94.0 = \underline{94\%}$$

EXAMPLE PROBLEM 11.8

The raw wastewater has a BOD concentration of 220 mg/L. If the BOD removal efficiency of the plant is 85%, what is the BOD of the plant effluent?

Given:

$BOD_i = 220$ mg/L $\quad BOD_e = ?$ $\quad PR = 85\%$

Solution:

BOD concentration in the effluent

$$BOD_e = BOD_i\left(1 - \frac{PR}{100\%}\right) = \frac{220\ mg}{L} \times \left(1 - \frac{85\%}{100\%}\right) = 33.0 = \underline{33\ mg/L}$$

EXAMPLE PROBLEM 11.9

A wastewater pollution control plant (WPCP) is designed to remove 90% of BOD. If the desired BOD of the plant effluent is 15 mg/L, what is the maximum BOD of the incoming wastewater that can be treated by this plant?

Given:

$BOD_i = ?$ $\quad BOD_e = 15$ mg/L $\quad PR = 90\%$

Solution:

BOD concentration in the raw wastewater

$$BOD_i = \frac{BOD_e}{(1 - PR/(100\%))} = \frac{15\ mg/L}{(1 - (90\%/100\%))} = 150.0 = \underline{150\ mg/L}$$

EXAMPLE PROBLEM 11.10

Primary clarification in a wastewater pollution control plant is designed to remove 45% of SS. If the desired SS in the plant effluent is not to exceed 110 mg/L, what is the maximum concentration of SS in the raw wastewater that can be treated by this plant?

Given:

$SS_i = ?$ $\quad SS_e = 110$ mg/L $\quad PR = 45\%$

Solution:

SS concentration in the raw wastewater

$$SS_i = \frac{SS_e}{(1 - PR/100\%)} = \frac{110\ mg/L}{(1 - 45\%/100\%)} = 200.0 = \underline{200\ mg/L}$$

EXAMPLE PROBLEM 11.11

The influent BOD of an activated sludge plant is 225 mg/L, and effluent BOD is 25 mg/L. What is BOD removal efficiency of the plant? If the BOD removal by primary clarification is 30%, find the BOD removal by the activated sludge process.

Given:

Primary	$BOD_i = 225$ mg/L	$BOD_e = ?$	$PR = 30\%$
Secondary	$BOD_i = ?$	$BOD_e = 25$ mg/L	$PR = ?$

Solution:

BOD removal by the plant

$$PR = \frac{(BOD_i - BOD_e)}{BOD_i} \times 100\% = \frac{(225 - 25)}{225} \times 100\% = 88.8 = \underline{89\%}$$

BOD of the primary effluent

$$BOD_e = BOD_i \times \left(1 - \frac{PR}{100}\right) = 225\frac{mg}{L} \times \left(1 - \frac{30\%}{100\%}\right) = 157 = 160 \, mg/L$$

BOD removal by the secondary treatment

$$PR = \frac{(BOD_i - BOD_e)}{BOD_i} \times 100\% = \frac{(157.5 - 25)}{157.5} \times 100\% = 84.12 = \underline{84\%}$$

EXAMPLE PROBLEM 11.12

Find the amount of iron and manganese removed per month (30 d) from a plant that treats an average of 12.0 MGD if the average iron concentration is 1.2 ppm. The removal efficiency for iron is 81%.

Given:

Fe = 1.2 ppm(8.34 lb/MG)   PR = 81%

Solution:

Iron removal per month

$$M = Q \times C \times PR = \frac{12 \, MG}{d} \times \frac{1.2 \times 8.34 \, lb}{MG} \times \frac{81\%}{100\%} \times \frac{30 \, d}{mo} = 2675 = \underline{2700 \, lb/mo}$$

PRACTICE PROBLEMS

PRACTICE PROBLEM 11.1

For a flow of 28 ML/d, the BOD concentrations in the influent and effluent are determined to be 220 mg/L and 150 mg/L, respectively. Calculate the tons of BOD removed daily as primary sludge. (2.0 t/d)

PRACTICE PROBLEM 11.2

Primary effluent flow is 15.6 ML/d containing 170 mg/L of BOD. If the secondary effluent BOD is 20 mg/L, how many kg of BOD are removed in the secondary treatment? (2340 kg/d)

PRACTICE PROBLEM 11.3

The flow to a primary clarifier is 0.80 MGD. If the BOD concentration in the influent and effluent is 190 mg/L and 120 mg/L, respectively, how many lb of BOD are removed by the primary treatment? (200 lb/d)

PRACTICE PROBLEM 11.4

The BOD removal efficiency of a primary clarifier is as high as 40%. For a daily wastewater flow of 10 ML/d with BOD concentration of 200 g/m^3, what BOD of the flow stream entering the aeration tank can be expected? (120 mg/L)

PRACTICE PROBLEM 11.5

The flow to a primary clarifier is 2.3 MGD. On average 40% of SS are removed when the raw wastewater contains 160 mg/L of SS. Estimate the volume of raw sludge produced assuming raw sludge contains 2.5% of dry solids. (5900 gal/d)

PRACTICE PROBLEM 11.6

The flow to a primary clarifier is 15 ML/d with solid content of 220 mg/L. If the average solids removal is 50%, find the daily volume of raw sludge produced assuming SS concentration in the sludge to be 3.0%. (55 m^3/d)

PRACTICE PROBLEM 11.7

For the data of the example problem, if the BOD removal by primary clarification is assumed to be 35%, what BOD removal is achieved by the secondary treatment? (91%)

PRACTICE PROBLEM 11.8

Raw wastewater has BOD of 180 mg/L. On average, the BOD removal efficiency of the plant is 90%. What is the BOD of the plant effluent? (18 mg/L)

PRACTICE PROBLEM 11.9

A sewage treatment plant (STP) is designed to remove 90% of BOD. If the desired BOD of the plant effluent is 20 mg/L, what is the maximum BOD of the incoming wastewater that can be treated by this plant? (200 mg/L)

PRACTICE PROBLEM 11.10

Secondary treatment of a wastewater pollution control plant is designed to remove 75% of BOD. If the desired BOD of the plant effluent is not to exceed 30 mg/L, what is the maximum BOD of the primary effluent that can be treated by this plant? (120 mg/L)

PRACTICE PROBLEM 11.11

The influent BOD of an activated sludge plant is 230 mg/L. Primary clarification is able to remove 35% of BOD. What should be the minimum removal by the secondary treatment to produce an effluent with BOD not exceeding 25 mg/L? (83%)

PRACTICE PROBLEM 11.12

Find the amount of manganese removed per annum from a plant that treats an average of 12.0 MGD if the average manganese concentration is 0.16 ppm. The removal efficiency for manganese is 69%. (4000 lb/a)

12 Molarity and Normality

Mass concentration of a solution is the mass of the chemical in a given volume of solution, for example, mg/L or g/m³. Mass of the chemical can also be expressed in number of moles of that substance. Concentration in moles per litre is called **molarity**. Molarity is one method devised to compare solution concentration. For example, a solution of one molarity would be 1 mole of the substance in 1 litre of solution.

12.1 MOLECULAR MASS

Every element shown on the **Periodic Table** has a corresponding atomic mass listed. This number indicates the relative mass of the atom of that element relative to a carbon atom, which is arbitrarily assigned a value of 12.00. This was necessary, as the absolute mass of atoms is extremely small. Depending on whether the mass density of a given element is lower or higher than that of carbon, its atomic mass will be less than or greater than 12.

A compound is made up of various elements, and thus the formula mass is the sum of all the atomic masses as indicated by the formula. A gram formula mass is the formula mass expressed in grams. For example, 12 grams of carbon would be 1 gram-mole or simply 1 mole of carbon. A mass of 100 g of calcium expressed as mole is:

Number of moles

$$\frac{Mass}{Formula\,mass} = 100\,g \times \frac{mol}{40.0\,g} = 2.5\,mol$$

To calculate the formula mass of a compound, first determine the total atomic mass represented by each element and then add them up. This is illustrated in the following example problems.

EXAMPLE PROBLEM 12.1

42 g of baking soda ($NaHCO_3$) is used in making up a 2.5-L solution. What is the molarity of the solution?

Given:

m = 42 g   V = 2.5 L   C = ?

DOI: 10.1201/9781003468745-12

Solution:

Element	g/mol	Atoms	Total
Na	23	1	23
H	1	1	1
C	12	1	12
O	16	3	48
		$\Sigma =$	84

Molar mass
$= (23 + 1 + 13 + 48) = 84 = 84 \ g/mol$

Molarity of the solution

$$Molarity = \frac{42\,g}{2.5\,L} \times \frac{mol}{84\,g} = 0.20 = \underline{0.2\ mol/L} = \underline{0.2\ M}$$

Note: M symbol indicates mol/L, and N indicates eq/L.

EXAMPLE PROBLEM 12.2

How many grams of $MgSO_4$ are required to make 5.0 L of 0.1 M solution?

Given:

$C = 0.1$ $M = 0.1$ mol/L $V = 5.0$ L $m = ?$

Solution:

Molar mass
$= 1 \times 24 \ g/mol + 1 \times 32 \ g/mol + 4 \times 16 \ g/mol = 120 \ g/mol$

$$Mass = \frac{0.10\,mol}{L} \times \frac{120\,g}{mol} = 60.0 = \underline{60\,g}$$

12.2 EQUIVALENT MASS OF AN ELEMENT

An atom is composed of protons, neutrons and electrons. The protons and neutrons make up the nucleus, and the mass of the atom is due to these particles. The electrons revolve around the nucleus and interact with electrons of other atoms to form new substances. The electrons in the outer orbit which interact are called **valence electrons**. Thus, valence indicates the number of electrons which an atom can exchange or share with other atoms to complete its orbit. Groups i, ii and iii in the periodic table of elements have 1, 2 and 3 valence electrons, respectively. By using valence information, that is, combining power and atomic mass, the equivalent mass or the combining mass can be calculated. Stated differently, it is the portion of the atomic mass of an element associated with each valence (combining) electron. For example, calcium has an atomic mass of 40.08 and a valence of 2. Thus, the equivalent mass is:

Equivalent mass

$$\boxed{\frac{Atomic\, mass}{Valence} = \frac{40.08\, g}{mol} \times \frac{mol}{2\, eq} = 20.04 = 20\, g/eq}$$

12.3 EQUIVALENT MASS OF A COMPOUND

Equivalent mass of a compound is the formula/molecular mass divided by the *net positive valence*. As discussed earlier, the molecular mass is the sum of the atomic masses of the combined elements and is the mass of 1 mole of a compound. For example, the molecular mass of $NaCl$ is 58.4 g/mol, while that of NH_3 gas is 17.0 g/mol.

EXAMPLE PROBLEM 12.3

Calculate the equivalent mass of ferric sulphate.

Given:

Formula: $Fe_2(SO_4)_3$

Solution:

Molar mass
$= 2 \times 55.8\, g/mol + 3 \times 32\, g/mol + 12 \times 16\, g/mol = 399.9 = 400\, g/mol$

The ferric (oxide iron) atom has a positive valence of 3; thus, a compound with 2 ferric atoms has a net positive valence of 2 × 3 = 6.

Equivalent mass
$$= \frac{399.9\, g}{mol} \times \frac{mol}{6\, eq} = 66.65\, g/eq = \underline{67\, g/eq}$$

Equivalent, or eq, is the equivalent mass in grams. Thus, milliequivalent (meq) represents equivalent mass in milligrams.

12.3.1 HARDNESS

The **hardness** of water is due to the presence of calcium and magnesium ions. Once the concentration of the ions in the water is known, the total hardness of the water can be found. However, before you can sum up the concentrations, they must be expressed in equivalent concentrations, that is, meq/L or eq/L. Hardness is generally expressed as equivalents or mg/L as $CaCO_3$. Conversion is made by knowing that $CaCO_3$ is 50 mg/eq.

12.3.2 ALKALINITY

Alkalinity is the buffering capacity (capacity to neutralize acid) of the water and is due to the presences of OH, HCO_3 and CO_3 ions. Alkalinity is also expressed in mg/L as $CaCO_3$.

EXAMPLE PROBLEM 12.4

The results of a water analysis are calcium 29.0 mg/L, magnesium 16.4 mg/L, sodium 23.0 mg/L, potassium 17.5 mg/L, bicarbonate 171 mg/L, sulphate 36.0 mg/L and chloride 24.0 mg/L. Calculate hardness as mg/L of $CaCO_3$ for this water sample.

Solution:

Cations

Component	mg/L	g/eq	meq/L
Ca	29.0	20.0	1.45
Mg	16.4	12.2	1.34
Na	23.0	23.0	1.00
K	17.5	39.1	0.45
Total			4.24

Anions

Component	mg/L	g/eq	meq/L
HCO_3	171	61.0	2.81
SO_4	36.0	48.0	0.75
Cl	24.0	35.5	0.68
		Σ =	4.24

Hardness of water analysed

$$Hardness = Ca + Mg = 1.45 + 1.34 = \frac{2.89\ meq}{L} \times \frac{50\ mg}{meq} = 144.5 = \underline{140\,mg/L}$$

Alkalinity and hardness are usually expressed as equivalent calcium carbonate.

12.4 NORMALITY

Like molarity, **normality** is the concentration expressed in equivalents per litre. For example: one normal (1.0 N) of NaOH (caustic) means 40 grams of sodium hydroxide in 1 litre of solution. When determining the chemical composition of a water sample, it is customary to report the concentration of each constituent in meq/L. This not only allows us to visualize the chemical composition but also provides a check on the accuracy of analyses for major ions. The sum of the meq/L of cations (positive ions) must equal the sum of anions (negative ions). In a perfect evaluation, they would be exactly the same, since water equilibrium is electrically balanced.

Equivalent mass represents the combining mass of an element or a compound. It means one equivalent of one substance will react with one equivalent of another substance. This will permit us to find the amount of chemical required (equivalent) to react with the chemical present in water. It follows that specific number of equivalents of one substance will react with the same number of equivalents of another substance.

Combining volume

$$N_1 \times V_1 = N_2 \times V_2$$

where N and V respectively represent the normality and volume of one substance required to completely react with another substance. This equation is very useful when working out the concentration by doing a titration, for example, an acid/base neutralization. Knowing the volume and normality of the titrant used to reach the end point (complete reaction) while titrating a given volume of water, the normality of the chemical in water reacting with the titrant can be calculated.

12.5 STANDARD SOLUTIONS

A **standard solution** is a solution whose exact concentration is known. A 1.0-N solution of NaOH requires 40 g of pure NaOH be weighed and made up to 1 L by adding distilled water. Occasionally a different quantity of reagent is used. To maintain the same concentration of solution, therefore, the corresponding quantity of distilled water to be added must be calculated.

EXAMPLE PROBLEM 12.5

How many litres of 0.5 N caustic (NaOH) is required to neutralize 50.0 L of waste with an acidity of 0.010 N as HCl?

Given:

Parameter	Caustic = 1	Acid = 2
Volume	?	50.0 L
Normality	0.50	0.010

Solution:

Volume of caustic required

$$V_1 = \frac{N_2}{N_1} \times V_2 = \frac{0.01\ eq/L}{0.5\ eq/L} \times 50.0\ L = 1.00 = \underline{1.0\ L}$$

EXAMPLE PROBLEM 12.6

To prepare a standard solution of NaOH, 50.0 g are to be dissolved to make up 1.0 L of solution. If 37.56 g are weighed out, how many mL of solution should be prepared?

Given:

$m_1 = 50.0$ g $V_1 = 1.0$ L $m_2 = 37.56$ g $V_2 = ?$

Solution:

Volume of solution

$$V_2 = \frac{m_2}{m_1} \times V_1 = 1.0\ L \times \frac{37.56\ g}{40.0} \times \frac{1000\ mL}{L} = 751.2 = \underline{751\ mL}$$

Normality of the solution

$$N_2 = \frac{50.0\ g}{L} \times \frac{eq}{40.0\ g} = 1.20 = \underline{1.2\ eq/L} = \underline{1.2\ N}$$

12.6   FRACTION OF AN ELEMENT IN A COMPOUND

In some situations, to add a given element, for example fluoride or copper, compounds of the elements like sodium fluoride and copper sulphate are added. In such cases, it is important to know that fraction of the element in the compound. This is done by knowing the molecular mass of the chemical compound.

EXAMPLE PROBLEM 12.7

What is the percent copper (Cu) in copper sulphate ($CuSO_4$)?

Solution:

Element	Number of Atoms	Atomic Mass, g/mole	Molecular Mass, g/mole
Cu	1	63.5	63.5
S	1	32	32
O	4	16	64
		$\Sigma =$	159.5

Fraction of copper

$$Cu = \frac{Cu}{CuSO_4} = \frac{63.5\,g/mol}{159.5\,g/mol} = 0.398 = \underline{40\%}$$

EXAMPLE PROBLEM 12.8

What is the percent aluminium (Al) in liquid alum $(Al_2(SO_4)_3.14.3H_2O$?

Element	Number of Atoms	Atomic Mass, g/mole	Molecular Mass, g/mole
Al	2	27	54
S	3	32	96
O	26.3	16	420.8
H	28.6	1	28.6
		$\Sigma =$	599.4

Solution:

Fraction of aluminium

$$Al = \frac{Al}{Al_2(SO_4)_3} = \frac{54\,g/mol}{599.4\,g/mol} = 0.090 = \underline{9.0\%}$$

PRACTICE PROBLEMS

PRACTICE PROBLEM 12.1

Determine the molarity of a 2.0 L solution containing 40 g of calcium carbonate. (0.2 M)

PRACTICE PROBLEM 12.2

Determine the mass in grams of NaCl (table salt) required to make 5.0 L of 0.5-M solution. (146 g)

PRACTICE PROBLEM 12.3

Calculate the equivalent mass of $CaCO_3$ in grams. (50.0 g/eq)

PRACTICE PROBLEM 12.4

Calculate the alkalinity as mg/L of $CaCO_3$ for the same water sample described in the example problem 12.4. (141 mg/L as $CaCO_3$)

PRACTICE PROBLEM 12.5

Laboratory tests (titration) on an industrial wastewater indicate that 100 mL of sample requires 11.3 mL of 0.5 N sulphuric acid (H_2SO_4) to lower the pH to 7.0 (neutralized). What is the normality of the wastewater? (0.0565 N)

PRACTICE PROBLEM 12.6

To prepare 1 L of 0.0192 N solution of $AgNO_3$, it is required to dissolve 3.27 g. If 3.14 g is weighed, how many mL of solution should be prepared? (960 mL)

PRACTICE PROBLEM 12.7

What is the fraction of Mn in $kMnO_4$? (35%)

PRACTICE PROBLEM 12.8

What is the percent fluoride (F) ion in fluorosilicic acid (H_2SiF_6)? (79%)

13 Basic Hydraulics

Hydraulics concepts are needed to understand most water system operations and maintenance. A pump is one of the most common hydraulic devices used. The intent is not to become an expert but to get the know the basics as applied to water and wastewater operations.

13.1 PRESSURE

In simple terms, **pressure** or pressure intensity is force per unit area. In water transportation systems, water may possess energy due to pressure. A pump is a device by virtue of which energy is imparted to the fluid.

The units of pressure are that of force divided by area units. In SI units, it is N/m^2, which is the same thing as one pascal (Pa). A pressure of 1 Pa is a very small; thus kPa is generally used to indicate pressure in water systems. For example, pressure in a water main is typically in the range of 300–400 kPa (40–50 psi).

13.1.1 Pressure and Force

For the same force applied, pressure is inversely proportional to area, and for the same area, pressure will vary directly with the force applied. For a piston-cylinder arrangement, pressure for the same force will generate more pressure in a smaller-diameter piston.

13.1.2 Pressure as a Liquid Column

Pressure due to a liquid column is directly proportional to its height, as shown in Figure 13.1. Thus, pressure is independent of the surface area or size of the liquid column. Whether it is a small-diameter tube or a large reservoir, if height of water is the same, pressure at the bottom will be the same.

Pressure and height

$$p = \gamma \times h \ \text{ or } \ h = \frac{p}{\gamma} = \frac{N}{m^2} \times \frac{m^3}{N} = m$$

Water column due to a pressure of 1.0 kPa

$$h = \frac{p}{\gamma} = 1\,kPa \times \frac{m}{9.81\,kPa} = 0.102 = 0.10\,m$$

Water column due to a pressure of 1.0 psi

$$h = \frac{p}{\gamma} = 1\,psi \times \frac{ft^3}{62.4\,lb} \times \frac{144\,in^2}{ft^2} = 2.31\,ft$$

DOI: 10.1201/9781003468745-13

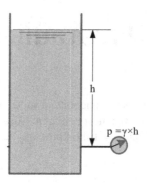

FIGURE 13.1   Pressure as a liquid column.

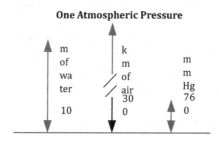

FIGURE 13.2   Atmospheric pressure as fluid column.

Thus 1 m of water column will result in a pressure of about 10 kPa. Saying it differently, a pressure of about 10 kPa will be able to support 1 metre of water column. You might have heard the weather woman saying the atmospheric pressure is 760 mm of mercury. She is referring to pressure as column of mercury. Knowing that the SG of mercury is 13.6, standard atmospheric pressure in kPa is:

Atmospheric pressure

$$p = \gamma \times h = SG \times \gamma_w \times h = 13.6 \times \frac{9.81\,kPa}{m} \times 760\,mm \times \frac{m}{1000\,mm} = 101.2 = \underline{101\,kPa}$$

The standard atmospheric pressure of 760 mm of mercury is equal to 101 kPa. This is the pressure exerted due to the weight of air (Figure 13.2). Warm air being lighter, falling atmospheric pressure indicates the arrival of a warm air mass, and atmospheric pressure at high altitudes is lower due to the decrease in weight of the air.

13.1.3   GAUGE AND ABSOLUTE PRESSURES

At this point, it is important to understand the difference between gauge pressure and absolute pressure. Water pressure in pipelines is generally indicated as gauge

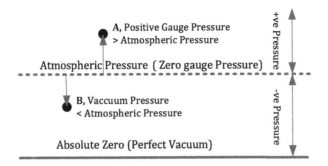

FIGURE 13.3 Absolute versus gauge pressure.

pressure, that is, pressure over and above atmospheric pressure. **Gauge pressure** is pressure with reference to atmospheric pressure. Therefore, a pressure below atmospheric pressure (vacuum) will be negative when expressed as gauge pressure. Until otherwise indicated, pressure is referred as gauge pressure. **Absolute pressure** is expressed with reference to absolute zero or perfect vacuum. Thus, absolute pressure is calculated by adding the atmospheric pressure on gauge pressure (Figure 13.3). As the name indicates, absolute pressure is always positive. Perfect vacuum corresponds to absolute zero pressure.

EXAMPLE PROBLEM 13.1

A pressure gauge attached at the bottom of a water tank reads 55 kPa. What is the height of water in the tank?

Given:

p = 55 kPa (kN/m²)   γ = 9.81 kPa/m   h =?

Solution:

Height of water in the tank

$$h = \frac{p}{\gamma} = 55\ kPa \times \frac{m}{9.81\ kPa} = 5.61 = \underline{5.6\ m}$$

EXAMPLE PROBLEM 13.2

A pressure gauge attached at the bottom of a water tank reads 15 psi. What is the height of water in the tank?

Given:

p = 15 psi   γ = 62.4 lb./ft³ = 2.31 ft/psi   h = ?

Solution:

Height of water in the tank

$$h = \frac{p}{\gamma} = 15 \ psi \times \frac{2.31 \ ft}{psi} = 34.65 = \underline{35 \ ft}$$

EXAMPLE PROBLEM 13.3

A cylindrical tank of 20 ft diameter is filled with 50,000 gal of water. What does the pressure gauge attached at the bottom of the tank read?

Given:

D = 20 ft   V = 50 000 gal   p, h = ?

Solution:

Height of water

$$h = \frac{V}{A} = 50000 \ gal \times \frac{4}{\pi (20 \ ft)^2} \times \frac{ft^3}{7.48 \ gal} = 21.27 = 21 \ ft$$

Hydrostatic pressure

$$p = \gamma \times h = \frac{0.433 \ psi}{ft} \times 21.27 \ ft = 9.21 = \underline{9.2 \ psi}$$

EXAMPLE PROBLEM 13.4

A 30-m-diameter cylindrical tank contains 6.0 ML of water. What does the pressure gauge attached at the bottom of the tank read?

Given:

D = 30 m   V = 6.0 ML = 6000 m³   p = ?

Solution:

Height of water

$$h = \frac{V}{A} = \frac{6.0 \ ML}{0.785 (30 \ m)^2} \times \frac{1000 \ m^3}{ML} = 8.49 = 8.5 \ m$$

Hydrostatic pressure

$$p = \gamma \times h = \frac{9.81 \ kPa}{m} \times 8.49 \ m = 83.2 = \underline{83 \ kPa}$$

EXAMPLE PROBLEM 13.5

The gauge on the suction side of a pump indicates a vacuum pressure of 25 cm of mercury. Express this pressure as gauge pressure and absolute pressure.

Given:

h = 25 cm of Hg   p =?   SG of Hg = 13.6

Solution:

Gauge pressure

$$p = \gamma \times h = SG \times \gamma_w \times h = 13.6 \times \frac{9.81\ kPa}{m} \times 25\ cm \times \frac{m}{100\ cm}$$

$$= 33.35 = 33(vacuum) = -33\ kPa$$

Absolute pressure

$$p_{abs} = -33\ kPa + 101\ kPa = 68\ kPa\ (abs)$$

EXAMPLE PROBLEM 13.6

A liquid chemical tank is 5.6 m in diameter and is filled to a height of 8.4 ft. If the SG of the liquid is 1.23, what is the hydrostatic pressure at the bottom of the tank?

Given:

h = 8.4 ft   SG = 1.23   p = ?

Solution:

Hydrostatic pressure

$$p = \gamma \times h = SG \times \gamma_w \times h = 1.23 \times \frac{psi}{2.31\ ft} \times 8.4\ ft = 4.47 = 4.5\ psi$$

EXAMPLE PROBLEM 13.7

A standpipe tank is 21 m in diameter and 8.5 m tall. What will the pressure be at the bottom when water held in storage is 2.0 ML?

Given:

h = 2.4 m   D = 21 m   V = 1.0 ML   p = ?

Solution:

Height of water in the tank

$$h_w = \frac{V}{A} = \frac{4}{\pi} \times \frac{3.0 \, ML}{(21 \, m)^2} \times \frac{1000 \, m^3}{ML} = 5.77 = 5.8 \, m$$

Pressure at the bottom

$$p = \gamma_w \times h_w = \frac{9.81 \, kPa}{m} \times 5.77 \, m = 56.6 = 57 \, kPa$$

EXAMPLE PROBLEM 13.8

A rectangular tank measures 8 ft by 12 ft. Water in the tank is 12 ft in depth. What is the pressure and force acting on the bottom of the tank?

Given:

A = 12 ft × 8.0 ft   d = 12 ft   p, F = ?

Solution:

Pressure at the bottom

$$p = \gamma \times d = \frac{0.433 \, psi}{ft} \times 12 \, ft = 5.19 = 5.2 \, psi$$

Force acting at the bottom

$$F = p \times A = \frac{5.19 \, lb}{in^2} \times \left(\frac{12 \, in}{ft} \right)^2 \times 8 \, ft \times 12 \, ft = 71829 = 72000 \, lb$$

EXAMPLE PROBLEM 13.9

What is the force acting at the bottom of a clarifier 25 m in diameter if the water level is 3.2 m deep?

Given:

D = 25 m   d = 3.2 m   F = ?

Solution:

Force acting at the bottom

$$F = \gamma \times d \times A = \frac{9.81 \, kN}{m^2} \times 3.2 \, m \times \frac{\pi}{4} \times (25 \, m)^2 = 15409 = 15\,000 \, kN = 15 \, MN$$

13.1.4 Lateral Pressure and Force

Calculating the force at the bottom of a tank is straightforward. The force at the bottom is expressed as:

Resultant force on the bottom

$$F_R = p \times A = \gamma \times h \times A$$

Where γ and h are weight density and height of the water in the tank open to atmosphere A is the area of the bottom.

Calculating the lateral force, that is, the force against the retaining wall or side of tank, is a bit different, as the pressure varies directly with height of the liquid. From the previous relationship, it can be said that pressure is zero at the water surface and maximum at the bottom, resulting in a triangular distribution, as shown in Figure 13.4. To calculate the total force, the average pressure occurring at half the depth should be used.

13.1.5 Average Pressure

Average pressure is the pressure acting midway or at half the depth of water.

$$p_{avg} = \frac{1}{2} \times \gamma \times d$$

Due to the triangular distribution, the centre of force is located along a line two thirds of the way down from the top surface, that is, the centroid of the triangle. This is because pressures in the lower half of the wall are greater.

Centre of force, $d_R = 2d/3$

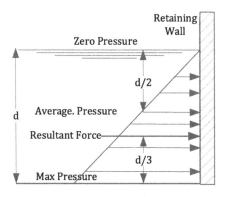

FIGURE 13.4 Lateral pressure on retaining wall.

13.1.6   RESULTANT FORCE

The resultant force is the average pressure multiplied by the lateral area, that is the face area of the wall.

Resultant force

$$F_R = p_{avg} \times A = \frac{1}{2} \times \gamma \times d^2 \times L$$

L refers to the length of the retaining wall.

EXAMPLE PROBLEM 13.10

Find the upward force on the bottom of an empty clarifier of diameter of 25 ft caused by groundwater depth of 8.5 ft above the tank bottom.

Given:

h = 8.5 ft   F = ?

Solution:

Upward force

$$F = p \times A = \gamma \times h \times A = \frac{62.4\ lb}{ft^3} \times 8.5\ ft \times \frac{\pi}{4} \times (25\ ft)^2 = 260359 = \underline{260\ 000\ lb}$$

EXAMPLE PROBLEM 13.11

Calculate the resultant force in kN and the centre of force on the side of a rectangular tank if the tank is 8.5 m long. The depth of water in the tank is 3.0 m.

Given:

L = 8.5 m   d = 3.0 m γ = 9.81 kN/m³ F_R = ?

Solution:

Resultant force

$$F_R = \frac{1}{2} \times \gamma \times d^2 \times L = \frac{1}{2} \times \frac{9.81\ kN}{m^3} \times (3.0\ m)^2 \times 8.5\ m = 750.4 = \underline{750\ kN}$$

Centre of force

$$d_R = \frac{2d}{3} = \frac{2}{3} \times 3.0\ m = \underline{1.0\ m}$$

13.2 CONTINUITY EQUATION

Based on the principle of **mass conservation**, under steady state conditions, the flow of water that enters the pipe is the same flow that exits the pipe. In mathematical form, this is known as the **continuity equation**. Water is relatively incompressible, so in terms of volume, the flow rate remains the same from one section to other. This is true so long as no additional flow is added to or exits the system. However, the velocity of the water would change if the area of the water section changes. At a given section, the flow velocity v is inversely proportional to flow area A. In other words, flow would slow down in a larger section, and flow would accelerate in a narrower section, as shown in Figure 13.5.

Continuity equation

$$\boxed{A_1 \times v_1 = A_2 \times v_2}$$

Flow velocity

$$\boxed{v = \frac{Q}{A} = \frac{4}{\pi} \times \frac{Q}{D^2}}$$

Velocity changes

$$\boxed{\frac{v_2}{v_1} = \frac{A_1}{A_2} \times \left(\frac{D_1}{D_2}\right)^2}$$

EXAMPLE PROBLEM 13.12

A flow of 4.5 ML/d is carried by a 200-mm-diameter pipe. What is the velocity of flow?

Given:

Q = 4.5 ML/d   D = 200 mm = 0.20 m

Solution:

$$v = \frac{Q}{A} = \frac{4.5\ ML}{d} \times \frac{1000\ m^3}{ML} \times \frac{d}{1440\ min} \times \frac{min}{60\ s} \times \frac{4}{\pi(0.20\ m)^2} = 1.66 = \underline{1.7\ m/s}$$

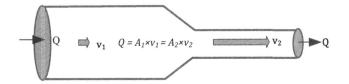

FIGURE 13.5   Continuity equation.

EXAMPLE PROBLEM 13.13

The velocity of flow in a 12-in-diameter pipe is 5.5 ft/s. If it discharges through a 3-in nozzle, what is the velocity of the jet issuing from the nozzle?

Given:

Parameter	Pipe = 1	Nozzle = 2
D, in	12	3
v, ft/s	1.6	?

Solution:

$$\frac{v_2}{v_1} = \frac{A_1}{A_2} = \left(\frac{D_1}{D_2}\right)^2 \quad or$$

$$v_2 = v_1 \times \frac{D_1^2}{D_2^2} = \frac{5.5 \ ft}{s} \times \left(\frac{12 \ in}{3 \ in}\right)^2 = 88.0 = \underline{88 \ ft/s}$$

EXAMPLE PROBLEM 13.14

Water is flowing at a velocity of 0.72 m/s in a 300-mm-diameter pipe. If the pipe changes from a 300-mm pipe to a 200-mm pipe, what will the velocity be in the 200-mm pipe? Also find the flow rate in L/s.

Given:

Parameter	Pipe = 1	Pipe = 2
D, mm	300	200
v, m/s	0.72	?

Solution:

Flow velocity in the smaller-diameter pipe

$$v_2 = v_1 \times \frac{D_1^2}{D_2^2} = \frac{0.72 \ m}{s} \times \left(\frac{300 \ mm}{200 \ mm}\right)^2 = 1.62 = \underline{1.6 \ m/s}$$

Water flow rate

$$Q = A_2 v_2 = \frac{\pi}{4} \times (D_2)^2 \times v_2 = \frac{\pi}{4} \times (0.20 \ m)^2 \times \frac{1.62 \ m}{s} \times \frac{1000 \ L}{m^3}$$

$$= 50.89 = \underline{51 \ L/s}$$

EXAMPLE PROBLEM 13.15

Water is flowing at a velocity of 4.2 ft/s in an 8.0-in-diameter pipe. If the pipe size changes from the 8.0-in pipe to a 10-in pipe, what will the velocity be in the 10-in pipe? What is the flow rate in gpm?

Given:

Parameter	Pipe = 1	Nozzle = 2
D, in	8	4.2
v, ft/s	10	?

Solution:

Flow velocity in the larger-diameter pipe

$$v_2 = v_1 \times \frac{D_1^2}{D_2^2} = \frac{4.2 \; ft}{s} \times \left(\frac{8 \; in}{10 \; in}\right)^2 = 2.68 = \underline{2.7 \; ft/s}$$

Water flow rate

$$Q = A_2 v_2 = \frac{\pi}{4} \times (D_2)^2 \times v_2 = \frac{\pi}{4} \times \left(\frac{8 \; ft}{12}\right)^2 \times \frac{2.68 \; ft}{s} \times \frac{7.48 \; gal}{ft^3} \times \frac{60 \; s}{min}$$

$$= 531.3 = \underline{530 \; gpm}$$

13.3 FORMS OF WATER ENERGY

The term **head** in hydraulics represents *energy per unit weight* of the liquid and has the same units as of length. As discussed earlier, the water pressure at a point can be represented by the equivalent column of water or other liquid.

Pressure head

$$h_p = \frac{E_{flow}}{w} = \frac{p \times V}{w} = \frac{p}{\gamma}$$

Energy in water can be due to pressure, position (elevation) and motion (velocity). As heads, they are called pressure head (h_p), elevation head (h_e) and velocity head (h_v). In mathematical terms, velocity head can be expressed as:

Velocity head

$$h_v = \frac{E_{kinetic}}{w} = \frac{1}{2} \times \frac{m \times v^2}{m \times g} = \frac{v^2}{2g}$$

Elevation head

$$h_e = \frac{E_{gravity}}{w} = \frac{m \times g \times z}{m \times g} = z$$

p = hydrostatic pressure, kPa   γ = weight density, 9.81 kN/m³
z = elevation or height from a datum,
v = flow velocity, m/s   g = acceleration due to gravity, 9.81 m/s²(32.2 ft/s²)

The total energy possessed by the water at a given point in the system is called total head, H. The total head is the sum of pressure, velocity and elevation heads.

Total head

$$H = h_p + h_v + h_e = \frac{p}{\gamma} + \frac{v^2}{2g} + z$$

Sometimes head terms are expressed as pressure. Velocity head when expressed as pressure is called **dynamic pressure**, elevation head expressed as pressure is called **hydrostatic pressure** and the pressure term is called **static pressure**.

EXAMPLE PROBLEM 13.16

What is the flow velocity head in a 6-in-diameter pipe carrying a flow of 500 gpm?

Given:

D = 6-in = 0.5 ft   Q = 500 gpm h_v = ?

Solution:

Flow velocity

$$v = \frac{Q}{A} = \frac{500\ gal}{min} \times \frac{ft^3}{7.48\ gal} \times \frac{4}{\pi(0.5\ ft)^2} \times \frac{min}{60\ s} = 5.67 = 5.7\ ft/s$$

Velocity head

$$h_v = \frac{v^2}{2g} = \frac{(5.67\ ft)^2}{s^2} \times \frac{s^2}{2 \times 32.2\ ft} = 0.499 = \underline{0.50\ ft}$$

EXAMPLE PROBLEM 13.17

Water is flowing at the rate of 50 L/s in a 150-mm-diameter water main. The water pressure in the pipe at a point 5.0 m above the datum is 350 kPa. Calculate the total head in m.

Given:

Q = 50 L/s   D = 150 mm = 0.150 m   p = 350 kPa   Z = 5.0 m

Solution:

Flow velocity

$$v = \frac{Q}{A} = \frac{50\ L}{s} \times \frac{m^3}{1000\ L} \times \frac{4}{\pi(0.15\ m)^2} = 2.82 = 2.8\ m/s$$

Velocity head

$$h_v = \frac{v^2}{2g} = \frac{(2.82\ m)^2}{s^2} \times \frac{s^2}{2 \times 9.81\ m} = 0.408 = \underline{0.41\ m}$$

Static head

$$h_p = \frac{P}{\gamma} = 350\ kPa \times \frac{m}{9.81\ kPa} = 35.67 = \underline{36\ m}$$

Total head

$$H = h_p + h_v + h_e = 35.67\ m + 0.408\ m + 5.0\ m = 41.08 = \underline{41\ m}$$

Note that kinetic head is negligiblely small as compared to other head values.

EXAMPLE PROBLEM 13.18

Water exits a fire nozzle with a flow velocity of 12 m/s. Express it as velocity pressure.

Given:

v = 12 m/s h_v = ?

Solution:

Velocity head

$$h_v = \frac{v^2}{2g} = \frac{(12\ m)^2}{s^2} \times \frac{s^2}{2 \times 9.81\ m} = 7.33 = 7.3\ m$$

Dynamic pressure

$$p = 7.33\ m \times \frac{9.81\ kPa}{m} = 72.0 = \underline{72\ kPa}$$

EXAMPLE PROBLEM 13.19

Water is flowing at the rate of 2.0 L/s in a 50-mm-diameter pipe. The water pressure in the pipe at a point 5.0 m above the datum is 300 kPa. What is the hydraulic head and total head?

Given:

Q = 2.0 L/s   D = 50 mm = 0.050 m   p = 300 kPa   Z = 5.0 m

Solution:

Flow velocity

$$v = \frac{Q}{A} = \frac{2.0 \, L}{s} \times \frac{m^3}{1000 \, L} \times \frac{4}{\pi (0.05 \, m)^2} = 1.02 = 1.0 \, m/s$$

Velocity head

$$h_v = \frac{v^2}{2g} = \frac{(1.02 \, m)^2}{s^2} \times \frac{s^2}{2 \times 9.81 \, m} = 0.0530 = 0.053 \, m$$

Dynamic pressure

$$p = \gamma \times h_v = 0.053 \times \frac{9.81 \, kPa}{m} = 0.519 = 0.52 \, kPa$$

Pressure head

$$h_p = \frac{p}{\gamma} = 300 \, kPa \times \frac{m}{9.81 \, kPa} = 30.58 = 31 \, m$$

Elevation head $h_e = z = 5.0 \, m$

Hydrostatic pressure

$$p_{hy} = \gamma \times h_e = 5.0 \, m \times \frac{9.81 \, kPa}{m} = 49.05 = 49 \, kPa$$

Hydraulic head
$$h = h_p + h_e = 30.58 \, m + 5.0 \, m = 35.58 = \underline{35.6 \, m}$$

Total head
$$H = h_p + h_v + h_e = 30.58 \, m + 0.053 \, m + 5.0 \, m = 35.63 = \underline{35.6 \, m}$$

Since velocity head is negligibly small, hydraulic head equals energy head.

EXAMPLE PROBLEM 13.20

A 12-in-diameter water main carries a flow of 750 gpm. The water pressure in the pipe at a point 50 ft above the datum is 45 psi. What is the hydraulic head and total head?

Given:

Q = 750 gpm   D = 12 in = 1.0 ft   p = 45 psi   Z = 50 ft

Solution:

Flow velocity

$$v = \frac{Q}{A} = \frac{750 \, gal}{min} \times \frac{ft^3}{7.48 \, gal} \times \frac{min}{60 \, s} \times \frac{4}{\pi (1.0 \, ft)^2} = 2.12 = 2.01 \, ft/s$$

Velocity head

$$h_v = \frac{v^2}{2g} = \frac{(2.12 \, ft)^2}{s^2} \times \frac{s^2}{2 \times 32.2 \, ft} = 0.070 = 0.070 \, ft$$

Pressure head

$$h_p = \frac{P}{\gamma} = 45 \, psi \times \frac{2.31 \, ft}{psi} = 103.9 = 104 \, ft$$

Hydraulic head

$$h = h_p + h_e = 103.9 + 50 \, ft = 153.9 = \underline{154 \, ft}$$

Total head

$$H = h + h_v = 153.95 + 0.07 \, ft = 154.0 = \underline{154 \, ft}$$

Again note that velocity head is negligibly small, hence hydraulic head and energy head have the same value.

PRACTICE PROBLEMS

PRACTICE PROBLEM 13.1

Calculate the pressure at a point 12 m below the water surface and express it as absolute pressure. (220 kPa)

PRACTICE PROBLEM 13.2

What is the pressure head in a pipeline carrying water at a pressure of 450 kPa? (46 m)

PRACTICE PROBLEM 13.3

A pipeline is designed on the basis of 20 cm of head loss for every 100 m length of pipe. How much pressure drop will be there over 1 km length of pipe? (20 kPa)

PRACTICE PROBLEM 13.4

What will be the total force acting on a retaining wall 5.0 m long when the depth of water is 2.0 m? (98 kN)

PRACTICE PROBLEM 13.5

What minimum diameter pipe is required to carry a flow of 2.0 L/s without exceeding the flow velocity of 1.5 m/s? (41 mm)

PRACTICE PROBLEM 13.6

What will be the flow carrying capacity of a 25-mm-diameter pipe when flowing at the rate of 2.0 m/s? How much flow can be carried by a 50-mm pipe (double the size) maintaining the same velocity of flow? (1.0 L/s, 4.0 L/s)

PRACTICE PROBLEM 13.7

A standpipe tank is 19 m in diameter and 8.5 m tall. How much water is held in storage when the pressure gauge attached to the bottom reads 65 kPa? (2.3 ML)

PRACTICE PROBLEM 13.8

What is the force acting at the bottom of a 15-ft-diameter alum storage tank if the level of the alum is 6.4 ft? The specific gravity of the alum is 1.23. (87,000 lb)

PRACTICE PROBLEM 13.9

A water rectangular tank measures 2.5 m by 4.5 m. The water in the tank is 2.1 m deep. What is the force at the bottom of the tank? (230 kN)

PRACTICE PROBLEM 13.10

Find the velocity head in a 8-in-diameter pipe carrying a flow of 650 gpm. (0.27 ft)

PRACTICE PROBLEM 13.11

For the data of Example Problem 13.8, calculate the total head at an elevation of 7.0 m where the pressure reading is 282 kPa (36 m).

PRACTICE PROBLEM 13.12

What is the maximum height a jet with flow velocity of 12 ft/s will rise in the air? (2.2 ft)

PRACTICE PROBLEM 13.13

Water is flowing at the rate of 32 L/s in a 150-mm-diameter pipe. The water pressure in the pipe at a point 15 m above the datum is 250 kPa. Find the total head. (41 m)

PRACTICE PROBLEM 13.14

Water is flowing at a velocity of 0.85 m/s in a 250-mm-diameter pipe. If the pipe changes from a 250-mm pipe to a 200-mm pipe, what will the velocity be in the 200-mm pipe? Also find the flow rate in L/s. (1.3 m/s, 42 L/s)

PRACTICE PROBLEM 13.15

Water is flowing at a velocity of 5.2 ft/s in a 10-in-diameter pipe. If the pipe size changes from the 10-in pipe to a 12-in pipe, what will the velocity be in the 12-in pipe? What is flow rate in gpm?(3.6 ft/s, 1300 gpm)

PRACTICE PROBLEM 13.16

A liquid chemical tank 18 ft in diameter is filled with a liquid chemical with SG of 1.34. A pressure gauge attached to the bottom of the tank reads 7.5 psi. What is the level of the liquid in the tank? (13 ft)

PRACTICE PROBLEM 13.17

A 200-mm-diameter water main is carrying water at the rate of 75 L/s. The water pressure in the main at a point 12 m above the datum is 280 kPa. Calculate the total head in m. (41 m)

PRACTICE PROBLEM 13.18

Water is coming out of a broken main at a flow velocity of 6.0 m/s. Express it as dynamic pressure. (18 kPa)

PRACTICE PROBLEM 13.19

In a 150-mm water main, water is flowing at the rate of 30 L/s. The water pressure main at a point 25 m above the datum is 290 kPa. What is the hydraulic head and total head? (55 m, 55 m)

PRACTICE PROBLEM 13.20

A 6-in-diameter water main carries a flow of 450 gpm. The water pressure in the pipe at a point 20 ft above the datum is 35 psi. Find the velocity head and hydraulic head. (0.32 ft, (100 ft)

14 Applied Hydraulics

14.1 BERNOULLI'S EQUATION

The principle of conservation of energy dictates that in a steady flow water system total energy or head remains the same; however, energy can change from one form to the other. A gain in elevation would cause drop in pressure and vice versa. Similarly, if the pipe diameter becomes larger resulting in a decrease in flow velocity, static pressure will increase. This principle of conservation of energy was put forward by Bernoulli. According to the principle of energy conservation in a given water system, the total energy remains the same.

According to **Bernoulli's principle**, the total head at various points of a water flow system remains the same. This is true if no fluid energy is added to or lost from the system between the two points in question. In the expanded form, this is known as Bernoulli's equation.

Bernoulli's equation

$$\frac{p_1}{\gamma} + Z_1 + \frac{v_1{}^2}{2\,g} = \frac{p_2}{\gamma} + Z_2 + \frac{v_2{}^2}{2\,g}$$

It is like saying you may have some money in different currencies. When you exchange currencies, you are converting one form of money into other. However, the buying power of your money remains the same. As seen in Figure 14.1, due to constriction at section 2, flow velocity increases by a factor of four, thus increasing the velocity head 16×, resulting in a drop in pressure. Since there is no change in elevation (horizontal position), any gain in kinetic energy causes the flow energy or pressure to drop in the same proportions.

14.1.1 HYDRAULIC GRADE LINE

Since hydraulic grade represents head due to elevation and pressure, at the constriction in Figure 12.1, there is a sudden drop in hydraulic grade line (HGL). If the pipe section reverts back to the original size, the pressure will recover, and it would cause the HGL to move back to its original position. In reality, this would never happen because there is always some loss of energy to overcome friction.

14.1.2 ENERGY GRADE LINE

In Figure 14.1, the energy grade remains horizontal to indicate the total head remains the same; only the form has changed. In real-life situations, energy grade line (EGL) will never be truly horizontal since there will be always some losses to overcome friction. Thus, the slope of the EGL would represent the rate at which friction losses

DOI: 10.1201/9781003468745-14

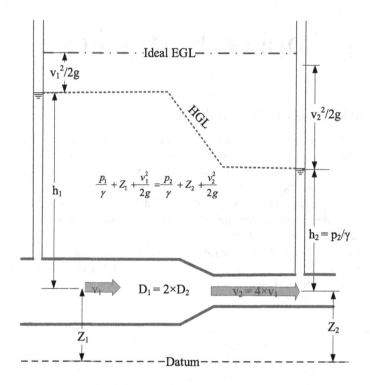

FIGURE 14.1 Bernoulli's principle.

take place. A steep EGL would mean high friction losses. When friction losses are considered, Bernoulli's equation is generalized and is called **general energy equation**.

EXAMPLE PROBLEM 14.1

Water is flowing from 1 to 2 in a given pipeline such that point 2 is 5.0 m above point 1. Determine the pressure in the line at point 2 if the water pressure at point 1 is 300 kPa.

Given:

$p_1 = 300$ kPa $(Z_2 - Z_1) = 5.0$ m $v_1 = v_2$

Solution:

Pressure downstream at point 2

$$p_2 = p_1 - \gamma(Z_2 - Z_1) = 300\,kPa - \frac{9.81\,kPa}{m} \times 5.0\,m = 251 = \underline{250\ kPa}$$

A gain in elevation causes a drop in pressure.

14.2 GENERAL ENERGY EQUATION

When there is motion (water is flowing), friction will come into play. Energy or head is lost from the fluid due to friction with the pipe surface (**major losses**) and other obstructions like valves and fittings (**minor losses**). These losses are not considered in Bernoulli's equation. They will limit the application of Bernoulli's equation to those systems in which the head losses due to friction (h_f) and minor losses (h_m) are not negligible. Furthermore, Bernoulli's equation is inapplicable to systems involving pumps required to add energy or head (h_a) to the water. Realizing that the real systems will involve the use of pumps and energy losses may be significant, Bernoulli's equation needs to be modified. When the terms for head added by the pump to water (h_a) and head loss (h_l) due to pipe friction and fittings are included, Bernoulli's equation becomes applicable to all real systems and thus is known as the **general energy equation**.

General energy equation

$$\boxed{\frac{p_1}{\gamma} + Z_1 + \frac{v_1^2}{2g} = h_a = \frac{p_2}{\gamma} + Z_2 + \frac{v_2^2}{2g} + h_l}$$

h_a = *head added by the pump*, h_l = *total head loss* = *major loss (h_f)* + *minor loss (h_m)*

Note each term in the general equation is that of *head* and should have the same units of length, like m or ft. It is illegal in the mathematical sense to add two

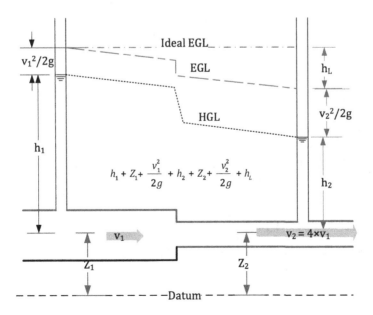

FIGURE 14.2 Energy concept.

quantities with different units. For example, 5 oranges and 5 apples don't make 10 apples or 10 oranges. By applying the energy equation between any two points in a water system, you can solve for one of the unknown quantity. If there is more than one unknown, it is not possible to solve the energy equation. Thus, the appropriate selection of the two points (sections) is very important. One of the sections should be such that pressure, velocity and elevation are all known quantities.

Not all the terms in the energy equation will be applicable to every water system. For example, if, at a given section, water is open to atmosphere, the pressure in terms of gauge pressure is zero. In larger tanks, the flow velocity is negligible and can be safely considered to be zero. If there is no pump between sections 1 and 2, the term h_a is zero. Depending on the length of the pipe between two sections and the number of accessories, head loss h_L can be estimated from tables or using a friction formula.

When applying the energy equation to solve for the unknown in a given hydraulic system, keep the following points in mind:

1. Pressure at a free surface (open to atmosphere) is zero.
2. Flow velocity in reservoirs and tanks is zero.
3. As indicated by the continuity equation, flow velocity is inversely proportional to the flow area. That is, velocity will be greater in smaller-diameter pipes. For example, if the diameter is reduced by half, water velocity increases four times.

$$\frac{v_1}{v_2} = \frac{A_1}{A_2} = \left(\frac{D_1}{D_2}\right)^2 = \frac{1}{2^2} = \frac{1}{4} \quad or \quad v_2 = 4 \times v_1$$

4. Head loss due to friction is relatively small if the length of the pipe between the two sections is small, and minor losses are negligible in water distribution systems when L/D >500.
5. Minor losses are usually estimated by multiplying velocity head by a loss coefficient.
6. The term h_a is zero when there is no pump between the two sections selected to write the energy equation.
7. Velocity head terms are usually negligible.
8. When applying the energy equation, first eliminate all the terms which are not applicable and then rearrange the terms to solve for the unknown.

EXAMPLE PROBLEM 14.2

The pressure drop observed over a pipeline length A to B is 40 kPa. What is the head loss in the line AB if point B is 1.5 m higher than point A?

Given:

$(p_A - p_B) = 40$ kPa $(Z_A - Z_B) = -1.5$ m $v_A = v_B$ $h_a = 0$ $h_l = ?$

Solution:

Head loss

$$h_l = Z_A - Z_B + \frac{p_A}{\gamma} - \frac{p_B}{\gamma} = -1.5 \ m + 40 \ kPa \times \frac{m}{9.81 \ kPa} = 2.57 = \underline{2.6 \ m}$$

EXAMPLE PROBLEM 14.3

A pump is pumping water from reservoir A to reservoir B. If the total head loss due to friction and minor losses is estimated to be 8.0 ft, determine the head added by the pump when the difference in elevation between the two water surfaces is 32 ft.

Given:

$(Z_2 - Z_1) = 32$ ft	$h_l = 8.0$ ft $h_a = ?$
$p_1 = p_2 = 0$ (Open)	$v_1 = v_2 = 0$ (Reservoir)

Solution:

Energy equation becomes
$$Z_1 + h_a = Z_2 + h_l$$

Head added
$$h_a = h_l + Z_2 - Z_1 = 8.0 \ ft + 32 \ ft = 40.0 = \underline{40 \ ft}$$

14.3   FLOW EQUATION

In the general energy equation, one of the terms is **head loss**. When applying the general energy equation, if the unknown is other than head loss h_l, then head losses must be estimated. A flow equation relates pipe characteristics including diameter, roughness, length to the flow rate and head loss due to friction, h_f. The three equations commonly in use are: Darcy–Weisbach, Hazen–Williams and Manning's equation. In addition to head losses due to friction, losses occur due to sudden changes in flow velocity as in the case of bends and valves. Such losses are called **minor losses, h_m**. Head losses are estimated using tables or the **equivalent pipe** concept. Equivalent length L_e is the length of straight pipe that will give the same head loss due to friction as the elbow or valve in question. This way, the same formula can be used by adding the equivalent length to the actual length of straight pipe.

14.3.1   Darcy–Weisbach Flow Equation

The Darcy–Weisbach flow equation is theoretical and can be used for any consistent units. Friction factor f in this equation depends on **Reynolds number** and **relative**

roughness. This equation yields more precise results and is applicable to all kinds of fluids and flow conditions.

Darcy–Weisbach equation (consistent units)

$$h_f = \frac{f}{1.23g} \times L \times \frac{Q^2}{D^5}, \quad Q = \sqrt{\frac{1.23g}{f} \times \frac{h_f}{L} \times D^5}, \quad D = \sqrt[5]{\frac{f}{1.23g} \times \frac{1}{S_f} \times Q^2}$$

In this equation, D, L are the diameter and length of the pipeline, and Q is the flow rate. The term h_f represents head loss due to friction, and friction factor f is a dimensionless number. It is based on two dimensionless parameters, relative roughness and Reynolds number.

$$Reynolds\ Number, N_R = \frac{velocity \times diameter}{kinematic\ viscosity} = \frac{vD}{v}$$

$$Relative\ Roughness, RR = \frac{Diameter}{Absolute\ roughness} = \frac{D}{\varepsilon}$$

$$Friction\ slope, S_f = \frac{Head\ loss}{Length\ of\ the\ conduit} = \frac{h_f}{L}$$

In turbulent flow conditions, the friction factor is dependent on the Reynolds number and the relative roughness of the pipe. The absolute roughness of a pipe is measured in terms of the average thickness of irregularities on the surface of the pipe and is denoted by the symbol ε (epsilon). To make it a dimensionless parameter, it is expressed as relative roughness, which is the ratio of the diameter of the pipe to the absolute roughness. Typical values of absolute roughness for commonly used pipe materials are shown in Table 14.1.

TABLE 14.1
Absolute Roughness

Pipe Material	Roughness, ε μm
Glass, plastic	Smooth
Copper, brass	1.5
Commercial steel	46
Cast iron, 1 uncoated	240
2 coated	120
Concrete	1200
Riveted steel	1800

14.3.1.1   Friction Factor f

The friction factor in the Darcy–Weisbach flow equation can be found using the following empirical relationship.

Friction factor (turbulent flow)

$$f = 0.0055 + 0.0055 \times \sqrt[3]{\frac{20000}{D/\varepsilon} + \frac{1000000}{N_R}}$$

EXAMPLE PROBLEM 14.4

What head loss can be expected when an 8-in water main (f = 0.02) is carrying a flow of 500 gpm? The length of the water main in question is 840 ft, and for losses due to fittings, assume an equivalent length of 50 ft.

Given:

D = 8-in = 0.667 ft   Q = 450 gpm $L + L_e$ = 840 ft + 50 ft = 890 ft h_l = ?

Solution:

Flow rate

$$Q = \frac{500\,gal}{min} \times \frac{min}{60\,s} \times \frac{ft^3}{7.48\,gal} = 1.114 = 1.1\,ft^3\,/\,s$$

Head loss

$$h_l = \frac{f(L+L_e)}{1.23g} \times \frac{Q^2}{D^5} = \frac{0.02}{1.23} \times \frac{s^2}{32.2\,ft} \times 890\,ft \times \left(\frac{1.114\,ft^3}{s}\right)^2 \times \frac{1}{(0.667\,ft)^5}$$

$$= 4.21 = \underline{4.2\,ft}$$

EXAMPLE PROBLEM 14.5

Find the head losses in the piping system of a pump station when the flow rate is 65 L/s. The 200-mm suction line is 4.5 m long and contains a bend and a gate valve reducer. The equivalent length for the accessories is estimated to be 12 m. The 150-mm discharge line is 10 m long and contains a check valve, a gate valve and three bends with total equivalent length of 25 m. Assume a friction factor of 0.022 for all pipes.

Given:

Line	Q, L/s	D, mm	L, m	L_e, m	$(L+L_e)$
Suction	65	200	4.5	12	16.5
Discharge	65	150	12	25	37

Solution:

Pumping rate

$$Q = \frac{65\,L}{s} \times \frac{m^3}{1000\,L} = 0.065\ m^3/s$$

Suction line

$$h_1 = \frac{f(L+L_e)}{1.23g} \times \frac{Q^2}{D^5} = \frac{0.022}{1.23} \times \frac{s^2}{9.81\,m} \times 16.5\ m \times \left(\frac{0.065\ m^3}{s}\right)^2 \times \frac{1}{(0.20\ m)^5}$$

$$= 0.397 = \underline{0.40\ m}$$

Discharge line

$$h_1 = \frac{f(L+L_e)}{1.23g} \times \frac{Q^2}{D^5} = \frac{0.022}{1.23} \times \frac{s^2}{9.81\,m} \times 37\ m \times \left(\frac{0.065\ m^3}{s}\right)^2 \times \frac{1}{(0.15\ m)^5}$$

$$= 3.75 = \underline{3.8\ m}$$

Note that head loss in the suction line is relatively small since the pipe diameter is one size larger than the discharge or delivery side. This is done by design to prevent cavitation.

EXAMPLE PROBLEM 14.6

For the data of Example Problem 14.5, find the head added, assuming that the static lift is 22 m. Based on overall efficiency of 71%, find the power input and daily cost of pumping at the rate of $0.25/kW.h. Assume this pump is operated 20 h/d.

Given:

$Q = 65\ L/s$ $\text{Lift} = Z_2\text{-}Z_1 = 12\ m$, $E_o = 71\%$, $\text{Rate} = \$0.25/kW\cdot h$ Operation: 20 h/d

Solution:

Head added

$h_a = h_l + Z_2 - Z_1 = 0.40\ m + 3.75\ m + 22\ m = 26.2\ m = \underline{26\ m}$

Power input

$$P_i = \frac{P_a}{E_o} = \frac{Q\gamma h_a}{E_o} = \frac{0.065\,m^3}{s} \times \frac{9.81\,kN}{m^3} \times \frac{26.2\,m}{0.71} \times \frac{kW.s}{kN.m} = 23.5 = \underline{24\ kW}$$

Cost of pumping

$$\text{Cost} = 23.5\ kW \times \frac{20h}{d} \times \frac{\$0.25}{kW.h} = 118 = \underline{\$120/d}$$

14.3.2 Hazen–Williams Flow Equation

The Darcy–Weisbach equation, when applied to solve for flow capacity, Q and similar problems, is cumbersome, and here the Hazen–Williams equation becomes handier. Therefore, empirical flow equations, which provide direct solutions, are commonly used. This equation relates the flow carrying capacity (Q) with the size (D) of the pipe, slope of the hydraulic gradient (S_f) and a coefficient of friction C which depends on the roughness of the pipe. However, it is important to note that the value of C is based on the judgement of the designer. Results based on this equation would be less accurate compared to the Darcy–Weisbach flow equation.

Hazen–Williams flow equation

$$Q = 0.278CD^{2.63} \times S_f^{0.54} \quad or \quad h_f = 10.7 \times \left(\frac{Q}{C}\right)^{1.85} \times \frac{L}{D^{4.87}} \quad - SI$$

Hazen–Williams flow equation

$$Q = 0.81CD^{2.63} \times S_f^{0.54} \quad or \quad h_f = 10.4 \times \left(\frac{Q}{C}\right)^{1.85} \times \frac{L}{D^{4.87}} \quad - USC$$

Q = Volume flow rate, m³/s, (gpm) C = Hazen–Williams friction coefficient
D = Diameter of pipe, m (in) S_f = friction slope = $h_f/L = \Delta h/L$

The Hazen–Williams flow equation is valid when the flow velocity is less than 3 m/s and the pipe diameter is larger than 2 cm. The values of the roughness coefficient C for selected pipe materials are given in Table 14.2. In water distribution hydraulics,

---

TABLE 14.2
Hazen–Williams Coefficient, C

Pipe Material	C
Asbestos cement	140
Cast iron	
-cement, lined	130–150
-new, unlined	130
-5-year-old, unlined	120
-20-year-old, unlined	100
Concrete	130
Copper	130–140
Plastic	140–150
Commercial steel	120
New riveted steel	110

the Hazen–Williams flow equation is commonly used. In this equation, the roughness of the pipe is indicated by the roughness coefficient C, which varies from 50 to 150, with higher values indicating smoother pipes.

Typical C values for water mains are in the range of 80–130 depending on age, degree of incrustation and tuberculation. Since this is an empirical relationship, the constant in the formula will change if the diameter of the pipe is in units other than m. The parameter friction slope is a dimensionless parameter and indicates head loss due to friction per unit length of the flow pipe. If instead of head loss, pressure loss is given, make sure to convert it to head units.

Friction slope/gradient

$$S_f = \frac{h_f}{L} = \frac{\Delta h}{L} = \frac{\Delta p}{\gamma} \times \frac{1}{L}$$

14.3.3 Flow Capacity Q

The flow carrying capacity of a given pipe size for a maximum permissible head loss or friction slope can be calculated using the flow equation. Note that in Hazen–Williams' equation, the exponent of the parameter diameter D is 2.63 to indicate an increase in pipe size results in a much greater increase in flow capacity.

Hazen–Williams equation (SI)

$$Q = 0.278\, C \times D^{2.63} \times S_f^{\,0.54} \quad D = m \;\; Q = m^3/s$$

Hazen–Williams equation (USC)

$$Q = 0.281\, C \times D^{2.63} \times S_f^{\,0.54} \quad D = in \;\; Q = gpm$$

14.3.4 Head Loss

For a given pipe size (D) and material (C) and length of the conduit, head loss due to friction can be found for various flow (Q) conditions. For the maximum flow, the expected pressure drop can be found.

Head loss (Hazen–Williams)

$$h_f = 10.7 \times \left(\frac{Q}{C}\right)^{1.85} \times \frac{L}{D^{4.87}} \quad - SI$$

$$h_f = 10.4 \times \left(\frac{Q}{C}\right)^{1.85} \times \frac{L}{D^{4.87}} \quad - D = in \;\; Q = gpm$$

EXAMPLE PROBLEM 14.7

Estimate the pressure loss due to friction in a 1.6-mile-long, 12-in-diameter water main with a coefficient C of 120 while carrying a flow of 750 gpm?

Given:

$C = 120$ $Q = 750$ gpm $D = 12$ in $L = 1.6$ mile $h_f = ?$

Solution:

Head loss

$$h_f = 10.4 \times \left(\frac{Q}{C}\right)^{1.85} \times \frac{L}{D^{4.87}} = 10.4 \times \left(\frac{750}{120}\right)^{1.85} \times \frac{1.6 \times 5280 \, ft}{12^{4.87}}$$

$$= 13.56 = 14 \, ft$$

Pressure loss

$$\Delta p = 13.56 \, ft \times \frac{0.433 \, psi}{ft} = 5.87 = \underline{5.9 \, psi}$$

EXAMPLE PROBLEM 14.8

Estimate the hydraulic gradient/frictional slope in a 300-mm-diameter water main with a friction factor of 120 and carrying a flow of 100 L/s?

Given:

$C = 120$ $Q = 100$ L/s $= 0.10$ m³/s $S_f = ?$

Solution:

Friction slope

$$S_f = 10.7 \times \left(\frac{Q}{C}\right)^{1.85} \times \frac{1}{D^{4.87}} = 10.7 \times \left(\frac{0.10}{120}\right)^{1.85} \times \frac{1}{0.30^{4.87}}$$

$$= 7.57 \times 10^{-3} = 7.57 \times 10^{-3} \times \frac{1000 \, m}{km}$$

$$= 7.57 = \underline{7.6 \, m/km}$$

EXAMPLE PROBLEM 14.9

Estimate the flow capacity of a 300-mm-diameter main for an allowable friction slope of 0.1%. Assume the Hazen–Williams coefficient of friction is 120, a relatively smooth pipe.

Given:

D = 300 m = 0.3 m $S_f = 0.1\% = 0.001$ C = 120

Solution:

Flow carrying capacity

$$Q = 0.278C \times D^{2.63} \times S_f^{0.54}$$
$$= 0.278 \times 120 \times 0.30^{2.63} \times 0.001^{0.54}$$
$$= 3.37 \times 10^{-2} = 3.4 \times 10^{-3} m^3 / s = \underline{34\ L/s}$$

EXAMPLE PROBLEM 14.10

Estimate the flow carrying capacity of an 8-in water main for an allowable friction loss of 6.0 ft/mile length of pipe. Assume the Hazen–Williams roughness coefficient to be 100.

Given:

D = 8 in S_f = 6.0 ft/mile   C = 100

Solution:

Friction slope

$$S_f = \frac{6.0\,ft}{mile} \times \frac{mile}{5280\,ft} = 0.113\%$$

Flow carrying capacity

$$Q = 0.281C \times D^{2.63} \times S_f^{0.54} = 0.281 \times 100 \times 8^{2.63} \times 0.00136^{0.54}$$
$$= 171.3 = \underline{170\,gpm}$$

PRACTICE PROBLEMS

PRACTICE PROBLEM 14.1

For the same data find the pressure at a point 5.0 m below the datum. (350 kPa)

PRACTICE PROBLEM 14.2

If the head loss in 1 km length of pipe is known to be 5.5 m, calculate the pressure drop assuming the pipe is horizontal. (54 kPa)

PRACTICE PROBLEM 14.3

For the data of Example Problem 14.3, if reservoir B is a pressurized tank such that the air pressure at the water surface is 150 kPa, how many m of head will be added by the pump? (26.9 m)

PRACTICE PROBLEM 14.4

Readings taken from the inlet and outlet pressure gauges of a pumping system, respectively, are 45 kPa and 450 kPa. Calculate TDH in m. (41 m)

PRACTICE PROBLEM 14.5

Find the head losses in the piping system of a pump for a flow rate of 48 L/s. The suction and discharge pipeline parameters are shown in the following table. Assume a Darcy friction factor of 0.020 for all pipes. (15 m)

Line	D, mm	L, m	L_e, m
Suction	150	5.0	10
Discharge	100	14	22

PRACTICE PROBLEM 14.6

For the data of Practice Problem 14.5, find the pumping head, assuming a lift of 25 m. Also determine the daily cost pumping at the rate of $0.22/kW.h. Assume the pump is operated 22 h/d, the electric motor is 92% efficient and the pump is operating at an efficiency of 65%. ($160/d)

PRACTICE PROBLEM 14.7

A pump operating against a head of 65 ft discharges at the rate of 850 gpm. If the overall operating efficiency is 70%, determine the water horsepower and input power. (20 hp)

PRACTICE PROBLEM 14.8

A pump is 82% efficient when pumping at the rate of 50 L/s against a head of 50 m. Calculate the power needed to supply to the motor, assuming the motor efficiency is 90%. (33 kW)

PRACTICE PROBLEM 14.9

The motor power requirement is calculated to be 35 kW. Based on average operation of 150 hours per week, calculate the power cost per week for the pump operation. Assume an energy rate of $0.045/unit. ($ 240/week)

PRACTICE PROBLEM 14.10

Water at the flow rate of 80 L/s is to be pumped against a pressure of 220 kPa. What is the power input required if the overall (wire to water) efficiency is 66%? What is the power cost for 16 hours of pumping if the unit cost is 8¢ per kW.h? (26.7 kW, $34)

15 Pump Performance

15.1 TYPES OF PUMPS

A wide variety of pumps are used to transport water in water and wastewater plant operations. As discussed earlier, pumps are hydraulic machines, which convert the mechanical power into waterpower. Two broad classes of pumps are positive displacement pumps and kinetic or velocity pumps. A brief description follows:

15.2 POSITIVE DISPLACEMENT PUMPS

These pumps are designed to deliver a fixed quantity of fluid for each revolution of the pump rotor. Therefore, except for minor slippage, *the delivery of the pump is unaffected by changes in the delivery pressure.* In general, these pumps will pump against high pressure, but their capacity is low. These pumps are well suited for pumping high-viscosity liquids and can be used for metering since output is directly proportional to rotative speed.

The capacity of positive displacement pumps is *independent of the delivery pressure.* At high pressures a small decrease in discharge may occur due to internal leakage. These characteristics make them suitable for use as chemical feed pumps. Since the capacity is practically constant at a given speed, the power required to drive the pump varies linearly with pressure. Therefore, it becomes necessary to protect positive displacement pumps with relief valves to prevent damage by overpressurization.

15.3 VELOCITY PUMPS

As the name indicates, this category of pump does add to the kinetic energy of the water, which is later converted to pressure energy. A centrifugal pump is the most commonly used velocity pump. One advantage of centrifugal pump is that the discharge valve can be closed without damaging the pump. Thus, if for some reason discharge becomes blocked (increase in discharge pressure), the pump will not be damaged. As in case of positive displacement pumps, it is not necessary to protect this pump with a relief valve. In fact, it is recommended to start centrifugal pumps with the discharge valve closed. As the speed is picked up, valve is slowly opened. The only effect of a blockage (shut-off head) is churning of the fluid, with a subsequent rise in the water temperature.

15.3.1 PUMPING HEAD

Pumps are used to add energy to water fluid. The amount of energy added (total dynamic head, TDH) can be calculated using the energy equation as follows:

Head added by the pump (TDH)

$$h_a = \frac{v_2^2}{2g} - \frac{v_1^2}{2g} + h_1 + Z_2 - Z_1 + \frac{p_2}{\gamma} - \frac{p_1}{\gamma}$$

DOI: 10.1201/9781003468745-15

The subscripts 1 and 2 refer to any two points in the direction of flow that the pump lies in between. Locations 1 and 2, respectively, are suction and discharge side of the pump.

15.4 TOTAL DYNAMIC HEAD

Total dynamic head is the head against which a pump must operate under the dynamic (pumping) conditions. This is also called the total head or the pumping head, as seen in Figure 15.1. In fact, this is the head added (h_a) to the water, as discussed earlier.

Head added by the pump, TDH

$$h_a = \frac{(v_2^2 - v_1^2)}{2g} + h_1 + (Z_2 - Z_1) + \frac{(p_2 - p_1)}{\gamma}$$

For most practical situations, term expressing difference in velocity head is negligibly small. Hence, for most practical applications, pumping head can be found by neglecting velocity head terms without losing accuracy.

Head added

$$h_a = h_1 + (Z_2 - Z_1) + \frac{(p_2 - p_1)}{\gamma}$$

In this equation, h_a head added is called the *pumping head* or *total dynamic head*. The pumping head is equal to the difference in the pressure head reading of suction, and delivery gauges are installed immediately before and after the pump. If the two

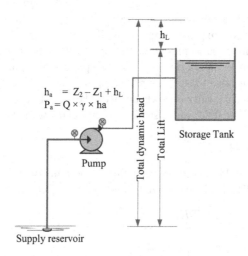

FIGURE 15.1 Pumping head.

gauges are read at the same elevation, $Z_2 - Z_1 = 0$. When the suction and discharge pipe diameters are the same, $v_1 = v_2$. Even when the sizes are different, the *velocity head term is relatively very small*. Under these conditions, the total dynamic head provided by the pump can be directly calculated from the pressure readings. The head loss in the pump body is considered part of pump efficiency.

Head added

$$h_a = \frac{p_2}{\gamma} - \frac{p_1}{\gamma} = \frac{(p_2-p_1)}{\gamma} = h_2 - h_1$$

The pressure reading on the suction side is usually negative. Only when the pumping water level is above the pump will p_1 be positive. When the gauge reads negative or vacuum pressure, the difference in pressure is obtained by adding the two readings. If the pressure is read as the height of another liquid, make sure to express it as column of water.

EXAMPLE PROBLEM 15.1

The pressure gauges attached to the inlet and outlet side of a centrifugal pump delivering water read 25 kPa vacuum and 150 kPa, respectively. Determine the total head added to water by the pump.

Given:

p_1 = 25 kPa vacuum = − 25 kPa p_2 = 150 kPa $z_2 - z_1 = 0$ $v_2 = v_1$

Solution:

Head added

$$h_a = \frac{(p_2-p_1)}{\gamma} = (150+25)kP_a \times \frac{m}{9.81kP_a} = 17.8 = \underline{18\ m}$$

EXAMPLE PROBLEM 15.2

During testing of a centrifugal pump, the suction and discharge gauge readings were recorded to be 28 cm of mercury vacuum and 98 kPa, respectively. What is the total head provided by the pump?

Given:

h_1 = −28 cm of Hg p_2 = 98 kPa   1 = suction   2 = delivery h_a =?

Solution:

In this case, suction pressure is read directly as head but as the height of mercury rather than water. Mercury is 13.6 times as heavy as water; thus, the equivalent column of water will be 13.6 times as much.

Suction head

$$h_a = -0.28 \ m \ Hg \times \frac{13.6 \ m \ water}{1 m \ of \ Hg} = -3.808 = \underline{-3.8 \ m}$$

Discharge head

$$h_2 = \frac{p_2}{\gamma} = 98 \ kPa \times \frac{m}{9.81 kPa} = 9.99 = \underline{10.m}$$

Head added

$$h_a = h_2 - h_1 = 9.99 \ m - (-3.808 \ m) = 13.8 = \underline{14 \ m}$$

EXAMPLE PROBLEM 15.3

The suction and discharge pressure gauge readings of a given pump respectively are 1.5 psi vacuum and 6.5 psi. The gauges are at the same elevation. What is the dynamic head of the pump?

Given:

Variable	Suction = 1	Discharge = 2
Pressure, psi	−1.5	6.5
Elevation, ft	$(Z_2 - Z_1) = 0$	
Pressure added, $p_{a,}$ psi	= 6.5 − (−1.5) = 8.0	

Solution:

Head added

$$h_a = \frac{p_2}{\gamma} - \frac{p_1}{\gamma} + Z_2 - Z_1 + h_1 = \frac{p_a}{\gamma} = 8.0 \ psi \times \frac{2.31 \ ft}{psi} = 18.48 = \underline{18 \ ft}$$

EXAMPLE PROBLEM 15.4

In a pumping system, the suction gauge reads 17 in of mercury column (vacuum), and the discharge gauge reads 45 psi. Assuming the discharge gauge is 1.5 ft above the suction gauge, find the pumping head.

Given:

Variable	Suction = 1	Discharge = 2
Pressure	$p_1/\gamma = -17 \ in \ Hg$	$p_2 = 45 \ psi$
Elevation	$(Z_2 - Z_1) = 1.5$	

Solution:

Head added

$$h_a = \frac{p_2}{\gamma} - \frac{p_1}{\gamma} + Z_2 - Z_1 + h_1 = \frac{p_2}{\gamma} - \frac{p_1}{\gamma} + Z_2 - Z_1$$

$$= \frac{45\,psi.\,ft}{0.433\,psi} - \left(-17in\,Hg \times \frac{13.6\,in\,H_2O}{1\,in\,Hg} \times \frac{ft}{12\,in} \right) + 1.5\,ft = 124 = \underline{120\,ft}$$

EXAMPLE PROBLEM 15.5

Water is being pumped from a water source with an elevation of 122.8 m to an elevation of 167.6 m. What is the total head if friction and minor head losses are 5.5 m?

Given:

$p_1 = p_2 = 0$ $Z_1 = 122.8$ m $Z_2 = 167.6$ m $h_1 = 5.5$ m

Solution:

Head added

$$h_a = \frac{p_2}{\gamma} - \frac{p_1}{\gamma} + Z_2 - Z_1 + h_1 = Z_2\_Z_1 + h_1 = (167.6 - 122.8)m + 5.5\,m$$

$$= 50.3 = \underline{50\,m}$$

15.5 PUMP POWER

Power is the rate of doing work or the rate at which energy is being spent.

Power and energy

$$\boxed{P = \frac{Energy}{t} = \frac{work}{t}}$$

Units of power are that of energy (J) divided by time (s), or J/s. Power of 1 J/s is also called 1 watt, named for James Watt, who invented the steam engine and used the term horsepower. The imperial unit of power is horsepower, which is equivalent to 550 ft.lb/s.

Power added to water, also called waterpower, can be determined by knowing the water pumping rate Q and the head added h_a, also called total dynamic head. Since head is the energy per unit weight, power added to the water can be calculated by multiplying h_a by the weight flow rate of water W.

Power added (waterpower)

$$\boxed{P_a = W \times h_a = Q \times \gamma \times h_a = Q \times p_a}$$

where p_a is the pressure added to water by the pump. Pressure added is the difference in pressure gauge readings of the pump inlet and outlet. The efficiency of a machine is the ratio of output to input. For example, to add a given power to water, the pump will need more power from the motor. This is because some power is lost due to slippage and friction of the shaft and other factors. The efficiency of a pump or electric motor is the ratio of output to input as power or energy.

Machine efficiency

$$E_{m/c} = \frac{OutnputPower}{InputPower}$$

In the case of a pump, output power is power added to water (P_a), and input power is the power given to the pump by the prime mover (P_1). Pump efficiency is measure of these losses; the higher the losses, the lower the pump efficiency. A typical range of pump efficiency is 50 to 85%.

Pump efficiency

$$E_p = \frac{Power\,added}{Shaft\,Power} = \frac{P_a}{P_m}$$

Motor efficiency

$$E_m = \frac{Pump\,power}{Power\,Supplied} = \frac{P_p}{P_1}$$

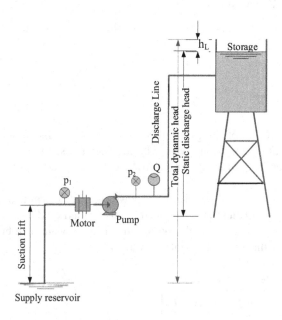

FIGURE 15.2 Simple pumping system.

If the motor and pump are not directly coupled, we should also consider the efficiency of the drive unit. Efficiency of the total system E_O will be the power output, that is, power delivered to the fluid divided by the power input to the system. Power input to the system is the power consumed by the system, as indicated by the watt meter.

Overall efficiency

$$\boxed{E_m = \frac{Power\ added}{Power\ Supplied} = \frac{P_a}{P_1}}$$

For example, a pumping system consisting of a pump with 60% efficiency and a 90% efficient electric motor will have an overall efficiency of 54%.

EXAMPLE PROBLEM 15.6

A centrifugal pump pumping water at the rate of 300 gpm adds a pressure of 35 psi. What power is added to water?

Given:

$p_a = 35$ psi $Q = 500$ gpm $P_a = ?$

Solution:

Power added

$$P_a = Q \times p_a = \frac{500\,gal}{min} \times \frac{35\,lb}{in^2} \times \frac{hp.s}{550\,lb.\,ft} \times \frac{144\,in^2}{ft^2} \times \frac{min}{60\,s} \times \frac{ft^3}{7.48\,gal}$$
$$= 10.2 = \underline{10.\,hp}$$

Hence the pump would require more than 10 hp to run.

EXAMPLE PROBLEM 15.7

Estimate the power added to the water (waterpower) by the pump when pumping at the rate of 3.5 L/s against a head of 35 m. Also find the pump power assuming the pump is 75% efficient.

Given:

$Q = 3.5$ L/s $h_a = 35$ m $P_a = ?$ $E_p = 75\%$ $P_p = ?$

Solution:

Power added

$$P_a = Q \times \gamma \times h_a = \frac{3.5\,L}{s} \times \frac{m^3}{1000\,L} \times \frac{9.81\,kN}{m^3} \times 35\,m \times \frac{kW.s}{kN.m} = \underline{1.2\,kW}$$

Pump Power

$$P_p = \frac{P_p}{E_p} = \frac{1.20\,kW}{75\%} \times 100\% = 1.60 = \underline{1.6\,kW}$$

EXAMPLE PROBLEM 15.8

A pump has a capacity of 350 gpm and lifts water 21 ft (total head). If the pump efficiency is 80%, what size (rated power) motor is required?

Given:

Q = 7500 gpm h_a = 21 ft E_p = 75% P_p = ?

Solution:

Power added

$$P_a = Q \times \gamma \times h_a = \frac{7500\,gal}{min} \times \frac{8.34\,lb}{gal} \times \frac{min}{60s} \times 21\,ft \times \frac{hp.s}{550\,ft.lb} = 39.8 = 40\,hp$$

Power to pump or shaft power

$$P_p = \frac{P_a}{E_p} = \frac{39.8}{75\%} \times 100\% = 53.07 = \underline{53\,hp}$$

EXAMPLE PROBLEM 15.9

A pumping unit draws 75 A at 220 V when pumping @ 35 L/s against a head of 31 m. What is the overall efficiency of the pumping unit?

Given:

I = 75 A   E = 220 V   Q = 35 L/s h_a = 31 m E_o = ?

Solution:

Power added

$$P_a = Q \times \gamma \times h_a = \frac{0.035m^3}{s} \times \frac{9.81\,kN}{m^3} \times 31m \times \frac{kW.s}{kN.m} = 10.6 = 11\,kW$$

Overall efficiency

$$E_O = \frac{Power\,added}{Input\,Power} = \frac{P_a}{p_i} = \frac{10.6\,kW}{220\,V \times 75A} \times \frac{V.A}{W} \times \frac{1000}{kW} \times 100 = 64.24 = \underline{64\%}$$

EXAMPLE PROBLEM 15.10

If the power supplied to a pumping unit is 10 kW, what is the brake power (pump) and waterpower (power added) given the motor is 90% efficient and the pump is 80% efficient?

Given:

$P_m = 10$ kW $\quad E_m = 90\%$ $\quad E_p = 80\%$ $\quad P_a = ?$ $\quad P_P = ?$

Solution:

Pump power

$$P_P = P_m \times E_m = 10\ kW \times \frac{90\%}{100\%} = 9.00 = \underline{9.0\ kW}$$

Power added

$$P_a = P_p \times E_P = 9.0\ kW \times 0.80 = 7.20 = \underline{7.2\ kW}$$

EXAMPLE PROBLEM 15.11

A pump operating against a head of 82 ft pumps at the rate of 1100 gpm. If the overall operating efficiency is 65%, determine the water horsepower and input power.

Given:

$Q = 1200$ gpm $\quad h_a = 82$ ft $\quad E_p = 65\%$ $\quad E_m = 90\%$ $\quad P_I = ?$

Solution:

Power added

$$P_a = Q \times \gamma \times h_a = \frac{1100\ gal}{min} \times \frac{min}{60s} \times \frac{8.34\ lb}{gal} \times 82\ ft \times \frac{hp.s}{550\ lb.\ ft} = 22.79 = \underline{23\ hp}$$

Power input

$$P_I = \frac{P_a}{E_O} = \frac{22.79\ hp}{65\%} \times 100\% = 35.07 = \underline{35\ hp}$$

EXAMPLE PROBLEM 15.12

A pump operating against a head of 21 m pumps at the rate of 65 L/s. If the pump operating efficiency is 75% and motor efficiency is 90%, determine the input power.

Given:

$Q = 65$ L/s $\quad h_a = 21$ m $\quad E_p = 75\%$ $\quad E_m = 90\%$ $\quad P_m = ?$

Solution:

Power added

$$P_a = Q \times \gamma \times h_a = \frac{65\ L}{s} \times \frac{m^3}{1000\ L} \times \frac{9.81\ kN}{m^3} \times 21\ m \times \frac{kW.s}{kN.m} = 13.39 = 13\ kW$$

Pump or shaft power

$$P_P = \frac{P_a}{E_p} = \frac{13.39\ kW}{75\%} \times 100\% = 17.85 = 18\ kW$$

Motor or input power

$$P_m = \frac{P_a}{E_m} = \frac{17.85\ kW}{90\%} \times 100\% = 19.84 = \underline{20.\ kW}$$

EXAMPLE PROBLEM 15.13

A 10-kW pump unit is rated to deliver 20 L/s with an efficiency of 72%. What is the operating pressure?

Given:

$Q = 20$ L/s $E_p = 72\%$ $P_a = 10$ kW $p_a = ?$

Solution:

Pump power

$$P_P = \frac{P_a}{E_p} = \frac{Q \times p_a}{E_P} \quad or \quad p_a = \frac{E_p \times P_P}{Q}$$

Pressure added

$$p_a = \frac{72\%}{100\%} \times 10.kW \times \frac{s}{0.02\ m^3} \times \frac{kPa.m^3}{kW.s} = 360.0 = \underline{360\,kPa}$$

EXAMPLE PROBLEM 15.14

A 10-hp pumping unit is rated to deliver 300 gpm with an efficiency of 70%. What is the operating pressure?

Given:

$Q = 300$ gpm $E_p = 70\%$ $P_b = 10$ hp $P_a = ?$

Solution:

Pump power

$$P_P = \frac{P_a}{E_P} = \frac{Q \times p_a}{E_P} \quad or \quad pa = \frac{E_P \times P_P}{Q}$$

Pressure added

$$p_a = \frac{70\%}{100\%} \times \frac{10.hp \times 550\ ft.lb}{hp.s} \times \frac{min}{300\ gal} \times \frac{7.48\ gal}{ft^3} \times \frac{60s}{min}$$

$$= \frac{5759.6}{ft^2} \times \frac{ft^2}{144\ in^2} = 39.9 = \underline{40\ psi}$$

EXAMPLE PROBLEM 15.15

A water pump is required to pump @ 15 L/s against a pressure of 350 kPa. If the pump is 65% efficient and the motor is 90% efficient, what is the power input required?

Given:

$Q = 15$ L/s $E_p = 65\%$ $E_m = 90\%$ $p_a = 350$ kPa $P_I = ?$

Solution:

Input power

$$P_I = = \frac{Q \times p_a}{E_p \times E_m} = \frac{154\ L}{s} \times \frac{m^3}{1000\ L} \times \frac{350\ kPa}{0.65 \times 0.90} \times \frac{kW.s}{kPa.m^3} = 8.97 = \underline{9.0\ kW}$$

EXAMPLE PROBLEM 15.16

Find the motor horsepower for a pump station with the following parameters: motor efficiency = 90%, total head = 186 ft, pump efficiency = 79%, discharge rate = 1200 gpm.

Given:

$Q = 1200$ gpm $E_p = 79\%$ $E_m = 90\%$ $ha = 186$ ft $P_m = ?$

Solution:

Power added to water

$$P_i = Q \times \gamma \times h_a = \frac{1200\ gal}{min} \times \frac{8.34\ lb}{gal} \times 186\ ft \times \frac{hp.s}{550\ ft.lb} \times \frac{min}{60s}$$

$$= 56.41 = \underline{56\ hp}$$

Power input to motor or motor power

$$P_m = \frac{P_a}{E_P \times E_M} = \frac{56.41\,hp}{0.79 \times 0.90} = 79.33 = \underline{79\ hp}$$

15.6 PUMPING COST

Since power is energy per unit time, energy spent by a pumping system will be equal to power supplied multiplied by hours of operation. Commonly used units of electric energy consumption are kW.h. One unit (kWh) of energy will be consumed when power is supplied at the rate of 1 kW for a period of 1 hour. Pumping cost can be determined by the power input (supplied) and hours of operation per unit (kW.h).

Energy cost

$$\boxed{Cost = E \times R = P \times t \times R}$$

where t = hours of operation, R = rate per kW.h

EXAMPLE PROBLEM 15.17

16 kW is required for a pumping application. If the cost of power is 5.3 ¢/kW.h, calculate the daily pump cost if the pump runs 20 hours a day.

Given:

P = 16 kW   t = 20 h/d   R = 5.3 ¢/kW.h

Solution:

Cost of pumping

$$= P \times t \times R = 16\ kW \times \frac{16\ kW}{d} \times \frac{\$0.053}{kW.h} = 16.9 = \underline{\$17/d}$$

EXAMPLE PROBLEM 15.18

Determine the cost to operate a 100-hp rated motor for 1 week if it runs an average of 14 h/d, is 83% efficient, and the electrical costs are $0.045/kW. h.

Given:

P = 100 hp   t = 20 h/d E_m = 83%   R = $0.045/kW.h

Solution:

Cost of pumping

$$Cost = P \times t \times R = \frac{100\,hp}{0.83} \times \frac{0.746\ kW}{hp} \times \frac{14\ h}{d} \times \frac{7d}{wk} \times \frac{\$0.045}{kW.h} = 396 = \underline{\$400\,/\,wk}$$

15.7 AFFINITY LAWS

The capacity of centrifugal pumps can be varied by varying the rotative speed N. Also, a given size pump casing can accommodate **impellers** of different diameters. The manner in which pump characteristics change with changes in either speed or diameter is determined by **affinity laws.** When speed N is changed from N_1 to N_2

Affinity laws

$$Q_2 = Q_1 \times \frac{N_2}{N_1}, \quad H_2 = H_1 \times \left(\frac{N_2}{N_2}\right)^2, \text{ and } P_2 = P_1 \times \left(\frac{N_2}{N_1}\right)^3$$

Increasing the rotative speed by 25%, the increase in pump capacity is 25%; increase in head is 56%; and increase in power is greatest, 95%.

EXAMPLE PROBLEM 15.19

A pump delivers 500 L/min at 1000 rpm against a total head of 45 m. Determine its performance at 1100 rpm.

Given:

$Q_1 = 500$ L/min $N_1 = 1000$ rpm $H_1 = 45$ m $N_2 = 1100$ rpm

Solution:

New pumping rate

$$Q_2 = Q_1 \times \frac{N_2}{N_1} = \frac{500\,L}{min} \times \left(\frac{1100}{1000}\right) = 550.0 = \underline{550\,L\,/\,min}$$

New pumping head

$$H_2 = H_1 \times \left(\frac{N_2}{N_1}\right)^2 = 45\,m \times \left(\frac{1100}{1000}\right)^2 = 54.450 = \underline{54\,m}$$

EXAMPLE PROBLEM 15.20

For a given centrifugal pump, if the speed of rotation of the impeller is cut in half, how does the capacity change?

Solution:

New pump discharge rate

$$Q_2 = Q_1 \times 0.5 = 0.5 Q_1 = \underline{50\% Q_1}$$

EXAMPLE PROBLEM 15.21

A pump with 12-in impeller discharges 550 gpm against a head 55 ft. What will be the pumping rate and head when the diameter of the impeller is trimmed by 1 inch?

Solution:

New discharge rate

$$Q_2 = Q_1 \times \frac{D_2}{D_1} = 550\, gpm \times \left(\frac{11}{12}\right) = 504 = \underline{500\, gpm}$$

New pumping head

$$H_2 = H_1 \times \left(\frac{D_2}{D_1}\right)^2 = 55\, ft \times \left(\frac{11}{12}\right)^2 = 46.2 = \underline{46\, ft}$$

EXAMPLE PROBLEM 15.22

A pump has a rated capacity of 40 L/s, develops a head of 25 m, and has a power requirement of 13 kW when operating at 1500 rpm. If the efficiency remains the same, what will be the operating characteristics if the speed drops to 1200 rpm?

Case	N (rpm)	Q (L/s)	H (m)	P (kW)
Old (1)	1500	40	25	13
New (2)	1200	?	?	?

Solution:

New discharge rate

$$Q_2 = Q_1 \times \left(\frac{N_2}{N_1}\right) = \frac{40\,L}{s} \times 0.8 = 32.0 = 32\, L/s$$

New pumping head

$$H_2 = H_1 \times \left(\frac{N_2}{N_1}\right)^2 = 25\, m \times 0.8^2 = 16.0 = \underline{16\, m}$$

New power required

$$P_2 = P_1 \times \left(\frac{N_2}{N_1}\right)^3 = 13\, kW \times 0.8^3 = 6.65 = \underline{6.7\, kW}$$

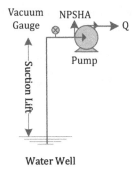

Vacuum Gauge   NPSHA   Q

Pump

Suction Lift

Water Well

FIGURE 15.3   Net positive suction head.

15.8   CAVITATION

The **vapour pressure** of a liquid refers to the *pressure at which liquid transforms into vapours*. For a given liquid, there is a definite relationship between the vapour pressure and the temperature. At normal atmospheric pressure, water starts boiling at 100°C. In other words, at 100°C, the vapour pressure of water is 1 atmosphere, or about 100 kPa (abs). When operating a pump, if the inlet pressure is allowed to drop to the extent that it approaches vapour pressure, bubbles of vapours will start forming. These bubbles suddenly collapse as they enter the high-pressure region in the pump body. This will create sudden noise and pitting of the metal surface due to the explosion of the bubbles. This phenomenon is called **cavitation**.

15.8.1   NET POSITIVE SUCTION HEAD

The basic measure to protect the pump against cavitation is to avoid pressures that are similar to the vapour pressure of the water. This is accomplished by designing and operating pumping systems so that total suction head of the liquid at the inlet is above the vapour pressure head. This is called **net positive suction head** (NPSH). Using the pump centreline as a reference, net positive suction head available is given by:

Net positive suction head available

$$h_{NSHA} = h_{atm} - h_{vap} - SL - h_L$$

SL = Suction lift = ΔZ   L = loss in the suction line
vap = Vapour pressure/head   atm = atmospheric

EXAMPLE PROBLEM 15.23

Determine the available NPSH for a pumping system pumping water at 20°C from a well in which the pumping water level is 2.5 m below the pump. The atmospheric pressure at 101 kPa and head losses due to friction in the pipe and fittings in the suction line are estimated to be 0.45 m.

Given:

$p_{atm} = 101$ kPa   SL $= Z_2$-$Z_1 = 2.5$ m $h_L = 0.45$ m $p_{vap} = 2.34$ kPa

Solution:

Net positive suction head

$$h_{NSHA} = h_{atm} - h_{vap} - (Z_2 - Z_1) - h_L$$

$$= (101 - 2.34) kPa \times \frac{m}{9.81 \ kPa} - 2.5m - 0.45 \ m$$

$$= 7.107 = \underline{7.1 \ m}$$

EXAMPLE PROBLEM 15.24

A pump is pumping under a suction head of 5.0 ft. (pump is sitting 5.0 ft above the pumping water level). Determine the available NPSH, assuming the atmospheric pressure is 14.5 psi and head losses due to friction in the pipe and fittings are estimated to be 2.5 ft. The vapour pressure of water at the prevailing temperature is 0.5 psi.

Given:

$p_{atm} = 14.5$ psi $Z_2 - Z_1 = -5.0$ ft (5.0 ft of suction head) $h_L = 2.5$ ft $p_{vap} = 0.50$ psi

Solution:

Net positive suction head available

$$h_{NSHA} = h_{atm} - h_{vap} - (Z_2 - Z_1) - h_L$$

$$= (14.5 - 0.5) psi \times \frac{ft}{0.433 \ psi} + 5.0 \ ft - 2.5 \ ft = 34.8 = \underline{35 \ ft}$$

This shows that pumping under suction head conditions, there is a minimal chance of cavitation. However, under suction lift conditions, it is necessary to check for NPSHA against NPSHR.

15.8.2 PERMISSIBLE SUCTION LIFT

Manufacturers specify the NPSH required for efficient operation of their pumps. As long as the machine is operated above this value, the operation will be satisfactory. Problems of cavitation can be avoided by providing positive head on the suction side of the pump. However, when the pumping level is below the pump (suction lift), it is important to check that NPSHA > NPSHR. To prevent cavitation, NPSHA should be greater than or equal to NPSHR. The maximum lift which can be allowed for a given pump can be found by knowing the NPSHR and vapour pressure of the liquid at the operating temperature. As NPSHA approaches NPSHR, suction lift approaches the maximum suction lift (MPSL) that could be afforded. We can increase the suction lift to the limit that NPSH approaches NPSHR. In other words,

Maximum permissible suction lift

$$\boxed{MPSL = h_{atm} - h_L - h_R - h_{vap}}$$

EXAMPLE PROBLEM 15.25

Find the permissible suction lift for a pump that requires 3.5 m of NPSH. Total head losses in the suction line are estimated to be 0.7 m. The temperature of water is 15°C, and the atmospheric pressure head is 10.2 m.

Given:

NPSHR = 3.5 m h_L = 0.7 m   T = 15°C h_{vap} = 0.21 m (Tables) p_{atm} = 10.2 m

Solution:

Maximum permissible suction lift

$$MPSL = h_{atm} - h_L - h_R - h_{vap} = 10.2 \ m - 0.7 \ m - 3.5 \ m - 0.21 \ m$$
$$= 5.69 = \underline{5.7 \ m}$$

PRACTICE PROBLEMS

PRACTICE PROBLEM 15.1

Due to positive pressure on the suction side, the pressure gauge on the suction side reads 25 kPa, and the discharge side gauge reads 175 kPa. Calculate the pump head. (15 m)

PRACTICE PROBLEM 15.2

Readings taken from the inlet and outlet pressure gauges of a pumping system respectively are 45 kPa and 450 kPa. Find total dynamic head. (41 m)

PRACTICE PROBLEM 15.3

Work out the pumping head for the data of Example Problem 5.3 assuming the discharge pressure gauge is located 2.0 ft above the suction gauge. (20 ft)

PRACTICE PROBLEM 15.4

Rework Example Problem 15.4 assuming the suction gauge reads 1.8 ft of mercury column and both gauges are read at the same elevation. (130 ft)

PRACTICE PROBLEM 15.5

Water is being pumped from a water source to a water tank. The elevation of water level at the source is 101.2 m, and that in the water tank elevation is 152.8 m. What is the total head if friction and minor head losses are 4.5 m? (57 m)

PRACTICE PROBLEM 15.6

A centrifugal pump when pumping water at the rate of 500 gpm adds a pressure of 25 psi. What power is added to water? (7.3 hp)

PRACTICE PROBLEM 15.7

What pump power is required to pump water @ 18 L/s against a head of 45 m assuming the pump is 65% efficient? (12 kW)

PRACTICE PROBLEM 15.8

A pump is pumping water at the rate of 950 gpm and lifts water 210 ft. If head losses are assumed to be 15 ft and the pump efficiency is 75%, what size (rated power) electric motor is required? (72 hp)

PRACTICE PROBLEM 15.9

35 kW of power is supplied to the motor. How many kW of power are added to water if the motor is 92% efficient and the pump is operating at an efficiency of 65%? (21 kW)

PRACTICE PROBLEM 15.10

A pump operating against a head of 65 ft discharges at the rate of 850 gpm. If the overall operating efficiency is 70%, determine the water horsepower and input power. (20 hp)

PRACTICE PROBLEM 15.11

A pump is 82% efficient when pumping at the rate of 50 L/s against a head of 50 m. Calculate the power needed to supply to the motor, assuming the motor efficiency is 90%. (33 kW)

PRACTICE PROBLEM 15.12

A 15-kW pump unit is rated to deliver 32 L/s with an efficiency of 75%. What is the operating pressure? (350 kPa)

PRACTICE PROBLEM 15.13

A 25-hp pumping unit is rated to deliver 1500 gpm with an overall efficiency of 65%. What is the operating pressure? (19 psi)

PRACTICE PROBLEM 15.14

Water at the flow rate of 35 L/s is to be pumped against a pressure of 320 kPa. What is the power input required if the overall (wire to water) efficiency is 70%? (16 kW)

PRACTICE PROBLEM 15.15

Find the motor horsepower for a pump station where the overall efficiency is 76%. Water is pumped at the rate of 1500 gpm, and the pumping head is 210 ft. (100 hp)

PRACTICE PROBLEM 15.16

The motor power requirement is calculated to be 35 kW. Based on average operation of 150 hours per week, calculate the power cost per week for the pump operation. Assume an energy rate of $0.045/unit. ($ 240/week)

PRACTICE PROBLEM 15.17

Water at the flow rate of 80 L/s is to be pumped against a pressure of 220 kPa. What is the power input required if the overall (wire to water) efficiency is 66%? What is the power cost for 16 hours of pumping if the unit cost is 8¢ per kW.h? (26.7 kW, $34)

PRACTICE PROBLEM 15.18

Determine the cost to operate a 75-hp motor for 1 month (assume 30 days) if it runs an average of 16 h/d, is 87% efficient, and the electrical costs are $0.055 per kW.h. ($1200/month)

PRACTICE PROBLEM 15.19

A pump delivers 500 L/min at 1000 rpm against a total head of 45 m. Determine its performance at 800 rpm. (400 L/min, 29 m)

PRACTICE PROBLEM 15.20

For a given centrifugal pump, if the speed of rotation of impeller is cut in half, how does the total head capability change? (Reduced 4×)

PRACTICE PROBLEM 15.21

A pump with 10-in impeller discharges 400 gpm against a head 65 ft. What will be the pumping rate and head when the diameter of the impeller is trimmed by 1 inch? (360 gpm, 53 ft)

PRACTICE PROBLEM 15.22

For the data of the example problem, determine the operating characteristics at 1350 rpm. (36 L/s; 20 m; 9.5 kW)

PRACTICE PROBLEM 15.23

Repeat the Example Problem15.23 assuming the temperature of water is 80°C. (2.5 m)

PRACTICE PROBLEM 15.24

A pump is pumping under a suction head of 4.0 ft. Determine the available NPSH assuming the atmospheric pressure is 14.7 psi and head losses due to friction in the pipe and fittings are estimated to be 3.5 ft. The vapour pressure of water at the prevailing temperature is 0.5 psi. (10 ft)

PRACTICE PROBLEM 15.25

For the data of Example Problem 13.4, find the permissible suction lift when: a) the temperature of the water is 60°C. (4.0 m, $\gamma = 9.78$ kN/m^3) b) The atmospheric pressure is 80 kPa (abs) due to high altitude (3.7 m)

16 Flow Measurement

The operation of water and wastewater treatment plants is dependent on knowledge of how much water is being processed. Both water and wastewater utilities need flow measurements to calculate chemical feed rates, hydraulic and organic loadings and flow-proportional composite sampling. It is necessary for scheduling of system maintenance and for sizing of pumps, pipes and metres. Flow records are needed to plan future expansions. Flow measurements can be based on flow rate or flow volume.

16.1 FLOW RATE

Flow rate can be expressed as L/s, gpm, m³/s, ft³/s, kL/h, MGD or ML/d. The time unit should not be mistaken to mean it is instant, hourly or daily flow. For example, hourly flow can be expressed as L/s or ML/d. This type of measurement is needed inside the plant in wastewater collection and water distribution.

16.1.1 FLOW VOLUME

Flow volume can be expressed in any units of volume, for example, m³ (kL) or dam³ (ML). A flow meter which gives the total volume of water passing through the meter is called a totalizer. A good example of this is the water meter in homes. The difference between readings at the beginning and end of a period is the volume of water used over that period. This difference divided by the time period will yield average flow rate over that period. There are several types of flow measurement devices available. Some of the most common are described subsequently.

EXAMPLE PROBLEM 16.1

A wet well measures 13 ft × 10 ft. The difference between minimum water level (pump shuts off) and the water level when the lead pump comes on is 2.0 ft. How much water flows into the well between two consecutive pumping cycles? If the time interval between two consecutive pumping cycles is 4 min and 40 s at 6:30 AM, calculate the inflow rate at 6:30 AM.

Given:

$A_s = 13$ ft × 10 ft $\Delta H = 2.0$ ft $\Delta t = 4$ min and 40 s = 280 s   Q =?

Solution:

Instantaneous flow rate

$$Q = \frac{\Delta V}{\Delta t} = \frac{A_S \times \Delta d}{\Delta t} = \frac{13\,ft \times 10\,ft \times 2.0\,ft}{280\,s} \times \frac{60\,s}{min} \times \frac{7.48\,gal}{ft^3} = 416 = \underline{420\,gpm}$$

16.2 VARIABLE HEAD METERS

This type of meter is more suitable for pressure flow conditions and measures instant flow rate. Under this category, venturi meter, orifice meter and flow nozzle are the important ones. The principle of operation is that a pressure differential is created across the unit as a *steady flow* of water passes through a constricted section—a venturi, orifice or nozzle. The increase in flow velocity at the constricted section causes pressure head to drop (Figure 16.1).

Flow through a venturi meter

$$Q = C_d A_2 \sqrt{\frac{2g(h_1 - h_2)}{1 - A_2^2 / A_1^2}} = C_c A_2 \sqrt{2g(h_1 - h_2)}$$

Q = *Volume flow rate*
A_2 = *Flow area of the constriction*
          (throat in case of venturi meter or nozzle and orifice opening)
C = *Coefficient of discharge, it is a empirically determined value*
Δh = *Pressure head drop across the meter* = $(p_1 - p_2)/\gamma = \Delta p/\gamma = h_1 - h_2$
$1, 2$ = *Upstream and downstream sections respectively*

Pressure head drop is usually measured by attaching a differential manometer to points 1 and 2.

Pressure head drop

$$\Delta h_p = \Delta h_g \left(SG_g - SG \right)$$

Δh_g = *Difference in heights of the gauge fluid*
SG_g = *Specific gravity of the gauge fluid*

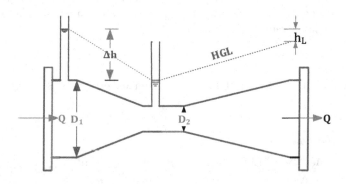

FIGURE 16.1   Venturi meter.

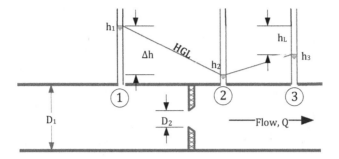

FIGURE 16.2   Orifice meter.

The pressure drop will change as the flow rate changes. This can be read by attaching manometers or pressure gauges across the meter. When observing pressure drop with the help of a differential manometer, the actual value read is the deflection of manometer fluid Δh_g.

EXAMPLE PROBLEM 16.2

A pressure gauge upstream of the venturi meter reads 420 kPa, and the one at the throat reads 280 kPa. The pipe diameter is 400 mm, and the diameter of throat section is 50 mm. Calculate the flow in L/s.

Given:

$p_1 = 420$ kPa $p_2 = 280$ kPa $D_2 = 50$ mm $= 5.0 \times 10^{-2}$ m $C = 0.95$

Solution:

Pressure head drop

$$\Delta h = \frac{\Delta p}{\gamma} = (420 - 280)\,kPa \times \frac{m}{9.81\ kPa} = 14.27\,m = 14\,m$$

Flow rate

$$Q = C_d A_2 \sqrt{2g(h_1 - h_2)} = 0.95 \times \frac{\pi (0.050\ m)^2}{4} \times \sqrt{\frac{19.62\,m}{s^2} \times 14.27\,m}$$

$$= \frac{3.119 \times 10^{-2}\,m^2}{s} \times \frac{1000\,L}{m^3} = 31.19 = \underline{31\ L/s}$$

EXAMPLE PROBLEM 16.3

What diameter orifice opening is required to discharge freely 16 L/s of water under a head of 8.5 m?

Given:

Q = 16 L/s = 0.016 m³/s h_1 = 8.5 m h_2 = 0 (free flow)

Solution:

Area of orifice opening

$$A_o = \frac{Q}{C_d\sqrt{2g(h_1 - h_2)}} = \frac{0.016\,m^3}{s} \times \frac{1}{0.62} \times \frac{1}{\sqrt{\frac{19.62\,m}{s^2} \times 8.5\,m}}$$

$$= 2.03 \times 10^{-3} = \underline{0.0020\,m^2}$$

Diameter of orifice opening

$$D_o = \sqrt{\frac{4A_o}{\pi}} = \sqrt{\frac{4 \times 2.03 \times 10^{-3}\,m^2}{\pi}} \times \frac{1000\,mm}{m} \times 50.8 = \underline{51\,mm}$$

EXAMPLE PROBLEM 16.4

A venturi meter with a throat diameter of 2 in is installed in 4-in-diameter pipe. The deflection in the differential manometer using mercury as the gauge fluid reads 2.2 in. Calculate the volume flow rate.

Given:

D_1 = 4 in	D_2 = 2 in	Δh_m = 2.2 in
SG = 13.6	C = 0.95	Q =?

Solution:

Pressure head drop

$$\Delta h = \Delta h_g (SG_g - SG) = 2.2\,in \times \frac{ft}{12\,in} \times (13.6 - 1) = 2.31\,ft$$

Discharge rate

$$Q = C_d A_2 \sqrt{2g\Delta h} = 0.95 \times \frac{\pi(2/12\,ft)^2}{4} \times \sqrt{2 \times \frac{32.2\,ft}{s^2} \times 2.31\,ft}$$

$$= \frac{0.225\,ft^3}{s} \times \frac{7.48\,gal}{ft^3} \times \frac{60\,s}{min} = \underline{113 - 110\,gpm}$$

16.3   PITOT TUBES

A pitot tube or gauge is used to measure the velocity of flow at a point. When a pitot tube is inserted in the flow stream, a pressure is created which is greater than the static pressure by the amount equal to velocity pressure head.

16.3.1   PITOT STATIC TUBES

A pitot-static tube facilitates the measurement of both the static pressure and total pressure in closed pipes. However, for calculating the flow velocity, the difference (velocity pressure) is needed. Generally, a differential manometer is used to measure the velocity head by reading the deflection of the gauge fluid. This is discussed in the previous section.

16.3.2   PITOT GAUGES

In water distribution systems, a pitot tube is used to measure system and fire flow. The quantity of discharge from a hydrant nozzle can be calculated from the pitot pressure reading. The kinetic energy of the water in the jet stream is converted into pressure. Under ideal flow conditions (no head loss and no contraction), flow velocity can be found as follows.

Ideal flow velocity

$$v = \sqrt{2gh_v} = \sqrt{2gp_v / \gamma}$$

The relationship yields the ideal or theoretical flow velocity. Introducing the coefficient of discharge C (typically 0.9), actual flow can be expressed in terms of pitot pressure.

Flow formula (pitot tube)

$$Q = Av_a = AC\sqrt{2gp / \gamma} = D^2\sqrt{p} \ (SI)$$

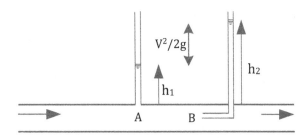

FIGURE 16.3   Pitot tube.

EXAMPLE PROBLEM 16.5

A static pitot tube is installed in a 100-mm-diameter pipe. The differential manometer connected to the pitot tube records the velocity head as 86 mm of water column. What is the velocity of flow in the pipe?

Given:

$h_v = 86$ mm $C = 0.95$ $v = ?$

Solution:

Flow velocity

$$v = c\sqrt{2gh_v} = 0.95\sqrt{\frac{2 \times 9.81\ m}{s^2} \times 86\ mm \times \frac{m}{1000\ mm}} = 1.23 = \underline{1.2\ m/s}$$

EXAMPLE PROBLEM 16.6

Pitot tube pressure measured in the discharge from a 60-mm nozzle of a fire hydrant is 70 kPa. Calculate the discharge in L/s.

Given:

$p = 70$ kPa $D = 60$ mm $= 0.06$ m $C = 0.9$ (assumed) $Q = ?$

Solution:

Discharge rate

$$Q = Av_a = AC\sqrt{2gp/\gamma}$$

$$= \frac{\pi(0.06\,m)^3}{4} \times 0.9 \times \sqrt{\frac{2 \times 9.81\,m}{s^2} \times 70\,kPa \times \frac{m}{9.81\,kPa}}$$

$$= 3.011 \times 10^{-2} = 3.0 \times 10^{-2}\,m^3/s = \underline{30\,L/s}$$

Alternatively,

$$Q = D^2\sqrt{p} = (0.06)^2 \times \sqrt{70} = 3.01 \times 10^{-2} = 3.0 \times 10^{-2}\,m^3/s$$

16.4   OPEN CHANNEL FLOW DEVICES

When flow is entirely due to gravity, the flow conditions are that of the **open channel flow** type. Under open channel flow conditions, the conduit flows partially full, and there is a free water surface open to atmosphere. There are many types of flow measurement devices available for open channel flow conditions.

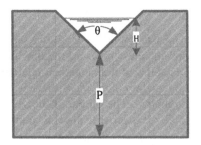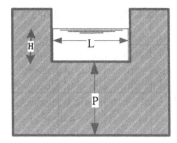

FIGURE 16.4   Sharp crested weirs: triangular weir, rectangular weir.

A primary flow measurement device is a human made flow control structure, which, when inserted into an open channel, creates a unique relationship between the depth of flow and the rate of flow. The depth of flow, referred to as head (H), can then be measured at the respective point area with a ruler or other staff gauge. When substituted into a formula which mathematically describes the relationship, the head measurement can be used to calculate the flow rate.

16.4.1   WEIRS

Weirs consist of a crest located across the width of an open channel (Figure 16.2). The primary factors which determine the relationship between head and discharge are the shape and size of the openings of the notches.

Rectangular weir

$$Q = 1.84 \times LH^{1.5} - SI \qquad Q = 3.32 \times LH^{1.5} - USC$$

Triangular weir

$$Q = 1.42 \tan(\theta/2) H^{2.5} - SI \qquad Q = 2.57 \tan(\theta/2) H^{2.5} - USC$$

L = Length of the weir
Θ = Angle of the weir
H = Head on the weir

16.4.2   FLUMES

Flumes are structures which force water through a narrow channel. They consist of converging section, a throat and a diverging much like a venturi meter. Figure 14.2 shows the most common type of flume and also provides formulas for calculating appropriate flow rates. Parshall flumes have fixed specifications relating to geometric shape. They vary only in throat width.

16.5   SECONDARY FLOW DEVICES

These are automated forms of flow rate measurement. Typically, a secondary device is used with a primary device to automatically measure the flow depth or head. This value is then processed, using established mathematical relationships to calculate flow rate and flow volume. Secondary flow measurement devices include floats, ultrasonic transducers and bubblers. The output of the secondary devices is transmitted to a display and/or totalizer to provide flow rate and volume information.

EXAMPLE PROBLEM 16.7

A 90°V-notch is to be installed in a 600-mm-diameter sewer to measure a flow of 35 L/s. What maximum head should be expected?

Given:

$Q = 35$ L/s $= 3.5 \times 10^{-2}$ m³/s $\theta = 90°$ H = ?

Solution:

Head over the weir

$$Q = 1.4 \tan(\theta/2) H^{2.5}$$

$$H = \left(\frac{Q}{1.42 \tan(\theta/2)}\right)^{0.40} = \left(\frac{1}{1.42} \times \frac{0.035\,m^3}{s} \times \frac{1}{\tan(90/2)}\right)^{0.40} = 0.227\,m = \underline{230\,mm}$$

EXAMPLE PROBLEM 16.8

A 120° V-notch is placed in a channel to measure flow. Calculate the discharge when the height of flow is 9.0 in as measured in a stilling well behind the weir.

Given:

$\theta = 120°$ H = 9.0 in = 0.75 ft

Solution:

Discharge rate

$$Q = 2.57 \tan(\theta/2) H^{2.5} = 2.57 \times \tan\left(\frac{120}{2}\right) \times (0.75)^{2.5}$$

$$= 2.16 = 2.2\ ft^3/s = \underline{1.4\,MGD}$$

EXAMPLE PROBLEM 16.9

A sharp rectangular weir is installed in a flow channel. The weir is 0.5 m wide and 0.4 m high from the bottom. If the walls of the weir extend to a height of 0.8 m above the bottom of the channel and the weir is flowing with level reaching full height, how much water is flowing over the weir?

Given:

H (max) = 0.8 m – 0.4 m = 0.4 m   L = 0.5 m

Solution:

Discharge rate

$$Q = 1.84 \times LH^{1.5} = 1.84 \times 0.5 \times (0.4)^{1.5} = 0.233\, m^3/s = \underline{200\, L/s}$$

EXAMPLE PROBLEM 16.10

Find the maximum flow a 60° V-notch can carry when operated at a head of 0.75 m.

Given:

θ = 60°   H = 0.75 m

Solution:

Discharge rate

$$Q = 1.42\, tan(\theta/2) H^{2.5} = 1.42 \times tan\left(\frac{60}{2}\right) \times (0.75)^{2.5}$$

$$= 0.399 = 4.0 \times 10^{-1} m^3/s = \underline{400\, L/s}$$

PRACTICE PROBLEMS

PRACTICE PROBLEM 16.1

A wet well measures 10 ft × 7.5 ft. The difference between the minimum water level (pump shuts off) and the water level when the lead pump comes on is 2.5 ft. How much water flows into the well between two consecutive pumping cycles? If the time interval between two consecutive pumping cycles is 6 min and 40 s, calculate the instantaneous inflow rate. (210 gpm)

PRACTICE PROBLEM 16.2

For the venturi meter described in the previous example, if the pressure drops 95 kPa, what is the flow rate? (26 L/s)

PRACTICE PROBLEM 16.3

If an orifice opening is 51 mm in diameter, what head is required to cause a free flow of 8.0 L/s? (2.0 m)

PRACTICE PROBLEM 16.4

A differential manometer attached across a 1-in-diameter sharp-edged orifice reads 1.2 in of mercury deflection. What is the flow rare? (14 gpm)

PRACTICE PROBLEM 16.5

For the data of Example Problem 16.5, calculate the water flow in L/s. (9.7 L/s)

PRACTICE PROBLEM 16.6

When another hydrant is opened, the pitot pressure reduces from 70 kPa to 55 kPa. Calculate the percent reduction in flow rate. (11%)

PRACTICE PROBLEM 16.7

What maximum flow can be measured with this arrangement if the flow depth is not to exceed 400 mm? (140 L/s)

PRACTICE PROBLEM 16.8

Solve the previous problem if the head is 20 cm. (15 L/s)

PRACTICE PROBLEM 16.9

If head over the weir is decreased to half of the full flow value, calculate the flow over the weir. (1.6 m³/s)

PRACTICE PROBLEM 16.10

Work the example problem assuming the angle of the notch is 90°. (690 L/s)

17 Data Analysis

Calculation of data without analysis, interpretation and use of results is a waste of time and money. Statistical analysis of data on a given sample is a means to determine how well your test results can be repeated (quality control). The same type of analysis performed on a group of samples allows us to estimate the variation over a given time period. Presentation of data in tables, graphs or charts make the information more usable by illustrating trends, variations and significant changes in raw water quality or treatment process efficiencies.

17.1 DATA TRANSFORMATION

Before data can be analysed and presented, frequently, it must be reduced or transformed in a usable form. For example, pressure gauge data read in kPa may have to be expressed in meters of water column and vice versa. Flow data, sometimes recorded in depths of flow in metres or feet needs to be transformed into flow rate by using the applicable formula for the flow measuring device, which may be a weir or a Parshall flume.

The transformation equation may be linear (straight-line relationship) or non-linear (curvilinear). For example, converting m of head into pressure in kPa is a linear relationship. However, converting head on a weir to flow is a non-linear relationship.

EXAMPLE PROBLEM 17.1

The general formula for a 90° triangular (V-notch) weir is $Q = 1.38\ H^{2.5}$, where Q is the flow rate in m³/s and H is the head in metres over the weir. Make a chart indicating the flow in ML/d for head over the weir ranging from 0.10 m to 1.0 m with an interval of 0.10 m.

Given:

$Q = 1.38\ H^{2.5}$ $H = 0.1$ m $Q = ?$

Solution:

Discharge rate

$$Q = 1.38 \times (0.10)^{2.5} = 4.364 \times 10^{-3} = 4.364 \times 10^{-3} m^3 / s$$

$$= \frac{4.364 \times 10^{-3} m^3}{s} \times \frac{ML}{1000\, m^3} \times \frac{3600\, s}{h} \times \frac{24\, h}{d} = 0.377 = \underline{0.38\, ML/d}$$

DOI: 10.1201/9781003468745-17

Following the same method, the rest of the values can be calculated

H (m)	0.10	0.20	0.30	0.40	0.50
Q (ML/d)	0.38	2.1	5.9	12	21

17.2   STATISTICAL ANALYSIS

Statistical analysis can be applied to study the *central tendency* (average) and *dispersion* (variation) in a given set of data. Arithmetic mean is commonly used to describe the average/centre of a distribution. Likewise, standard deviation is usually used to express the spread in a given distribution.

17.2.1   CENTRAL TENDENCY

The commonly used parameters used to describe the central tendency are arithmetic mean, geometric mean, median and mode.

17.2.2   ARITHMETIC MEAN

Arithmetic mean is the most commonly used way of describing the average or central tendency of a data series or set.

$$Mean, \bar{x} = \frac{\sum x}{n}$$

The units of arithmetic mean are the same as those of individual values in the data.

17.2.3   GEOMETRIC MEAN

In some cases, geometric mean is a better indicator of the central tendency of a variable. Geometric mean can be found using the following formula.

$$\bar{x}(geometric) = \sqrt[n]{x_1.x_2.....x_n} = (x_1.x_2.....x_n)^{1/n}$$

The units of geometric mean are the same as those of individual values in the data.

17.2.4   MEDIAN

Median divides the data in two halves. Median is the middle measurement when measurements are ranked in ascending or descending order (may fall between two measurements).

17.2.5   MODE

Mode is that value which occurs most frequently. There may be more than one mode or no mode in a given set of data.

EXAMPLE PROBLEM 17.2

The BOD determined each day is based upon a composite sample. For the 2-week data given, calculate the arithmetic mean, median and geometric mean.

150, 155, 155, 155, 160, 145, 160, 160, 165, 160, 160, 160, 165, 170, 180

Solution:

Sum of data values

$$\sum x = 150 + 155 + \cdots + 180 = 2240$$

Arithmetic mean

$$\bar{x} = \frac{\sum x}{n} = \frac{2240}{14} = 160.0 = \underline{160\,mg/L}$$

Geometric mean

$$\bar{x}(geometric) = \sqrt[n]{x_1.x_2.....x_n} = \sqrt[14]{150 \times 155 \times .....180} = 159.44 = \underline{159\,mg/L}$$

EXAMPLE PROBLEM 17.3

The turbidity of raw water at a filtration plant in NTU for a 2-week period is summarized as follows: 2.3, 1.9, 2.1, 2.9, 2.7, 2.6, 3.5, 3.7, 36, 3.3, 2.7, 2.9, 2.5, 2.7. Find the median and mode of this data series

Solution:

Arrange the values in ascending order, as shown in the following table.

Rank	1	2	3	4	5	6	7	8	9	10	11	12	13	14
NTU	1.9	2.1	2.3	2.5	2.6	2.6	2.7	2.7	2.9	2.9	3.3	3.5	3.6	3.7

The median is the middle value. With an even number of data points, the median is the mean of the middle two values. In this case, the middle two values are both 2.7 NTU; hence, the median of this data series is 2.7 mg/L.

The mode is the data value that occurs most often or has the highest frequency. The value of 2.7 NTU occurs three times. Thus, the mode in this case is also 2.7 NTU.

17.2.6 DISPERSION

Common parameters used to describe dispersion include range, average edition, variance and standard deviation. The units of measurement of dispersion, including range and standard deviation, are the same as those of individual values in the data.

17.2.7 RANGE

Range is the difference between the largest and smallest value.

17.2.8 STANDARD DEVIATION

Standard deviation is a measure of the average variability of a set of data from its mean. The formula is as follows:

Standard deviation

$$S_x = \sqrt{\frac{SS}{n-1}} = \sqrt{\frac{n\sum x^2 - \left(\sum x\right)^2}{n(n-1)}}$$

The second form is more convenient to use. Variance is the square of standard deviation. The term SS represents the sum of squared deviations from the mean.

Sum of squared deviations

$$SS = \frac{n\sum x^2 - \left(\sum x\right)^2}{n} = \sum x^2 - \frac{\left(\sum x\right)^2}{n}$$

17.2.9 COEFFICIENT OF VARIATION

Coefficient of variation is standard deviation divided by the arithmetic mean. It has no units and hence can be expressed as decimal fraction or percentage.

$$Coeff \ of \ variation, C_v = \frac{S_x}{\bar{x}}$$

EXAMPLE PROBLEM 17.4

For the data of Example Problem 17.2, calculate the range and standard deviation.

Solution:

$$Range = max - min = 180 - 145 = \underline{35 \ mg/L}$$

$$\sum x^2 = 150^2 + 155^2 + ... + 180^2 = 359350$$

Standard deviation

$$S_x = \sqrt{\frac{n\sum x^2 - (\sum x)^2}{n(n-1)}} = \sqrt{\frac{14 \times 359350 - 2240^2}{14(14-1)}} = 8.548 = \underline{8.55 \ mg/L}$$

EXAMPLE PROBLEM 17.5

The BOD of plant effluent determined each day is based upon composite sample. For the 2-week data given, calculate the coefficient of variation.

Day	1	2	3	4	5	6	7	8	9	10	11	12	13	14
mg/L	19	21	23	25	26	26	27	27	29	29	33	3.5	36	37

Solution:

Arithmetic mean

$$\bar{x} = \frac{\sum x}{n} = \frac{393}{14} = 28.07 = 28.1 \ mg/L$$

Sum of squared data values

$$\sum x^2 = 19^2 + 21^{22} + ... + 37^2 = 11427$$

Standard deviation

$$S_x = \sqrt{\frac{n\sum x^2 - (\sum x)^2}{n(n-1)}} = \sqrt{\frac{14 \times 11427 - 393^2}{14(14-1)}} = 5.511 = 5.51 \ mg/L$$

Coefficient of variation

$$C_v = \frac{S_x}{\bar{x}} = \frac{5.511}{28.1} = 0.196 = \underline{0.20 = 20\%}$$

17.2.10 MOVING AVERAGE

In water and wastewater treatment, there are number of parameters which vary from day to day or even hour to hour, such as turbidity, BOD and sludge volume index. Operational control of the treatment system is most often accomplished based on trends in data rather than individual data points or values. A trend is more accurately described by the moving average. In wastewater treatment, moving averages are commonly 7-day moving averages, which allows each day of the week to be included in the average.

To calculate the moving average, simply find the arithmetic mean for the days being considered, for example, 5 or 7. To complete the next moving average, repeat the procedure by replacing the oldest data value with the new value.

Moving average

$$\frac{(previous\ sum + new\ value - oldest\ value)}{n}$$

EXAMPLE PROBLEM 17.6

The results of the sludge volume index (SVI) test for an activated sludge plant for a period of 2 weeks were as follows. Calculate the 7-day moving averages.

Day	1	2	3	4	5	6	7	8	9	10	11	12	13	14
SVI	120	115	120	120	125	110	100	100	90	95	100	110	115	120

Solution:

Day	1	2	3	4	5	6	7	8	9	10	11	12	13	14
SVI							116	111	109	106	104	103	104	108

Sample of calculations

Summation

$$\sum x = 120 + 115 + 120 + 120 + 125 + 110 + 100 + 100 = 810$$

Moving average (8th day)

$$\bar{x} = \frac{(\sum x + x_8 - x_1)}{7} = \frac{810 + 100 - 120}{7} = 11.4 = \underline{111}$$

17.3   GRAPHS AND CHARTS

For easy interpretation of data, graphs are very helpful. Graphs show the characteristic tendency or average and dispersion) or trend in a relatively compact form.

17.3.1   BAR GRAPH

The grouped data can be represented in a graphical form by means of a *histogram*. A histogram consists of a set of rectangles or bars. The width of the base of each rectangle represents the class (group) interval, and the height of each rectangle equals the number of data values (frequency) falling into that class. Steps for grouping data:

1. Determine the range of data, the difference between the largest and smallest values.
2. Divide the range into a convenient number of classes (interval or groups), anywhere from 5 to 20, depending on the data.
3. Determine the number of data values (frequency) falling into each class.

4. Find the cumulative frequency by adding up the individual frequencies.
5. Calculate the relative frequency by dividing the individual frequency by the total frequency.
6. Plot the data by selecting suitable scale.

17.3.2 LINE GRAPHS

A line graph is a simple plot of data values you are interested in (e.g., flow and BOD). A line graph will reveal the trends. A line graph of flow data is called a **flow hydrograph**. If the values of flow are cumulative (totalized), the slope (steepness) of line will indicate the flow rate. Slope is the ratio of the rise to run for the time period in question.

17.4 LINEAR REGRESSION

The mathematical form of a linear relationship between two variables (x,y) is:

$$y = a + bx$$

x and y = independent and dependent variables, respectively
a, b = intercept and slope of the straight line

The graphical form of a linear relationship is a straight line, as shown in Figure 17.1. In spectrophotometry, the absorbency bears a linear relationship with the concentration. By measuring the absorbency of the standards (known concentration), a calibration equation can be developed. The equation, once developed, can be used to predict the concentrations by measuring the absorbency. This procedure is commonly used to

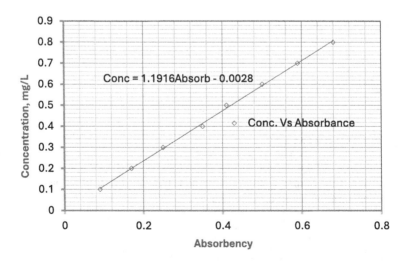

FIGURE 17.1 Straight-line relationship.

determine the concentration of iron manganese, fluoride, phosphates or nitrates in a given water sample.

In wastewater treatment, BOD data cannot be used to control the process, as the test takes 5 days. Knowing that COD and BOD are linearly related for a given wastewater, the COD test, which takes couple of hours, can be used to control activated sludge processes. Needless to say, it requires sufficient data on comparative values of BOD and COD on a sample with wide range of values. This will ensure prediction of BOD with more accuracy and confidence. Fitting the equation to a set of data consisting of paired values of x and y can be done graphically (eyeball line method) or statistically using the least square method.

17.4.1   Graphical Method

The data is plotted on a graph paper. In most cases, not all the points will lie exactly on a straight line. A best fit line is drawn, and the intercept (a) and slope (b) are read from the plot.

a = intercept with the y-axis = value of y when x = 0
b = slope of the straight line = rise/run = $\Delta y / \Delta x$

17.4.2   Least Square Method

The best fit line or equation for a set of data can be found using the following formulas to evaluate the constants a, b:

Slope of the straight line

$$\boxed{slope, b = \frac{n\left(\sum xy\right) - \left(\sum x\right)\left(\sum y\right)}{n\left(\sum x\right)^2 - \left(\sum x\right)^2} \qquad Intercept, a = \frac{\sum y - b\sum x}{n}}$$

Line passing through origin (0, 0) or no intercept, a= 0

Slope of the line (intercept=0)

$$\boxed{b = \frac{\left(\sum xy\right)}{\left(\sum x\right)^2}, when\, a = 0}$$

EXAMPLE PROBLEM 17.7

Using standards of total phosphorus ranging from 0.1 mg/L to 0.8 mg/L absorbency was recorded using stannous chloride method. Develop the calibration equation to predict concentration based on absorbance.

Conc. mg/L	0.1	0.2	0.3	0.4	0.5	0.6	0.7	0.8
Absorbency	0.09	0.17	0.25	0.35	0.41	0.50	0.59	0.68

Solution:

Table of computations

Y	X	Y²	X²	XY
0.1	0.09	0.01	0.008	0.009
0.2	0.17	0.04	0.028	0.034
0.3	0.25	0.09	0.063	0.075
0.4	0.35	0.16	0.0123	0.140
0.5	0.41	0.25	0.168	0.205
0.6	0.50	0.36	0.250	0.300
0.7	0.59	0.49	0.348	0.413
0.8	0.68	0.64	0.410	0.544
3.6	3.04	2.04	1.45	1.726

Parameters of the fitted equation

$$b = \frac{n\left(\sum xy\right) - \left(\sum x\right)\left(\sum y\right)}{n\left(\sum x^2\right) - \left(\sum x\right)^2} = \frac{8 \times 1.726 - 3.04 \times 3.6}{8 \times 1.45 - \left(3.04\right)^2} = 1.19$$

$$a = \frac{\sum y - b \sum x}{n} = \frac{3.6 - 1.19 \times 3.04}{8} = 0.004$$

Small value of the intercept shows that theoretically, intercept must be zero

PRACTICE PROBLEMS

PRACTICE PROBLEM 17.1

Calculate the weir flow when the head is 45 cm. (16 ML/d)

PRACTICE PROBLEM 17.2

The turbidity in the NTU of raw water over a period of 2 weeks is recorded as follows:

2, 3, 5, 17, 14, 13, 11, 9, 9, 8, 6, 4, 3, 3.

Calculate the mean and geometric mean over the 14-day period. (7.64, 6.26 both in NTU)

PRACTICE PROBLEM 17.3

For the data of Practice Problem 17.2, calculate the median and mode. (3 NTU, 7 NTU)

PRACTICE PROBLEM 17.4

Calculate the range and standard deviation of the secondary effluent BOD in mg/L over the 14-day period. (15 NTU, 4.7 NTU)

Day	1	2	3	4	5	6	7	8	9	10	11	12	13	14
mg/L	29	21	33	25	21	18	27	33	19	29	23	25	26	27

PRACTICE PROBLEM 17.5

For the data of Practice Problem 17.4, calculate the coefficient of variation. (19%)

PRACTICE PROBLEM 17.6

Calculate the 7-day moving average for the effluent BOD from the conventional activated sludge plant for the second week. (34, 34, 33, 30, 28, 26, 24)

Day	1	2	3	4	5	6	7	8	9	10	11	12	13	14
SVI	25	23	38	41	32	35	37	29	23	31	24	17	19	24

PRACTICE PROBLEM 17.7

COD standards ranging from 10 to 1000 were prepared. Following the standard procedure for COD determination, samples were prepared, and absorbency was recorded. The data is as follows:

Conc., mg/L	10	20	40	80	100	200	400	800	1000
Absorbance	0.005	0.009	0.018	0.039	0.050	0.101	0.201	0.399	0.493

Using the least square method, determine the regression equation to predict concentration based on absorbance. (CONC = 2000 ABS)

18 Water Sources

18.1 STORAGE CAPACITY

The storage capacity of a small reservoir (pond or lake) can be estimated by multiplying the surface area A_s by the average depth d. For a small reservoir, the average depth is typically about 40% of the maximum depth.

Storage capacity

$$V_{Storage} = A \times d_{avg}$$

18.2 CHEMICAL DOSING

Copper sulphate is used for algae control in lakes and ponds. Copper ions are a very effective algaecide. The chemical dose of copper sulphate pentahydrate ($CuSO_4.5H_2O$) depends upon the characteristics of water, particularly the alkalinity, turbidity and temperature. At high levels of alkalinity, citric acid may have to be added to prevent precipitation of copper.

18.2.1 SMALL RESERVOIRS

In order to calculate the dosage of copper sulphate, the following quantities must be known:

Chemical dosage rate

$$M_{Chemical} = V_{Storage} \times C$$

V = the storage capacity (volume) of the reservoir
C = the desired dosage of copper sulphate

18.2.2 LARGE RESERVOIRS

For large reservoirs, sometimes only a depth of perhaps 6 m or a depth down to the thermocline may be used in the calculation of volume of water to be treated.

The desired copper sulphate dosage may also be expressed as mass of the chemical per unit surface area, for example, kg/ha. This format is generally used for waters with a relatively high alkalinity. Under such conditions, algae control is limited to the upper volume due to the interference of other ions.

18.2.3 DOSING COPPER

When the dosage is expressed as mg/L of copper, we must know the fraction of copper in copper sulphate.

DOI: 10.1201/9781003468745-18

$$Molar\ mass = (63.5 + 32 + 4 \times 16 + 5 \times 18) = 249.5\ g/mol$$

$$Fraction\ of\ Cu = \frac{63.5\ g/mol}{249.5\ g/mol} \times 100\% = 25.4 = \underline{\underline{25\%}}$$

When dose is expressed as Cu, it may be changed to the equivalent copper sulphate pentahydrate. Note that one unit of the copper sulphate is 25% (1/4th) copper.

EXAMPLE PROBLEM 18.1

A small lake has an average length of 210 m and an average width of 95 m. The mean depth of the lake is 2.7 m. If the desired dosage rate of copper is 2.5 mg/L, how many kg of copper sulphate are required?

Given:

L = 210 m   W = 95 m   d = 2.7 m   C = 2.5 mg/L   m = ?

Solution:

Copper dosage

$$m = 210\,m \times 95\,m \times 2.7\,m \times \frac{2.50\,g\,Cu}{m^3} \times \frac{kg}{1000\,g} \times \frac{100\%\,salt}{25\%\,Cu} = 538 = \underline{\underline{540\ kg}}$$

EXAMPLE PROBLEM 18.2

The desired copper dosage in a reservoir is 0.5 mg/L for algae control. The reservoir has an estimated volume of 420 ha·cm. How many kg of copper sulphate (25% available copper) will be required?

Given:

C = 0.5 mg/L (Cu)   V = 420 ha·cm = 42000 m³ = 42 ML

Solution:

Quantity of copper sulphate

$$m = 42ML \times \frac{0.50\,kg}{ML} \times \frac{100\%\,salt}{25\%\,Cu} = 84.0 = \underline{\underline{84\,kg}}$$

EXAMPLE PROBLEM 18.3

One bag containing 80 lb of copper sulphate is applied to a lake. Calculate the dose of copper sulphate in lb/acre if the water surface is approximately a circle of diameter 800 ft.

Given:

D = 800 ft   m = 80 lb   Application rate = ?

Solution:

Application rate

$$= \frac{m}{A} = 80\,lb \times \frac{4}{\pi\left(800\,ft\right)^2} \times \frac{43560\,ft^2}{acre} = 6.93 = \underline{6.9\,lb/acre}$$

18.3   WATER WELLS

In ground water supplies, well is the most important structure. Well efficiency needs to be monitored on a continuous basis. In determining efficiency of the well, you need to measure the drawdown in the well.

18.3.1   Drawdown (s)

Ground water moves into the well by gravity. During pumping, the water level in the well is lowered. The term drawdown is a measure of the lowering of the water level due to pumping. To determine drawdown, the level of water before and after pumping must be known.

Drawdown in a well

$$\boxed{s = DWL - SWL}$$

Static water level = SWL
Dynamic water level = DWL

Drawdown will vary with pumping time and pumping rate. After continuous pumping for longer periods, water levels are almost stabilized (equilibrium). When indicating drawdown in the well, pumping time and capacity should be specified.

18.3.2   Water Level Measurement

The water level in a well may be directly measured by a graduated tape. However, for accurate measurement, the tape is connected to an electric sounder. As the tape touches the water surface, an electrical circuit is completed, and the reading on the tape is observed.

Pumping wells in municipalities and large installations are generally equipped with an airline. In this case, air pressure is applied to the airline until the pressure reading becomes constant. The amount of pressure is directly proportional to the depth of water level above the bottom of the line. For a given length of the air line (L), the water level (WL) can be calculated by reading the pressure (Figure 18.1):

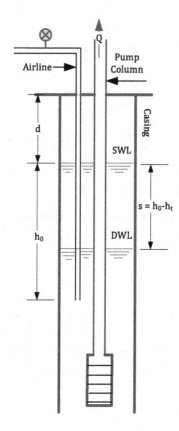

FIGURE 18.1   Water level measurement in a well.

Depth to WL

$d = L - p/\gamma$

Depth to SWL   *Depth to PWL*

$$d_0 = L - p_0 / \gamma \qquad d_t = L - p_t / \gamma$$

Drawdown

$$s = d_t - d_0 = \frac{(p_o - p_t)}{\gamma}$$

Change in drawdown

$$\Delta s = d_2 - d_1 = \frac{(p_1 - p_2)}{\gamma}$$

γ = *Weight density of water* = *9.81 kN/m³ (62.4 lb/ft³)*
WL = *Depth of water level with respect to ground surface*
L = *Length of air line*
d_o = *depth water surface at time t=0 (static conditions = no pumping)*
d_t = *depth to water surface at time t after pumping started*
p_0 = *airline gauge reading before pumping started (zero time)*
p_t = *airline gauge reading after pumping started (time t)*

18.3.3 WELL YIELD

Well yield is essentially the capacity of the well to produce water. It is often expressed as L/s. It may also be expressed as m³/s, m³/h or m³/d. The yield of a well is basically determined by the hydraulic characteristics of both the aquifer and the well. After constructing the well, pumping tests are performed to determine the well yield. This information is provided by the contractor. However, well operators are required to monitor the yield to indicate performance of the well.

18.3.4 SPECIFIC CAPACITY

Well yield is not significant unless the corresponding drawdown is compared. The term specific yield or capacity is the capacity of the well per unit of drawdown.

Specific capacity

$$SC = \frac{Capacity}{Well\,Drawdown} = \frac{Q}{s_w}$$

For most wells, specific capacity ranges from 0.20 L/s.m in a poor aquifer to as high as 20 L/s.m for a properly developed well in a highly permeable aquifer. Knowing the specific capacity of a well, the operator can predict the drawdown at various pumping rates.

EXAMPLE PROBLEM 18.4

The specific capacity of a well is 5.0 gpm/ft. What is the maximum yield of a well if the maximum available drawdown is 45 ft?

Given:

SC = 5.0 gpm/ft $\qquad$ s_w = 45 ft $\qquad$ Q =?

Solution:

Specific capacity of the well

$$SC = \frac{Q}{s_w} \quad or \quad Q = SC \times s_w = \frac{5.0\,gpm}{ft} \times 45\,ft = 225 = 230\,gpm$$

EXAMPLE PROBLEM 18.5

Before pumping started, the static water level in a well is observed to be at a depth of 64 feet. After 22 hours of continuous pumping @ 25 L/s, the water level stabilizes at a depth of 115 ft. Determine the specific capacity of the well in L/s.m.

Given:

$$SC = ? \quad s_w = 115-64 = 51 \text{ ft} \quad Q = 25 \text{ L/s}$$

Solution:

Specific capacity of the well

$$SC = \frac{Q}{s_w} = \frac{25 L}{s \times 51\,ft} \times \frac{3.28\,ft}{m} = 1.607 = 1.6\,L/s.m$$

18.3.5 WELL EFFICIENCY

Well production capacity may decrease with pumping time due to slow plugging of the intake portion of the well. This is indicated by increased drawdown for the same pumping rates. Well efficiency E_w can be defined as follows:

Well efficiency

$$\boxed{E_w = \frac{SC_{current}}{SC_{theretical}} = \frac{s_{act}}{s_{new}}}$$

Theoretical specific capacity refers to the ideal conditions when there is no well loss. For practical purposes, the maximum specific capacity when the well is new can be used for ideal specific capacity. Drawdown in the pumped well s is typically for a set period, usually 1 day.

EXAMPLE PROBLEM 18.6

Before pumping started, the pressure reading on the airline was 22 psi. What is the depth of the static water level if the length of the airline is 100 ft?

Given:

$$SWL = ? \quad L = 1000 \text{ ft} \quad p = 22 \text{ psi}$$

Solution:

Depth to static water level

$$d_o = L - p_o/\gamma = 100\,ft - \frac{22\,psi.ft}{0.433\,psi} = 49.2 = \underline{49\,ft}$$

EXAMPLE PROBLEM 18.7

The static water level of a well is 32.3 m below the ground surface. The pumping (dynamic) water level is determined using the airline method. After 10 hours of continuous pumping at the rate of 32 L/s, the water level stabilizes and the pressure reading on the air line is read to be 55 kPa. The length of the airline is 50.0 m. Calculate: a) the dynamic water level, b) drawdown in the pumping well and c) specific yield of the well.

Given:

SWL = 32.3 m   L = 50 m   p = 55 kPa

Solution:

Depth to dynamic water level

$$d_t = L - p_t/\gamma = 50.0\,m - \frac{55\ kPa.m}{9.81\ kPa} = 44.39 = \underline{44\,m}$$

Drawdown

$$s = d_t - d_0 = 44.39\ m - 32.3\ m = 12.09 = \underline{12\ m}$$

Specific capacity

$$SC = \frac{Q}{s} = \frac{32\ L}{s} \times \frac{1}{12.09\ m} = 2.64 = \underline{2.6\ L/s.m}$$

EXAMPLE PROBLEM 18.8

A pressure gauge attached to a 100-ft airline installed in a municipal well reads a pressure of 10 psi before pumping starts. After 1 day of continuous pumping, the gauge reads 7.2 psi. Calculate the static water level and drawdown in the well.

Given:

L = 100 ft p_0 = 10 psi p_t = 7.2 psi

Solution:

Depth to static water level

$$d_0 = L - \frac{p_0}{\gamma} = 100\,ft - 10\,psi \times \frac{ft}{0.433\ psi} = 76.9 = \underline{77\,ft}$$

Drawdown

$$s = \frac{(p_0 - p_t)}{\gamma} = (10 - 7.2)\,psi \times \frac{ft}{0.433\ psi} = 6.46 = \underline{6.5\,ft}$$

EXAMPLE PROBLEM 18.9

A well is pumped at a rate of 25 L/s. Before starting the pump, the gauge on the airline reads 320 kPa. After continuous pumping for 20 hours, the gauge reading stabilizes at 205 kPa. During 20 hours of pumping, 1900 m³ of water was pumped. Calculate the specific capacity in L/s.m.

Given:

$p_0 = 320$ kPa $p_t = 205$ kPa $t = 20$ h   V = 1900 m³

Solution:

Drawdown

$$s = \frac{(p_0 - p_t)}{\gamma} = (320 - 205) kPa \times \frac{m}{9.81 kPa} = 11.72 = 11.7 m$$

Specific capacity

$$SC = \frac{Q}{s} = \frac{1900\ m^3}{20\ h} = \frac{1}{11.72\ m} \times \frac{h}{3600\ s} \frac{1000\ L}{m^3} = 2.25 = \underline{2.3\ L/s.m}$$

> *Any change in pumping conditions will affect the operating efficiency of the pumping unit.*
>
> *A drop in well efficiency below 80% should signal caution and indicate a need to consider well rehabilitation.*

EXAMPLE PROBLEM 18.10

The pumping water level for a submersible pump is 55 m below the discharge header. The discharge pressure measured at the pump discharge header is 460 kPa. If the pump flow rate is 30. L/s, what is the water horsepower (power added to water), assuming the velocity head and head loss in the pipe are negligible?

Given:

Variable	Intake	Discharge
Pressure, kPa	0	460
Velocity, m/s	0	0
Elevation, m	$Z_2 - Z_1 = 55$ m	
Head Loss, m	0	

Solution:

Head added

$$h_a = \frac{p_2}{\gamma} - \frac{p_1}{\gamma} + \frac{v_2^2}{2g} - \frac{v_1^2}{2g} + Z_2 - Z_1 + h_1 = \frac{460\,kPa.m}{9.81\,kPa} + 55\,m = 101.8 = 102\,m$$

Power added

$$P_a = Q \times \gamma \times h_a = \frac{30\,L}{s} \times \frac{9.81\,N}{L} \times 101.8\,m \times \frac{W.s}{N.m} \times \frac{kW}{1000\,W} = 29.9 = \underline{30.\,kW}$$

EXAMPLE PROBLEM 18.11

The static water level is 30 m, and the drawdown in the pumped well is 9.5 m when pumping at a rate of 15 L/s. The discharge pressure is 420 kPa, and the head loss in the pump column and discharge line up to the pressure gauge is estimated to be 1.5 m. A submersible pump is set at a depth of 50.0 m below the pressure gauge. Calculate the total dynamic head and efficiency of the pumping unit if the input power is 11 kW.

Given:

SWL = 30 m, s = 9.5 m, DWL = 30 + 9.5 = 39.5 m, $Z_2 - Z_1$ = 50.0 − 39.5 = 10.5 m
p_1 = 0, p_2 = 420 kPa, v_1 = 0, v_2 = small, Q = 15 L/s, P_I = 11 kW, h_L = 1.5 m

Solution:

Head added

$$h_a = \frac{p_2 - p_1}{\gamma} + Z_2 - Z_1 + h_1 = \frac{420\,kPa.m}{9.81\,kPa} + 10.5\,m + 1.5\,m = 54.8 = \underline{55\,m}$$

Power added

$$P_a = Q \times \gamma \times h_a = \frac{15\,L}{s} \times \frac{9.81\,N}{L} \times 54.8\,m \times \frac{W.s}{N.m} \times \frac{kW}{1000\,W} = 8.06 = 8.1\,kW$$

Overall efficiency

$$E_0 = \frac{P_a}{P_i} = \frac{8.06\,kW}{11\,kW} \times 100\% = 73.2 = \underline{73\%}$$

EXAMPLE PROBLEM 18.12

The volume of water in a well casing is determined to be 220 gal. How many quarts (1/4 gallon) of sodium hypochlorite with 4.0% available chlorine are required to disinfect the well at a chlorine dosage of 75 ppm?

Given:

Parameter	Chemical = 1	Water = 2
Concentration	4.0%	75 mg/L
Volume	?	220 gal

Solution:

Volume of hypochlorite solution

$$V_1 = V_2 \times \frac{C_2}{C_1} = 220\,gal \times \frac{75\,lb}{10^6\,lb} \times \frac{100\%}{4.0\%} \times \frac{4\,Quart}{gal} = 1.65 = \underline{1.7\,Quarts}$$

EXAMPLE PROBLEM 18.13

An existing well has a total casing length of 76 m. The upper casing has a diameter of 30 cm, and the lower 15 m of the casing is of 20 cm diameter. How many kg of 70% HTH will be required to disinfect the well at a chlorine dosage of 100 mg/L? The static water level is 4 m from the ground.

Given:

Parameter	Larger Section	Smaller Section
Length, m	76–15 = 61	15
Diameter, m	0.30	0.20
Height of water, m	61–4 = 57	15 m

Solution:

Volume of water held in the casing

$$V = \frac{\pi(0.30\,m)^2 \times 57\,cm}{4} + \frac{\pi(0.20\,m)^2 \times 15\,m}{4} = 4.49 = 4.5\,m^3$$

Quantity of HTH

$$m = V \times C = 4.49\,m^3 \times \frac{100\,g}{m^3} \times \frac{kg}{1000\,g} \times \frac{kg\,HTH}{0.70\,kg} = 0.642 = \underline{0.64\,kg}$$

EXAMPLE PROBLEM 18.14

A water well has a depth of 315 ft. The upper casing of the well consists of 100 ft of 12-in pipe, and the lower casing is of 10-in diameter and 216 ft long. What is the volume of water held in the casing when the depth to the water level is 84 ft?

Solution:

Since the depth of the water is 84 ft, the height of water in the well is 315−84 = 231 ft. That is, all of the smaller casing is completely filled with water, and 231−216 = 15 ft of the larger casing is filled with water.

Parameter	Casing	
	1	2
Length, ft	100	216
Height of water, ft	15	216
Diameter, in	12	10

Volume of water in the casing

$$V = V_1 + V_2 = \frac{\pi}{4}\left[1.0\,ft^2 \times 15\,ft + \left(\frac{10}{12}ft\right)^2 \times 216\,ft\right] = 129.59 = \underline{130\,ft^3}$$

EXAMPLE PROBLEM 18.15

For the data of Example Problem 18.14, how many gallons of sodium hypochlorite (5.25% available chlorine) are required to disinfect the well to apply a free chlorine dosage of 75 mg/L?

Solution:

Strength of hypochlorite

$$C_1 = \frac{5.25\%}{100\%} \times \frac{kg}{L} \times \frac{1000g}{L} = 52.5\,g\,/\,L$$

Volume of hypochlorite

$$V_1 = V_2 \times \frac{C_2}{C_1} = 129.59\,ft^3 \times \frac{7.48\,gal}{ft^3} \times \frac{75\,mg}{L} \times \frac{L}{52.5\,g} \times \frac{g}{1000\,mg}$$

$$= 1.38 = \underline{1.4\,gal}$$

PRACTICE PROBLEMS

PRACTICE PROBLEM 18.1

A small lake surface is estimated to be 340 m × 180 m. The maximum depth is 5.0 m. If the average depth is assumed to be 40% of the maximum depth, calculate the kg of copper sulphate required to dose the lake at the rate of 2 mg/L. (240 kg)

PRACTICE PROBLEM 18.2

If 240 kg of copper sulphate pentahydrate with 25% available copper is dosed into a reservoir with a capacity of 150 ML, what dosage of Cu is achieved? (0.4 mg/L)

PRACTICE PROBLEM 18.3

If the desired dose of copper sulphate is 6.0 kg/ha, how many kg of the chemical will be required to dose a lake measuring approximately 350 m × 215 m? (45 kg)

PRACTICE PROBLEM 18.4

The specific capacity of a well is 7.0 gpm/ft. What is the maximum yield of a well if the maximum available drawdown is 50 ft? (350 gpm)

PRACTICE PROBLEM 18.5

The static water level in a well is at a depth of 75 feet. After continuous pumping @ 20 L/s, the water level stabilizes at a depth of 115 ft. Determine the specific capacity of the well in gpm/ft. (7.9 gpm/ft)

PRACTICE PROBLEM 18.6

The length of airline in a water well is 120 ft. With the help of a depth meter, the depth of the static water level is observed to be 55 ft. Assuming the airline gauge reads correctly, what should be the pressure reading? (28 psi)

PRACTICE PROBLEM 18.7

During the spring runoff period, a well was not operated for 2 days. A pressure gauge reading on a 50-m airline was noted to be 102 kPa. Pumping was started, and drawdown stabilized after 20 hours of pumping. The constant pumping rate was recorded to be 25 L/s, and the airline gauge reading dropped by 6.2 kPa. Calculate the drawdown and specific capacity of the well. (4.5 L/s.m)

PRACTICE PROBLEM 18.8

A well is equipped with a 40-m-long airline. Before pumping started, the pressure gauge on the air line was read as 310 kPa. After 1 hour of continuous pumping, the pressure dropped to 255 kPa. Calculate the drawdown in the well. (5.61 m)

PRACTICE PROBLEM 18.9

Before starting the pump, the gauge on the airline read 300 kPa. After continuous pumping for 20 hours, the gauge reading stabilized at 200 kPa. During 20 hours of pumping, 1500 m³ of water was pumped. Find the specific capacity. (2.5 L/s.m)

PRACTICE PROBLEM 18.10

The pumping water level for a submersible pump is 52 m below the discharge header. The discharge pressure measured at the pump discharge header is 410 kPa. If the pumping rate is 35 L/s, what is the water horsepower (power added to water), assuming the velocity head and head loss in the pipe are negligible? (32 kW)

PRACTICE PROBLEM 18.11

The static water level is 33 m, and the drawdown in the pumped well is 11 m when pumping at a rate of 22 L/s. The discharge pressure is 390 kPa, and the head loss in the pump column and discharge line up to the pressure gauge is estimated to be 2.5 m. A submersible pump is set at a depth of 55 m below the pressure gauge. Calculate the total dynamic head and efficiency of the pumping unit if the input power is 14 kW. (53 m, 82%)

PRACTICE PROBLEM 18.12

The volume of water in an open well is determined to be 850 gal. How many litres of sodium hypochlorite with 5% available chlorine are required to disinfect the well at a chlorine dosage of 100 ppm? (6.4 L)

PRACTICE PROBLEM 18.13

A new well is to be disinfected with chlorine at a dosage of 60 mg/L. How many grams of 70% HTH will be required to disinfect a well casing with a diameter of 20 cm and length of 54 m? The water level in the well is 4.0 m from the ground. (130 g)

PRACTICE PROBLEM 18.14

Repeat Example Problem 18.14, assuming the diameter of the larger casing is 16 in and that of the lower casing it is 12 in. (190 ft³)

PRACTICE PROBLEM 18.15

For the data of Practice Problem 18.14, what volume of 12% bleach is required to apply a dosage of 100 mg/L? (1.2 gal)

19 Coagulation and Flocculation

The purpose of coagulation and flocculation is to remove turbidity. Non-settleable particles in water are removed by the use of coagulating chemicals. These chemicals cause the non-settleable particles to clump together, forming floc (Figure 19.1). These flocs become heavy and large as the small flocs clump together. These heavy clumps can be removed by the following process of sedimentation and filtration.

Metallic salts, primarily those of aluminium (alum), iron (ferrous or ferric sulphate) and synthetic organic polymers, are commonly used as coagulation chemicals in water treatment. Anionic and non-ionic polymers have also proven to be effective as coagulation and flocculation aids.

19.1 CHEMISTRY OF COAGULATION

Coagulants like alum react with the alkalinity of the water and form insoluble flocs. Sufficient chemicals must be added to water to exceed the solubility limit of the metal hydroxide. The optimum pH range for alum is between 5 and 8. There should be sufficient natural alkalinity present to react with the coagulating chemical and to serve as a buffer. If the source water is low in alkalinity, it may be increased by the addition of lime or soda ash.

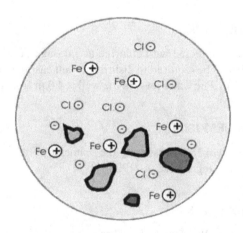

**COLLOIDAL SOLIDS COAGULATED
WITH FERRIC CHLORIDE FeCl$_3$**

FIGURE 19.1 Flocculation.

DOI: 10.1201/9781003468745-19

19.1.1 Chemical Reactions

Coagulation reactions are quite complex. The chemical reactions presented in the following are hypothetical but are useful in estimating the quantities of reactants and products. The chemical reaction of alum with alkalinity is shown as follows:

Alum reacting with natural alkalinity

$$Al_2(SO_4)_3.14.3H_2O + 3Ca(HCO_3)_2 = 2Al(OH)_3 \downarrow +3CaSO_4 + 6CO_2 + 14.3H_2O$$

Molar mass of alum

$$= 2 \ mole \ Al \times \frac{27 \ g}{mol} + 3 \ mole \ SO_4 \times \frac{96 \ g}{mol} + 14.3 \ mol \ H_2O \times \frac{18 \ g}{mol}$$

$$= 599.4 = 600 \ g/mol$$

Stoichiometric requirements

$$\frac{Alkalinity}{Alum} = \frac{3 \ mol \times 100 \ g/mol}{1 \ mol \times 600 \ g/mol} = 0.50 \ or$$

$$Alk \ Consumed = 0.5 \times Alum \ dose$$

Per stoichiometry, the molar ratio of alum to alkalinity is 1 to 3. As shown, one unit of alum reacts with half a unit of the alkalinity expressed as $CaCO_3$. Based on the chemical reaction, alum reacts with alkalinity to produce carbon dioxide, and sulphate ions are added to the finished water. Thus, the addition of alum results in reducing the pH, the alkalinity or both. In addition, a permanent harness as calcium sulphate is added, and production of carbon dioxide makes water corrosive. Despite these drawbacks, alum is still the most-used coagulant. Alum is available both in solid and dry forms. Filter alum is the name used for solid alum. Liquid alum comes in three strengths. The strongest liquid alum is about 48.5%, with a specific gravity of 1.32.

19.1.2 Lime Dose Required

When alum is used for coagulation in the solids contact clarification process, lime or soda ash is sometimes added to provide adequate alkalinity for coagulation and precipitation of the solids. To determine the lime dose that must be added to raw water being treated, you need to know the following:

1. The alkalinity of the raw water (present)
2. The alkalinity that must be present to ensure complete precipitation of the alum and desirable alkalinity in the finished water (residual).
3. The amount of alkalinity that reacts with alum (consumed).

$$Alkalinity \ added = required - present$$
$$alkalinity \ required = consumed + residual$$

As in any algebraic equation, each term should represent the same quantity. The units must be the same (e.g., mg/L or g/m³) and refer to the same chemical. When the natural alkalinity may be as bicarbonates (HCO_3) and lime is used to add alkalinity, it is necessary to express each term as equivalent. For example, 1 mg/L of lime as $Ca(OH)_2$ is not equal to 1 mg/L of alkalinity as $CaCO_3$ because the equivalent masses of the two chemicals in question are different. In other words, one equivalent of alum will react with one equivalent of alkalinity. In terms of equivalence, the chemical form of alkalinity is disregarded.

1. Express the alum dose in alkalinity present and residual alkalinity in equivalent terms. This is done by dividing the concentration of a given reactant or product by its equivalent mass.
2. After completing step 1, calculate the alkalinity in equivalents required to be added.
3. Step 2 will yield the value of alkalinity to be added in equivalents meq/L or eq/m³. To convert this back into lime dosage, multiply the value from step 2 by the equivalent mass of lime or whatever chemical you are going to use to add alkalinity.

EXAMPLE PROBLEM 19.1

Chemical tests indicate that 20 mg/L of alkalinity as $CaCO_3$ is required to be added along with alum for coagulating surface water. Calculate the feed rate in lb/d of CaO (80% pure) for treating a flow of 4.3 MGD.

Given:

Alkalinity added = 20 mg/L as of $CaCO_3$

Solution:

Dosage of CaO for adding alkalinity

$$C_{CaO} = \frac{20\ mg}{L} \times \frac{meq}{50\ mg} \times \frac{28\ mg\ CaO}{meq} \times \frac{1\ mg}{0.80\ mg} = 14\ mg/L$$

Dosage rate

$$M_{CaO} = Q \times C = \frac{4.3MG}{d} \times \frac{14\ mg}{L} \times \frac{8.34\ lb/MG}{mg/L} = 502 = \underline{500\ lb/d}$$

EXAMPLE PROBLEM 19.2

The optimum alum dosage is 50 mg/L, as determined by jar testing. Calculate natural alkalinity as $CaCO_3$ consumed by alum.

Given:

Alum dose = 50 mg/L   Natural alkalinity = ?

Solution:

Alum dosage

$$C_{alum} = \frac{50\ mg}{L} \times \frac{meq}{100\ mg} = 0.50 = 0.5\ meq/L$$

Alkalinity consumed

$$C_{alk} = \frac{0.5\ meq}{L} \times \frac{50\ mg}{meq} = \underline{25\ mg/L}$$

EXAMPLE PROBLEM 19.3

Based on jar testing, the optimum alum dosage for coagulation is 40 mg/L. If it is desired that only 10 mg/L of natural alkalinity be consumed, what amount of alkalinity needs to be added?

Given:

Alum dose = 40 mg/L   Alkalinity consumed = 10 mg/L   Alkalinity added = ?

Solution:

Alum dosage

$$C_{alum} = \frac{40\ mg}{L} \times \frac{meq}{100\ mg} = 0.40 = 0.4\ meq/L$$

Alkalinity consumed by alum

$$C_{alum} = \frac{0.4\ meq}{L} \times \frac{50\ mg}{meq} = 20.0 = 20\ mg/L$$

Alkalinity to be added

$$C_{add} = 20 - 10 = 10\ mg/L$$

EXAMPLE PROBLEM 19.4

Based on jar testing, the optimum dosage of alum is 60 mg/L. If it is desired that only 10 mg/L of natural alkalinity be consumed during coagulation with alum, what dose of 85% lime as CaO is required?

Given:

Alum dose = 60 mg/L   Natural alkalinity consumed = 10 mg/L   85% lime dose = ?

Solution:

Alkalinity consumed by alum

$$C_{cons} == 0.5 \times Alum = 0.5 \times 60 \ mg/L = 30 \ mg/L$$

Alkalinity to be added

$$C_{add} = 30 - 10 = 20 \ mg/L$$

Lime dosage

$$C_{lime} = \frac{20 \ mg}{L} \times \frac{meq}{50 \ mg} \times \frac{28 \ mg \ CaO}{meq} \times \frac{1}{0.85} = 13.2 = \underline{13 \ mg/L}$$

EXAMPLE PROBLEM 19.5

The optimum dosage for coagulating river water is 80 mg/L. If it is desired that only 20 mg/L of natural alkalinity be consumed during coagulation, what dosage rate of 92% soda ash is required to treat a water flow of 3.8 MGD?

Given:

Alum dose = 80 mg/L   Natural alkalinity consumed = 20 mg/L
Q = 3.8 MGD   92% soda-ash dose = ?   Dosage rate, M = ?

Solution:

Alkalinity consumed

$$C_{cons} = 0.5 \times Alum = 0.5 \times 80 \ mg/L = 40 \ mg/L$$

Alk to be added = 40 − 20 = 20 mg/L

Dosage of soda ash

$$C_{soda} = \frac{20 \ mg}{L} \times \frac{meq}{50 \ mg} \times \frac{53 \ mg \ Na_2CO_3}{meq} \times \frac{1}{0.92} = 23.0 = \underline{23 \ mg/L}$$

Dosage rate of soda ash

$$M_{soda} = \frac{3.8 \ MG}{d} \times \frac{23 \ mg}{L} \times \frac{8.34 \ lb/MG}{mg/L} = 730.0 = \underline{730 \ lb/d}$$

EXAMPLE PROBLEM 19.6

Surface water is coagulated by adding 50 mg/L of alum and an equivalent dosage of lime. How many kg of coagulant are used to treat 1 million litres of water? How many kg of lime are required at a purity of 88% CaO?

Given:

Alum = 50 mg/L   Lime as CaO = equivalent dosage Purity = 88%

Solution:

Alum dosage

$$C_{alum} = \frac{50\,mg}{L} \times \frac{meq}{100\,mg} = 0.5\,meq/L$$

Lime dosage

$$C_{lime} = \frac{0.50\,meq}{L} \times \frac{56\,mg\,CaO}{meq} \times \frac{1}{0.88} = 31.8 = 32\,mg/L = 32\,kg/ML$$

It is recommended to adjust the strength of the feed solution so that the feeder setting is about in the middle range (40% to 60%). This allows it to go up and down and ensures feed pump operation under optimal conditions.

EXAMPLE PROBLEM 19.7

A dose of 40 mg/L of alum is used in coagulating turbid surface water. Calculate a) natural alkalinity consumed and b) changes in the SO_4 and HCO_3 content of water.

Solution:

Alum dosage

$$C_{alum} = \frac{40\,mg}{L} \times \frac{meq}{100\,mg} = 0.40 = 0.4\,meq/L$$

Alkalinity consumed by alum

$$C_{cons} = \frac{0.4\,meq}{L} \times \frac{50\,mg}{meq} = 20.0 = 20\,mg/L$$

Increase in sulphate content

$$C_{sulphate} = \frac{0.4\,meq}{L} \times \frac{48\,mg}{meq} = 19.2 = 19\,mg/L$$

Reduction in bicarbonate content

$$C_{carbo} = \frac{0.4\,meq}{L} \times \frac{61\,mg}{meq} = 24.4 = \underline{24\,mg/L}$$

EXAMPLE PROBLEM 19.8

The raw water of the Salman River has an alkalinity of 35 mg/L as $CaCO_3$. The jar test indicated an optimum alum dosage of 30 mg/L to reduce the turbidity to less than 1 NTU. To ensure complete precipitation, a minimum of 30 mg/L of residual alkalinity is recommended. Work out the dosage of alkalinity to be added to maintain the required level of residual alkalinity. Also find the dose of 85% hydrated lime in mg/L that will be needed to complete the reaction. What should be the setting on the lime feeder in kg/d when the daily flow is 10 ML/d? Also work out the production of carbon dioxide.

Given:

Raw alk = 35 mg/L   Residual alk = 30 mg/L
Alum dose = 30 mg/L   Q = 10 ML/d

Solution:

Alum dosage

$$C_{alum} = \frac{30\,mg}{L} \times \frac{mmole}{600\,mg} = 0.05\,mmol/L$$

Alkalinity consumed by alum

$$C_{cons} = 0.50 \times Alum = 0.5 \times 30\,mg/L = 15\,mg/L$$

$$C_{add} = (15 + 30 - 35)\frac{mg}{L} \times \frac{mmol}{100\,mg} = 0.10\,mmol/L$$

Lime dosage

$$C_{lime} = \frac{0.1\,mmol}{L} \times \frac{74\,mg}{mmol} = 7.40 = \underline{7.4\,mg/L}$$

Lime dosage rate

$$M_{lime} = \frac{7.4\,kg}{ML} \times \frac{10\,ML}{d} \times \frac{100\%}{85\%} = 87.05 = \underline{87\,kg/d}$$

Increase in CO_2 content

$$C_{CO_2} = \frac{6.0\,mmol\,CO_2}{1\,mmol\,alum} \times \frac{0.05\,mmol\,alum}{L} \times \frac{44\,mg}{mmol\,CO_2} = 13.2 = \underline{13\,mg/L}$$

19.2 MIXING AND FLOCCULATION

Mixing allows uniform distribution of the chemical in water. This is accomplished by flash mixing. Detention time is usually not a critical factor in flash mixing. Typically, a mixing time of 1 min is considered adequate.

19.2.1 FLOCCULATION

Detention time is quite important in flocculation basins, as more detention time means more opportunity for growth of floc. The detention time recommended for flocculation ranges from 5 to 30 minutes.

EXAMPLE PROBLEM 19.9

A water treatment plant treats a flow of 9.5 ML/d. The flash mix chamber is 75 cm square, and the depth of water is 90 cm. Calculate the detention time in seconds.

Given:

Dimensions: 75 cm × 75 cm × 90 cm $Q = 9.5$ ML/d

Solution:

Capacity of the flash chamber

$$V = 75\,cm \times 75\,cm \times 90\ cm \times \left(\frac{m}{100\,cm}\right)^3 = 0.506 = 0.51\,m^3$$

Detention time

$$t_d = \frac{V}{Q} = 0.506\ m^3 \times \frac{d}{9.5\,ML} \times \frac{ML}{1000\ m^3} \times \frac{1440\ min}{d} \times \frac{60\ s}{min} = 4.60 = \underline{4.6\ s}$$

EXAMPLE PROBLEM 19.10

A water filtration plant treats a flow of 2300 gpm. What must be the diameter of the flash mix chamber with a depth of 2.8 ft to achieve a mixing time of 5.0 s?

Given:

Dimensions: 75 cm × 75 cm × 90 cm $Q = 9.5$ ML/d

Solution:

Detention time

$$t_d = \frac{V}{Q} = \frac{\pi D^2}{4} \times \frac{d}{Q} \quad or \quad D = \sqrt{\frac{4Qt_d}{\pi d}}$$

Diameter of the flash mixer

$$D = \sqrt{\frac{4}{\pi} \times \frac{2300\, gal}{min} \times \frac{5.0\, s}{2.8\, ft} \times \frac{min}{60\, s} \times \frac{ft^3}{7.48\, gal}} = 3.41 = \underline{3.4\, ft}$$

19.3　CHEMICAL FEED RATES

The amount (dosage) and type of chemical (coagulant) change depending on the characteristics of raw water. This determination is made by performing a jar test, which is discussed in the laboratory section. Based on the jar test results, we select the optimum dosage and set the chemical feeders accordingly. The chemical feed pump rate needs to be fed to water for chemical treatment. The chemical feed pump rate depends on the characteristics of the raw water, strength of the chemical and volume of flow being treated.

Feed pump rate

$$\boxed{Q_1 \times C_1 \times Q_s \times C_s}$$

Q = *water flow/pump feed rate*
C = *dosage/strength of solution*

Concentration can be expressed either as mass of the chemical per unit volume or mass of chemical in a given mass of solution. The second form is commonly used for concentrated solutions (liquid chemicals). The relationship between two types of concentrations is as follows:

Chemical strength and concentration

$$\boxed{C = C_{m/m} \times \rho = C_{m/m} \times SG \times \rho_w}$$

ρ = *density of the liquid chemical*
ρ_w = *density of water = 1 kg/L= 1.0 g/mL*

EXAMPLE PROBLEM 19.11

Jar tests indicate that 40 mg/L is the optimum alum dose for treating Rainy River water. If 30 mg/L of residual alkalinity expressed as $CaCO_3$ is required to promote complete precipitation, what is the total alkalinity required for chemical coagulation?

Given:

Alum dose = 40 mg/L　　　Residual alkalinity = 30 mg/L

Solution:

Alum dosage

$$C_{alum} = \frac{40\, mg}{L} \times \frac{meq}{100\, mg} = 0.40 = 0.4\, meq/L$$

Alkalinity consumed by alum

$$C_{cons} = \frac{0.4\,meq}{L} \times \frac{50\,mg}{meq} = 20.0 = 20\,mg/L$$

Alkalinity required

$$Alk = 20.0\,mg/L + 30.0\,mg/L = 50.0 = \underline{50\,mg/L}$$

EXAMPLE PROBLEM 19.12

The dosage of alum in an alum–lime coagulation of water is 60 mg/L. It is desired that alum react with only 10 mg/L of natural alkalinity. What dosage of lime as 85% CaO is required?

Given:

Natural alkalinity = 35 mg/L as $CaCO_3$ Alum dose = 50 mg/L
Residuaalkalinity = 30 mg/L as $CaCO_3$ Lime dose = ?

Solution:

Alkalinity consumed by alum

$$C_{cons} = 0.50 \times (60\,mg)/L = 30.0 = 30\,mg/L$$

Natural alkalinity consumed = 10 mg/L

$$C_{add} = 30\,mg/L - 10\,mg/L = 20\,mg/L$$

Lime dosage

$$C_{lime} = \frac{20\,mg}{L} \times \frac{meq}{50\,mg} \times \frac{28\,mg\,CaO}{meq} \times \frac{mg\,commercial}{0.85\,mg\,Cao} = 13.1 = \underline{13\,mg/L}$$

EXAMPLE PROBLEM 19.13

The natural alkalinity in Kaddy River water is found to be 35 mg/L as $CaCO_3$. The alum dose required, as determined by jar tests, is 50 mg/L. Calculate the lime dose required in mg/L as 80% CaO if 30 mg/L of alkalinity as $CaCO_3$ is required for complete precipitation.

Given:

Natural alkalinity = 35 mg/L as $CaCO_3$ Alum dose = 50 mg/L
Residual alkalinity = 30 mg/L as $CaCO_3$ Lime dose = ?

Solution:

Alkalinity consumed by alum

$$C_{cons} = \frac{50\ mg}{L} \times \frac{meq}{100\ mg} \times \frac{50\ mg}{meq} = 25.0 = 25\ mg/L$$

Alkalinity required

$$C_{reqd} = 25.0\ mg/L + 30.0\ mg/L = 55.0 = 55\ mg/L$$

Alkalinity to be added

$$C_{add} = 55.0\ mg/L - 35.0\ mg/L = 20.0 = 20\ mg/L$$

Lime dosage

$$C_{lime} = \frac{20\ mg}{L} \times \frac{meq}{50\ mg} \times \frac{28\ mg\ Cao}{meq} \times \frac{Commer}{0.8} = 14.0 = \underline{14\ mg/L}$$

EXAMPLE PROBLEM 19.14

A dose of 40 mg/L of alum is used in coagulating turbid surface water. Calculate a) the natural alkalinity consumed and b) changes in the SO_4 and HCO_3 content of water.

Solution:

Alum dosage

$$C_{alum} = \frac{40\ mg}{L} \times \frac{meq}{100\ mg} = 0.40 = 0.4\ meq/L$$

Alkalinity consumed

$$C_{cons} = \frac{0.4\ meq}{L} \times \frac{50\ mg}{meq} = 20.0 = \underline{20\ mg/L}$$

Increase in sulphate content

$$C_{sulphate} = \frac{0.4\ meq}{L} \times \frac{48\ mg}{meq} = 19.2 = \underline{19\ mg/L}$$

Reduction in bicarbonate content

$$C_{carbo} = \frac{0.4\ meq}{L} \times \frac{61\ mg}{meq} = 24.4 = \underline{24\ mg/L}$$

EXAMPLE PROBLEM 19.15

The optimum alum dose as determined by jar testing is 10 mg/L. The chemical feed pump has a maximum capacity of 4.0 gal/h at a setting of 100% capacity. The liquid alum delivered to the plant is 48.5%, with a SG of 1.35. Determine the setting on the liquid alum feeder for treating a flow of 4.5 MGD. How long will 100 gal of liquid alum last?

Given:

Parameter	Liquid Alum = 1	Treated Water = 2
Concentration	48.5%	10 mg/L
SG	1.35	1.0
Pump rate	?	4.5 MGD (mil gal/d)

Solution:

Strength of commercial liquid alum

$$C = C_{m/m} \times \rho = \frac{48.5\%}{100\%} \times \frac{1.35 \times 8.34 \ lb}{gal} = 5.46 = 5.5 \ lb/gal$$

Alum feed pump rate

$$Q_1 = Q_2 \times \frac{C_2}{C_1} = \frac{4.5 \ MG}{d} \times \frac{10 \ mg/L \times 8.34 \ lb/MG}{mg/L} \times \frac{gal}{5.46 \ lb}$$

$$= \frac{68.73 \ gal}{d} \times \frac{d}{24 \ h} = 2.86 = 2.9 \ gal/h$$

Percentage setting

$$S_2 = S_1 \times \frac{Q_1}{Q_2} = 100\% \times \frac{2.9 \ gal/h}{4.0 \ gal/h} = 71.5 = \underline{72\%}$$

Time period to last 100 gal of alum

$$t = \frac{V}{Q} = \frac{1000 \ gal.h}{2.86 \ gal} \times \frac{h}{24 \ h} = 14.56 = \underline{15 \ d}$$

For liquid chemicals with relatively low concentrations, the density can be safely assumed the same as that of water. However, some liquid chemicals may have densities quite different from that of water. As noted earlier, the relative density or specific gravity compares the density of the liquid with that of water. For example, a specific gravity of 1.5 means the liquid is 50% heavier than water, or each litre of liquid weighs 1.5 kg.

PRACTICE PROBLEMS

PRACTICE PROBLEM 19.1

For the data of Example Problem 19.1, calculate the lb/d of 95% pure soda ash (Na_2CO_3) to be added. (800 lb/d)

PRACTICE PROBLEM 19.2

River water requires 60 mg/L of alum dose as found by jar testing. If a residual alkalinity of 30 mg/L must be present in water to ensure complete precipitation of alum added, what is the total alkalinity required? (60 mg/L)

PRACTICE PROBLEM 19.3

The optimum alum dosage for coagulation is 35 mg/L. If it is desired that only 5.0 mg/L of alkalinity be consumed, what amount of alkalinity needs to be added? (13 mg/L)

PRACTICE PROBLEM 19.4

The optimum dosage of alum is 75 mg/L. If it is desired that only 15 mg/L of natural alkalinity be consumed during coagulation with alum, what dose of 85% lime as CaO is required? (15 mg/L)

PRACTICE PROBLEM 19.5

Jar testing of a river water indicates the optimum dosage of alum is 60 mg/L. If it is desired that only 15 mg/L of natural alkalinity be consumed during coagulation, what dosage rate of 88% CaO is required to treat a water flow of 4.5 MGD? (360 lb/d)

PRACTICE PROBLEM 19.6

Given the following data, calculate the lime dose as 80%CaO in mg/L to add the required alkalinity. Alum dose = 60 mg/L, natural alkalinity = 40 mg/L, residual = 40 mg/L. (21 mg/L as 80% CaO)

PRACTICE PROBLEM 19.7

A 30-mg/L dosage of alum and a stoichiometric amount of soda ash are added in coagulation of given surface water. Calculate the amount of soda ash applied per day when water is being treated at the rate of 18 ML/d. What amount of sulphate ions expressed in mg/L is added as a result of this chemical treatment? (290 kg/d, 14 mg/L)

PRACTICE PROBLEM 19.8

The raw water of the Simmon River water has an alkalinity of 40 mg/L as $CaCO_3$. The jar test indicated an optimum alum dosage of 30 mg/L to reduce the turbidity to less than 1 NTU. To ensure complete precipitation, a minimum of 30 mg/L of residual alkalinity is recommended. Work out the dosage of alkalinity to be added to maintain the required level of residual alkalinity. Also find the dose of 85% CaO in mg/L that will be needed to complete the reaction. What should be the setting on the lime feeder in kg/d when the daily flow is 15 ML/d? (3.3 mg/L, 49 kg/d)

PRACTICE PROBLEM 19.9

The flocculation basin for the same water plant is a rectangular tank of size 15 m × 5 m with a water depth of 2.5 m. Calculate the detention time in min. (28 min)

PRACTICE PROBLEM 19.10

Rework the example problem 19.10 assuming the flash mixer is square shaped. (3.0 ft)

PRACTICE PROBLEM 19.11

For removal of turbidity, the optimum alum dose is found to be 40 mg/L. Alkalinity present in water is 120 mg/L expressed as $CaCO_3$. Calculate how much alkalinity will be left in the water after sedimentation. (100 mg/L)

PRACTICE PROBLEM 19.12

The natural alkalinity in Kaddy River water is 35 mg/L as $CaCO_3$. The alum dose required, as determined by jar tests, is 50 mg/L. Calculate the lime dose required in mg/L as 80% CaO if 30 mg/L of alkalinity as $CaCO_3$ is required for complete precipitation. (13 mg/L)

PRACTICE PROBLEM 19.13

Given the following data, calculate the lime dose as 80%CaO in mg/L to add the required alkalinity. Alum dose = 60 mg/L, natural alkalinity = 40 mg/L, residual = 40 mg/L. (21 mg/L as 80% CaO)

PRACTICE PROBLEM 19.14

A dose of 50 mg/L of alum is used in coagulating turbid surface water. Calculate

a) the natural alkalinity consumed (25 ng/L) and
b) changes in the SO_4 and HCO_3 content of water. (24 mg/L↑, 31 mg/L↓)

PRACTICE PROBLEM 19.15

Determine the settings in percent stroke on a chemical feed pump to apply a dosage of 2.5 mg/L. The water is pumped at the rate of 500 gpm, and the strength of the chemical solution fed is 5.0%. The chemical feed pump has a maximum capacity (100% setting) of 5.0 gal/h. (30%)

20 Sedimentation

Sedimentation, also called clarification, is the removal of particulates, chemical flocs and precipitation by gravity settling. The parameters commonly used to express hydraulic loading include detention time, overflow rates and weir loading and, with rectangular tanks, horizontal velocity or mean flow velocity. A schematic of rectangular sedimentation tank is shown in Figure 20.1.

20.1 HYDRAULIC PARAMETERS

Detention time, also called fill time, is the theoretical time for which water stays in the tank. The actual time may be less than this due to short circuiting. Usually expressed in hours, it is calculated by dividing the basin volume by the average daily flow.

Hydraulic detention time

$$HDT = \frac{V}{Q}\,(2-3\,h)$$

20.1.1 OVERFLOW RATE/SURFACE LOADING

Overflow rate is the flow per unit surface area. Basically, overflow rate is the upward velocity with which water rises in the tank. It is important that the settling velocity be more than the overflow rate.

Overflow rate

$$SOR = v_0 = \frac{Flow\ Rate}{surface\ area} = \frac{Q}{A_S}\,(20-30\,m^3/m^2.\,d,500-800\,gpd/ft^2)$$

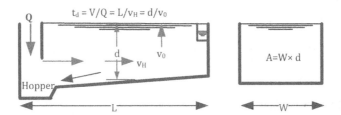

FIGURE 20.1 Rectangular clarifier.

DOI: 10.1201/9781003468745-20

20.1.2 MEAN FLOW VELOCITY

Main flow velocity is the average horizontal velocity of water as it travels through a rectangular basin. The calculation is important since excessive horizontal velocities can cause scouring of the sludge. Mean flow velocity is calculated by dividing the flow by the cross-sectional area of the sedimentation basin.

Horizontal flow velocity

$$v_H = \frac{Flow\ Rate}{Sectional\ area} = \frac{Q}{A_X}(2-2.5\ mm/s,\ 0.5\ ft/min)$$

$Ax = Area\ perpendicular\ to\ flow = W \times d$

20.2 WEIR LOADING

This parameter is computed by dividing the water flow rate by the weir length,

L_w *Weir loading rate*

$$WLR = \frac{Flow\ Rate}{Weir\ Length} = \frac{Q}{L_W}(< 250\ m^3/m.d, < 20000\ gpd/ft)$$

Low weir loading ensures slow uniform water movement with a minimum of short circuiting. Excessive weir loading can cause carryover of solids.

EXAMPLE PROBLEM 20.1

A water treatment plant treats a flow of 6.0 ML/d. The dimensions of the rectangular sedimentation basin as read from the plant design drawings are 25 m long, 8 m wide and water depth of 4 m. Calculate the theoretical detention time in hours.

Given:

Basin: L = 25 m × 8 m × 4 m   Q = 6.0 ML/d   HDT = ?

Solution:

Hydraulic detention time

$$t_d = \frac{V}{Q} = 25\ m \times 8.0\ m \times 4.0\ m \times \frac{d}{6.0\ ML} \times \frac{ML}{1000\ m^3} \times \frac{24\ h}{d} = 3.20 = \underline{3.2\ h}$$

EXAMPLE PROBLEM 20.2

A sedimentation basin is 70 ft long and 20 ft wide and has an effective water depth of 10 ft. What maximum flow can be treated to provide a minimum hydraulic detention time of 2.0 hours?

Given:

L = 70 ft   W = 20 ft   d = 10 ft   HDT = 2.0 h   Q = ?

Solution:

Maximum flow

$$Q = \frac{V}{t} = \frac{(70 \; ft \times 20 \; ft \times 10 \; ft)}{2.0 \; h} \times \frac{7.48 \; gal}{ft^3} \times \frac{h}{60 \; min} = 872 = \underline{870 \; gpm}$$

EXAMPLE PROBLEM 20.3

A water treatment plant has two parallel trains of water treatment units. Each train consists of a rapid mixing chamber, flocculation tank and sedimentation tank. The sedimentation tank is 300 ft long, 150 ft wide and 16 ft in depth. The length of the effluent weir along the four channels at the end of the tank is 1250 ft. Calculate horizontal flow velocity, overflow rate and weir loading for producing a daily flow of 40 MGD.

Given:

Q = 40 MGD   Q/train = 20 MGD   Tank A_s = 300 ft × 150 ft   d = 16 ft

Solution:

Overflow rate

$$v_O = \frac{Q}{A_S} = \frac{20 \times 10^6 \; gal}{d} \times \frac{1}{(300 \; ft \times 150 \; ft)} = 444 = \underline{440 \; gal/ft^2 \cdot d}$$

Horizontal flow velocity

$$v_H = \frac{Q}{A_X} = \frac{20 \times 10^6 \; gal}{d} \times \frac{1}{(150 \; ft \times 16 \; ft)} \times \frac{d}{1440} = 0.774 = \underline{0.77 \; ft/min}$$

Weir loading rate

$$WLR = \frac{Q}{L_W} = \frac{40 \times 10^6}{d} \times \frac{1}{1250 \; ft} = \underline{32000 \; gal/ft.d}$$

EXAMPLE PROBLEM 20.4

A sedimentation basin in a water plant is 23 m square with a liquid depth of 3.6 m and 90 m of effluent weir. Calculate detention time, overflow rate, mean flow velocity and weir loading based on a flow of 11 ML/d.

Given:

Q = 11 ML/d   L × W × d = 23 m × 23 m × 3.5 m L_w = 90 m

Solution:

Hydraulic detention time

$$HDT = \frac{V}{Q} = 23 \ m \times 23 \ m \times 3.5 \ m \times \frac{d}{11ML} \times \frac{ML}{1000 \ m^3} \times \frac{24 \ h}{d} = 4.03 = \underline{4.0 \ h}$$

Overflow rate

$$v_O = \frac{Q}{A_S} = \frac{11 \ ML}{d} \times \frac{1}{(23 \ m \times 23 \ m)} \times \frac{1000 \ m^3}{ML} = 20.7 = \underline{21 \ m^3/m^2.d}$$

Weir loading rate

$$WLR = \frac{Q}{L_W} = \frac{11 \ ML}{d} \times \frac{1}{90 \ m} \times \frac{1000 \ m^3}{ML} = 122.2 = \underline{120 \ m^3/m.d}$$

20.3   FLOCCULATORS

As discussed in the previous chapter, flocculation tanks precede sedimentation tanks. For these tanks, recommended flow through velocity 0.5 ft/min $< v_H <$ 1.5 ft/min (2.5–7.5 mm/s) and minimum detention time of 30 min. Agitators should be driven by variable-speed drives, with the peripheral speed of paddles ranging from 0.15–0.90 m/s (0.5 to 3.0 ft/s).

20.3.1   FLOCCULATOR CLARIFIERS

Flocculator clarifiers, also referred to as solid contact units or upflow tanks, combine the processes of mixing, flocculation and sedimentation in a single compartmented tank. In the beginning, solid contact units were used for removing hardness only, but now they are considered equally good for removing turbidity as well. Reduced space requirements and less costly installation are the major factors favouring their use. However, such a system is less flexible and more suited for treating waters with uniform characteristics.

EXAMPLE PROBLEM 20.5

A water treatment has two parallel trains of water treatment units. Each train consists of a rapid mixing chamber, flocculation tank and sedimentation tank. The flocculation tank is 150 ft wide and 50 ft long and has aside water depth of 15 ft. Calculate horizontal flow velocity and flocculation time based on a daily plant flow of 40 MGD.

Given:

Q = 40 MGD Q = 20 MGD/train A_s = 150 ft × 50 ft d = 15 ft

Solution:

Horizontal flow velocity

$$v_H = \frac{Q}{A_X} = \frac{20 \times 10^6 \; gal}{d} \times \frac{1}{(150 \; ft \times 15 \; ft)} \times \frac{ft^3}{7.48 \; gal} \times \frac{d}{1440} = 1.54 = \underline{1.5 \; ft/min}$$

Hydraulic detention time

$$t_d = \frac{V}{Q} = \frac{L \times W \times d}{Q} = \frac{150 \; ft \times 50 \; ft \times 15 \; ft.d}{20 \times 10^6 \; gal} \times \frac{1440 \; min}{d} = 60.6 = \underline{61 \; min}$$

EXAMPLE PROBLEM 20.6

Calculate the detention time for each of four flocculation basins measuring 15 m × 3.5 m with a water depth of 3.2 m and a sedimentation basin that is 90 m long, 22 m wide and has a water depth of 3.0 m. The flow is 42 ML/d.

Given:

Q = 42 ML/d   Sedimentation: 90 m × 22 m × 3.0 m
Flocculation: 15 m × 3.5 m × 3.2 m   # = 4

Solution:

Detention time (flocculation tanks)

$$t_d = \frac{V}{Q} = \frac{4 \times L \times W \times d}{Q} = \frac{4 \times 15 \; m \times 3.5 \; m \times 3.2 \; m.d}{42 \; ML} \times \frac{ML}{1000 \; m^3} \times \frac{1440 \; min}{d}$$
$$= 23.0 = \underline{23 \; min}$$

Detention time (sedimentation tank)

$$t_d = \frac{90 \; m \times 22 \; m \times 3.0 \; m.d}{42 \; ML} \times \frac{ML}{1000 \; m^3} \times \frac{24 \; h}{d} = 3.39 = \underline{3.4 \; h}$$

EXAMPLE PROBLEM 20.7

For the sedimentation and flocculation units discussed in Example Problem 20.6, find the horizontal flow velocity in both of these units and overflow rate in the sedimentation tank.

Given:

Q = 42 ML/d   Flocculation: A_X = 3.5 m × 3.2 m   # = 4 v_H = ?
Sédimentation: A_X = 22 m × 3.0 m A_S = 110 m × 22 m v_H, v_O = ?

Solution:

Horizontal flow velocity (flocculation tank)

$$v_H = \frac{Q}{A_x} = \frac{42\ ML}{d} \times \frac{1}{4} \times \frac{1}{(3.5\ m \times 3.2\ m)} \times \frac{1000\ m^3}{ML} \times \frac{d}{1440 \times 60\ s} \times \frac{1000\ mm}{m}$$
$$= 1.09 = \underline{1.1\ mm/s}$$

Horizontal flow velocity (sedimentation tank)

$$v_H = \frac{42\ ML}{d} \times \frac{1}{(22\ m \times 3.0\ m)} \times \frac{1000\ m^3}{ML} \times \frac{d}{1440 \times 60\ s} \times \frac{1000\ mm}{m} = 7.37 = \underline{7.4\ mm/s}$$

Overflow rate

$$v_O = \frac{Q}{A_S} = \frac{42\ ML}{d} \times \frac{1}{(110\ m \times 22\ m)} \times \frac{1000\ m^3}{ML} = 17.4 = \underline{17\ m^3/m^2.d}$$

EXAMPLE PROBLEM 20.8

In a water treatment plant, each of the flocculation tanks is fitted with four horizontal shafts, each supporting four paddles. Each paddle is 13 m × 0.20 m and is centred 1.5 m from the shaft. Calculate the peripheral velocity when rotated at 1.5 rpm.

Given:

Paddle: A = 13 m × 0.20 m   r = 1.5 m   N = 1.5 rpm v_p = ?

Solution:

Peripheral flow velocity

$$v_p = N \times 2\pi r = \frac{1.5\ rev}{min} = \frac{2\pi \times 1.5\ m}{rev} \times \frac{min}{60\ s} \times \frac{1000\ mm}{m} = 236 = \underline{240\ mm/s}$$

20.3.2   SETTLEABILITY TEST

In the operation of solids contact clarification units, the settling test provides the operator an indication of the settleability of the slurry or sludge in the sludge blanket. The slurry is allowed to settle for 10 minutes, and the volume of the settled slurry on the bottom of the graduated cylinder is measured and recorded.

Percent settleable solids

$$\boxed{PSS = \frac{V_{SS}}{V_{sample}} \times 100\%}$$

V_{ss} = *settled sludge volume*
V_{sample} = *sample volume, usually 100 mL*

20.3.3 DETENTION TIME

Detention time in the various zones of a solids contact unit is calculated the same way as any other unit that is volume per unit flow rate. The recommended flocculation and mixing time (reaction) should be more than 30 min with a minimum sedimentation time of 2 hours.

20.3.3.1 Alum Floc (Sludge)

In the hypothetical coagulation equation, aluminium floc is written as $Al(OH)_3$. The quantity of sludge produced as $Al(OH)_3$ can be estimated as follows:

$$\frac{Alum\ Floc}{Alum} = \frac{2\ mol \times 78\ g/mol}{1\ mol \times 600\ g/mol} = 0.26 \quad or \quad alum\ floc = Alum\ dose \times 0.26$$

Experiments have shown that actual production of alum floc is twice as much. In addition to aluminium hydroxide, turbidity will be removed. The total solids produced in alum coagulation can be estimated by using the following empirical relationship.

Total solids produced during coagulation flocculation
Total SS $= 0.44 \times Alum\ dose + 0.74 \times Turbidity\ removed\ in\ NTU$

For a known concentration of dry solids produced as sludge, the volume of sludge can be estimated. Remember, sludge with solids content of 1% will contain only 10 g of dry solids in every litre of sludge. The remaining 99% is liquid. The consistency of alum sludge ranges from 1–3%.

EXAMPLE PROBLEM 20.9

A water plant is to process water at the rate of 12 MGD. It is planned to have two rectangular flocculation tanks, each with an operating depth of 10 ft. Select the size of each tank to provide a minimum detention time of 20 min. Assume the length is twice the width.

Given:

Q = 12 mil gal/d $\quad$ d = 10 ft $\quad$ L = 2 × W $\quad$ t_d = 20 min

Solution:

Detention time

$$t_d = \frac{V}{Q} = \frac{L \times W \times d}{Q} = \frac{2W \times W \times d}{Q} = \frac{2W^2 \times d}{Q} \quad or$$

Width of the tank

$$W = \sqrt{\frac{t_d \times Q}{2d}} = \sqrt{\left(20\ min \times \frac{d}{1440\ min} \times \frac{12 \times 10^6\ gal}{d} \times \frac{ft^3}{7.5\ gal} \times \frac{1}{2 \times 10\ ft}\right)}$$

$$= 33.3 = 33\ ft \quad hence \quad L = 2 \times 33.3 = 66.7 = 67\ ft$$

EXAMPLE PROBLEM 20.10

100 mL of slurry from a solid contact unit is tested for settleability. After 10 minutes of settling time, a total of 21 mL of sludge settled to the bottom of the graduated cylinder. What is the percent of settled sludge of the sample?

Given:

$$V_{ss} = 21\ mL \qquad V_{sample} = 100\ mL$$

Solution:

Percent of settleable solids

$$PSS = \frac{V_{ss}}{V_{sample}} \times 100\% = \frac{21\ mL}{100\ mL} 100\% = 21.0 = \underline{21\%}$$

EXAMPLE PROBLEM 20.11

Jar tests indicate that 60 mg/L is the optimum alum dose for treating Murky River water. Estimate the production of chemical sludge as $Al(OH)_3$ when treating a flow of 15 ML/d. Assume the solids concentration in the sludge is 2.0%.

Given:

Alum dose = 60 mg/L Q = 15 ML/d SS_{sl} = 2.0% = 20 g/m^3

Solution:

SS as alum floc

$$SS = \frac{60\ mg}{L} \times 0.26 = 15.6 = 16\ mg/L = 16\ kg/ML$$

Volume of chemical sludge

$$V_{sl} = \frac{Q \times SS}{SS_{sl}} = \frac{15\ ML}{d} \times \frac{15.6\ kg}{ML} \times \frac{L}{20\ g} = 11.7 = \underline{12\ m^3/d}$$

EXAMPLE PROBLEM 20.12

An alum dose of 40 mg/L is applied to treat water at the Geni water filtration plant. Estimate the production of chemical sludge when treating a flow of 5.0 MGD. Assume a solids concentration in the sludge of 2.0%.

Given:

Alum dose = 40 mg/L $\quad$ Q = 5 MGD $\quad$ SS_{sl} = 2.0%

Solution:

SS as alum floc

$$SS = \frac{40\ mg}{L} \times 0.26 \times \frac{8.34\ lb/MG}{mg/L} = 86.7\ lb/MG$$

Production of chemical sludge

$$V_{sludge} = \frac{Q \times SS}{SS_{sl}} = \frac{5.0\ MG}{d} \times \frac{86.7\ lb}{MG} \times \frac{1}{0.02} \times \frac{gal}{8.34\ lb} = 2598 = \underline{2600\ gal/d}$$

EXAMPLE PROBLEM 20.13

A river water treatment plant coagulates raw water with a turbidity of 12 NTU by applying an alum dosage of 40 mg/L. Estimate the sludge solids production based on the empirical formula for processing a water flow of 4.0 MGD. Calculate the volume of sludge from the bottom of the settling basin based on a solid concentration of sludge of 1.5%.

Given:

Alum = 40 mg/L $\quad$ Q = 4.0 MG/d $\quad$ SS_{sl} = 1.5% $\quad$ Turbidity = 12 NTU

Solution:

SS as alum floc

$$SS = \frac{40\ mg}{L} \times 044 + 0.74 \times 12 = 26.48 = 26\ mg/L(ppm)$$

Production of chemical sludge

$$V_{sludge} = \frac{Q \times SS}{SS_{sl}} = \frac{4.5 \times 10^6\ gal}{d} \times \frac{26.48\ lb}{10^6\ lb} \times \frac{100\%}{1.5\%} = 79440 = \underline{79\ 000\ gal/d}$$

EXAMPLE PROBLEM 20.14

A surface water treatment plant coagulates raw water by applying an alum dosage of 50 mg/L. the concentration of SS in raw water and settled water, respectively, are 37 mg/L and 12 mg/L. Estimate the sludge production for processing a water flow of 12 ML/d. Assume the sludge has a consistency of 2.5% solids.

Given:

Alum dose	Flow, Q	SS_{sl}	SS_{raw}	$SS_{settled}$
50 mg/L	12 ML/d	2.5% = 25 kg/ML	37 mg/L	12 mg/L

Solution:

SS conc. of the chemical sludge

SS = removed as hydroxide + removed as solids

$$= 0.26 \times 50 \ mg/L + ((37-12) \ mg)/L = 38 \ mg/L$$

Volume of chemical sludge

$$V_{sl} = \frac{Q \times SS}{SS_{sl}} = \frac{12 \ ML}{d} \times \frac{38 \ kg}{ML} \times \frac{kL}{25 \ g} = 18.2 = \underline{18 \ kL/d}$$

EXAMPLE PROBLEM 20.15

A water treatment plant has two parallel trains of water treatment units. Each train consists of a rapid mixing chamber, flocculation tank and sedimentation tank. The flocculation tank is 140 ft wide, 58 ft long and 15 ft liquid depth. The sedimentation tank is 140 ft wide, 280 ft long and 17 ft in depth. Calculate the major parameters used in sizing these units based on a flow of 40 MGD to each of the trains.

Given:

Flocculation: A_X = 140 ft × 15 ft A_S = 140 ft × 58 ft $t_d, v_H =?$

Sedimentation: A_X = 22 m × 3.0 m A_S = 110 m × 22 m $t_d, v_H, v_O = ?$

Solution:

Horizontal flow velocity (flocculation tank)

$$v_H = \frac{Q}{A_X} = \frac{40 \times 10^6 \ gal}{d} \times \frac{1}{(140 \ ft \times 15 \ ft)} \times \frac{ft^3}{7.48 \ gal} \times \frac{d}{1440 \ min} = 1.77 = \underline{1.8 \ ft/min}$$

Detention time (flocculation tank)

$$t_d = \frac{V}{Q} = \frac{(140\ ft \times 58\ ft \times 15\ ft).d}{40 \times 10^6\ gal} \times \frac{7.48\ gal}{ft^3} \times \frac{1440\ min}{d} = 32.8 = \underline{33\ min}$$

Horizontal flow velocity (sedimentation tank)

$$v_H = \frac{Q}{A_X} = \frac{40 \times 10^6\ gal}{d} \times \frac{1}{(140\ ft \times 17\ ft)} \times \frac{ft^3}{7.48\ gal} \times \frac{d}{1440\ min} = 1.56 = \underline{1.6\ ft/min}$$

Detention time (sedimentation tank)

$$t_d = \frac{V}{Q} = \frac{(140\ ft \times 280\ ft \times 17\ ft).d}{40 \times 10^6\ gal} \times \frac{7.48\ gal}{ft^3} \times \frac{24\ h}{d} = 2.99 = \underline{3.0\ h}$$

Overflow rate (sedimentation tank)

$$v_O = \frac{Q}{A_S} = \frac{40 \times 10^6\ gal}{d} \times \frac{1}{(280\ ft \times 140\ ft)} = 1020 = \underline{1000\ gal/ft^2.d}$$

PRACTICE PROBLEMS

PRACTICE PROBLEM 20.1

A water treatment plant has a circular clarifier for a sedimentation basin. The clarifier has a diameter of 65 ft and a side water depth of 10 ft. What is the hydraulic detention in hours for the basin when the flow is 2.0 MGD? (3.0 h)

PRACTICE PROBLEM 20.2

A clarifier is 17 m in diameter and has a water depth of 3.5 m. At what flow rate would the detention time be 1.5 hours? (13 ML/d)

PRACTICE PROBLEM 20.3

A water treatment plant has two parallel trains of water treatment units. Each train consists of a rapid mixing chamber, flocculation tank and sedimentation tank. The sedimentation tank is 250 ft long, 130 ft wide and 15 ft in depth. The length of the effluent weir along the four channels at the end of the tank is 1100 ft. Calculate horizontal flow velocity, overflow rate and weir loading based on a design flow of 30 MGD. (0.71 ft/min, 460 gal/ft²·d, 14000 gal/ft·d)

PRACTICE PROBLEM 20.4

Two rectangular clarifiers, each 27 m long, 5.0 m wide and 3.8 m deep, settle coagulated water at an average daily flow rate of 6.0 ML/d. The total weir length is 50 m. Calculate the detention time, overflow rate, mean flow velocity and weir loading. (4.1 h, 22 m³/m²·d, 1.8 mm/s, 120 m³/m·d)

PRACTICE PROBLEM 20.5

A water treatment plant has two parallel trains of water treatment units. Each train consists of a rapid mixing chamber, flocculation tank and sedimentation tank. The flocculation tank is 70 ft wide and 45 ft long and has a side water depth of 15 ft. Calculate horizontal flow velocity and flocculation time based on a design flow of 30 MGD. (1.3 ft/min, 34 min)

PRACTICE PROBLEM 20.6

In a water treatment plant, there are four flocculation basins, each measuring 20 m wide and 4.0 long, with a water depth of 3.1 m. The sedimentation basin is 90 m long, 22 m wide and has a water depth of 3.2 m. Find the detention time for each unit based on a flow of 44,000 m³/d. (33 min, 3.5 h)

PRACTICE PROBLEM 20.7

For the sedimentation and flocculation units discussed in Practice Problem 20.6, find the horizontal flow velocity for each unit. For the sedimentation tank, also find the overflow rate. (2.1 mm/s, 7.2 mm/s, 22 m/d)

PRACTICE PROBLEM 20.8

In a water treatment plant, each of the flocculation tanks is fitted with four horizontal shafts, each supporting four paddles. Each paddle is 13 m × 0.20 m and is centred 140 cm from the shaft. Calculate the peripheral velocity when rotated at 2.0 rpm. (290 mm/s)

PRACTICE PROBLEM 20.9

A water plant treats a flow of 10 MGD. It is planned to have two rectangular flocculation tanks, each with an operating depth of 11 ft. Select the size of each tank to provide a minimum detention time of 0.4 h. Assume the length is twice the width. (46 ft × 23 ft)

PRACTICE PROBLEM 20.10

A graduated cylinder is filled to the 100 mL mark with a slurry sample. After 10 minutes of settling, the volume of settled sludge is observed to be 19 mL. Calculate the percent settled sludge of the sample studied. (19%)

PRACTICE PROBLEM 20.11

Jar testing of Muddy River water indicates that 55 mg/L is the optimum alum dose for removal of turbidity. Estimate the production of chemical sludge as $Al(OH)_3$ when treating a flow of 18 ML/d. Assume the solids concentration of the sludge is 1.5%. (17 m^3)

PRACTICE PROBLEM 20.12

The optimum dose of alum is 50 mg/L for treating Murky River water. Estimate the production of chemical sludge as $Al(OH)_3$ when treating a flow of 12 ML/d. Assume solids concentration in the sludge is 1.2% (13 m^3/d).

PRACTICE PROBLEM 20.13

Estimate the sludge solids production based on the empirical formula for processing a daily flow of 5.8 MGD. A river water treatment plant coagulates raw water with a turbidity of 16 NTU by applying an alum dosage of 50 mg/L. Calculate the volume of sludge from the bottom of the settling basin assuming the solid concentration of sludge is 1.5%. (13,000 gal/d)

PRACTICE PROBLEM 20.14

A surface water treatment plant coagulates raw water by applying alum at the rate of 35 mg/L. The concentrations of SS in raw water and settled water, respectively, are 32 mg/L and 11 mg/L. Estimate the sludge production for processing a water flow of 15 ML/d. Assume the sludge has a consistency of 2.0% solids. (23 m^3/d)

PRACTICE PROBLEM 20.15

A water treatment plant has two parallel trains of water treatment units. Each train consists of a rapid mixing chamber, flocculation tank and sedimentation tank. The flocculation tank is 130 ft wide, 60 ft long and 14 ft liquid depth. The sedimentation tank is 130 ft wide, 260 ft long and 16 ft in depth. Calculate the major parameters used in sizing these units based on a flow of 37 MGD to each of the trains. (32 min, 1.9 ft/min, 2.6 h, 1.7 ft/min, 1100 gal/ft^2.d)

21 Filtration

The granular media gravity filter is the most common type of filter. The purpose of filtration is to remove nonsettleable floc remaining after the chemical coagulation and sedimentation. Direct filtration, which does not include sedimentation, can be used to treat surface waters with low turbidity and colour. A schematic showing the main components of a gravity filter is shown in Figure 21.1.

In the operation of the filtration process, you need to determine the following: flow rate, filtration rate, backwash rate, water production, unit filter run volume (UFRV) and percent of water production used to backwash filters.

21.1 FILTRATION RATE

Filtration rate is defined as flow per unit surface area of the filter and is a measure of the hydraulic loading rate. A typical filtration rate is 3.5 L/s.m^2 or 5 gpm/ft^2.d.

Filtration rate/velocity

$$v_F = \frac{Flow\ Rate}{Filter\ surface} = \frac{Q}{A_F}$$

This equation is based on the principle of continuity. Thus, the filtration rate basically indicates the velocity of the flow (v_F) through the filter. It can be directly observed by noting the water drop rate after closing off the influent valve. Higher filtration rates can be afforded in multimedia filters.

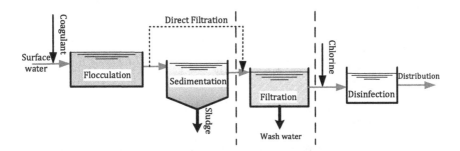

FIGURE 21.1 Conventional water treatment process scheme.

DOI: 10.1201/9781003468745-21

Velocity of filtration

$$v_F = \frac{Drop\ in\ water\ level}{Time\ for\ drop} = \frac{\Delta h}{\Delta t}$$

EXAMPLE PROBLEM 21.1

A 30 ft × 24 ft dual media sand filter treats 3.0 MGD. Calculate the filtration rate in ft/h and gpm/ft^2.

Given:

$A_F = 30\ ft \times 24\ ft$ $Q = 3.0\ MGD$ $v_F = ?$

Solution:

Filtration rate

$$v_F = \frac{Q}{A_F} = \frac{3 \times 10^6\ gal}{d} \times \frac{ft^3}{7.48\ gal} \times \frac{1}{30\ ft \times 24\ ft} \times \frac{d}{24\ h} = 23.21 = \underline{23\ ft/h}$$

$$v_F = \frac{3 \times 10^6\ gal}{d} \times \frac{1}{30\ ft \times 24\ ft} \times \frac{d}{1440\ min} = 2.89 = \underline{2.9\ gpm/ft^2}$$

EXAMPLE PROBLEM 21.2

A filter measuring 14 m × 7 m produces a total of 72 ML during a 72-hour filter run. What is the average filtration rate in L/s.m^2?

Given:

$V = 72\ ML/72\ h$ $A_F = 14\ m \times 7\ m = 98\ m^2$ $v_F = ?$

Solution:

Filtration rate

$$v_F = \frac{Q}{A_F} = \frac{72\ ML}{72\ h} \times \frac{1}{98\ m^2} \times \frac{10^6\ L}{ML} \times \frac{h}{3600s} = 2.83 = \underline{2.8\ L/m^2.s}$$

EXAMPLE PROBLEM 21.3

After closing the influent valve, it is observed that it took 6 min and 30 seconds for the water level to drop by 0.50 m. What is the filtration rate in mm/s and flow through the filter in ML/d if the filter measures 8.0 m × 8.0 m?

Given:

$\Delta H = 0.50$ m $\Delta t = 6$ min 30 s $A_F = 8.0$ m × 8.0 m

Solution:

Filtration rate

$$v_F = \frac{\Delta h}{\Delta t} = \frac{0.50\,m}{6.5\,min} \times \frac{min}{60\,s} \times \frac{1000\,mm}{m} \times 1.28 = \underline{1.3\,mm\,/\,s}$$

Flow through the filter

$$Q = A_F \times v_F = 8.0\,m \times 8.0\,m \times \frac{1.28\,L}{m^2.s} = 81.92 = \underline{82\,L/s}$$

$$= \frac{81.92\,L}{s} \times \frac{m^3}{1000\,L} \times \frac{ML}{1000\,m^3} \times \frac{3600\,s}{h} \times \frac{24\,h}{d} = 7.07 = \underline{7.1\,ML/d}$$

The filtration rate calculated by measuring the drop rate will correspond to the time when the drop rate is observed. For a given filter it may vary depending upon operation time after the backwash cycle. It makes sense to expect higher values in the beginning and drop off towards the end of the filter run due to increased head loss. Filtration rate is one measure of filter production. Along with filter run time, it provides valuable information for the operation of the filter. Problems can develop when design filtration rate values fall outside the rate of 2–6 L/s·m²(3–9 gal/ft²·d).

EXAMPLE PROBLEM 21.4

Six slow sand filter beds are used to treat a maximum flow of 15 ML/d with a filtration rate of 5.0 m³/m²·d. What should be the size of the rectangular filter box of length twice the width? Also assume that one unit out of six is kept as standby.

Given:

Q = 15 ML/d $v_F = 5.0$ m/d   L = 2B

Solution:

Area of the filter surface

$$A_F = \frac{Q}{v_F} = \frac{15\,ML}{d} \times \frac{d}{5.0\,m} \times \frac{1000\,m^3}{ML} = 3000.0 = 3000\,m^2$$

Since one unit is to be kept as standby, 5 units should provide the required surface. Width of the filter surface

$$W = \sqrt{\frac{A_F}{2 \times 5}} = \sqrt{\frac{3000\,m^2}{10}} = 17.3 = \underline{17.5\,m\,say}$$

Six filter units each measuring 35 m × 17.5 m will meet the requirements.

EXAMPLE PROBLEM 21.5

A water treatment plant has six filters with an average filtration rate of 5.9 gal/min·ft². If the plant flow rate is 61 ft³/s, what is the surface area of each filter? If the filter surface is a square, find the length of each filter.

Given:

v_F = 5.9 gal/min.ft²   Q = 61 ft³/s   # = 6 A_F, L = ?

Solution:

Total area of the filter surface

$$A_F = \frac{Q}{v_F} = \frac{61\,ft^3}{s} \times \frac{ft^2.min}{5.9\,gal} \times \frac{60\,s}{min} \times \frac{7.48\,gal}{ft^3} \times \frac{1}{6\,units} = 773\,ft^2\,/unit$$

Length of the filter of the filter surface

$$L = \sqrt{A_F} = \sqrt{773\,ft^2} = 27.8 = \underline{28\,ft}$$

21.1.1   VOLUME OF WATER FILTERED (V_{WF})

In the calculation of filtration rate, it is required to know the flow rate, which can be directly read from the flow meter. The average rate of filtration can be determined if the total volume of water filtered per filter run is known. This is done by obtaining the total flow volume produced and dividing it by the length of the filter run (time).

Avg. rate

$$\boxed{\bar{v} = \frac{V_{WF}}{t_{FR}}}$$

If the flow rate is known, the expected volume of water filtered can be estimated by multiplying it by the length of filter run. It is important to note that this equation will yield the average values over the filter run period.

21.1.2   UNIT FILTER RUN VOLUME (UFRV)

This parameter is a measure of filter performance and is used to compare and evaluate filter runs. UFRV is determined by obtaining the total volume of water filtered between filter runs and dividing this by the surface area of the filter, A_F. UFRV falls in the range of 200–400 m³/m² (650–1300 ft³/ft²) of filter area.

Unit filter run volume

$$UFRV = \frac{V_{WF}}{A_F} = \frac{Q \times t_{FR}}{A_F} = v_F \times t_{FR}$$

v_F used in these calculations should be more appropriately the average filtration rate over the filter run period. Depending on the mode of operation, the actual rate of filtration may vary over the filter run period.

- *Due to the fact that some flow area is occupied by the media particles, the actual velocity of water through the media is more than the filtration rate or velocity.*
- *A filtration rate of 1 L/s·m² is equivalent to a filtration velocity of 1 mm/s.*

EXAMPLE PROBLEM 21.6

Calculate the unit filter run volume for a filter 8.0 m long and 5.0 m wide if the volume of water filtered between successive backwash cycles (filter run) is 8800 m³.

Given:

$A_F = 8.0 \text{ m} \times 5.0 \text{ m}$ $V_{WF} = 8800 \text{ m}^3$ UFRV = ?

Solution:

Unit filter run volume

$$UFRV = \frac{V_{WF}}{A_F} = \frac{8800 \, m^3}{8.0 \, m \times 5.0 \, m} = 220.0 = 220 \, m^3/m^2$$

EXAMPLE PROBLEM 21.7

A multimedia filter with a media surface of 40 ft × 25 ft is operated maintaining a filtration rate of 5.0 gpm/ft². What should be the minimum filter run to get a UFRV of 2500 ft³/ft²? Also find the volume of water that is filtered in one run.

Given:

$A_F = 25 \text{ ft} \times 40 \text{ ft} = 1000 \text{ ft}^2$ t_{FR} = ?   UFRV = 2000 ft³/ft²

Solution:

Filter run

$$t_{FR} = \frac{UFRV}{v_F} = \frac{2500\,ft^3}{ft^2} \times \frac{min.\,ft^2}{5.0\,gal} \times \frac{7.48\,gal}{ft^3} \times \frac{h}{60\,min} = 60.3 = \underline{62\,h}$$

Volume of filter filtered per run

$$V_{WF} = UFRV \times A_F = \frac{2500\,ft^3}{ft^2} \times 1000\,ft^2 \times \frac{7.48\,gal}{ft^3} = 1.496 \times 10^7 = \underline{15\,MG}$$

EXAMPLE PROBLEM 21.8

The average filtration rate for a filter is determined to be 1.6 L/s.m². Calculate UFRV if the filter is backwashed after operating for 71 hours.

Given:

v_F = 2.2 L/s.m² t_{FR} = 71 h   UFRV = ?

Solution:

Unit filter run volume

$$UFRV = v_F \times t_{FR} = \frac{1.6\,L}{m^2.s} \times 71\,h \times \frac{m^3}{1000\,L} = \frac{3600\,s}{h} = 408.9 = \underline{410\,m^3/m^2}$$

EXAMPLE PROBLEM 21.9

Determine the UFRV and total volume filtered per run for a 56 m² filter if the average filtration rate was 1.8 mm/s during a 55-hour filter run.

Given:

v_F = 1.8 mm/s t_{FR} = 55 h   UFRV = ?

Solution:

Unit filter run volume

$$UFRV = v_F \times t_{FR} = \frac{1.8\,L}{m^2.s} \times 55\,h \times \frac{m^3}{1000\,L} = \frac{3600\,s}{h} = 356 = \underline{360\,m^3/m^2}$$

Volume of filter filtered per run

$$V_{WF} = UFRV \times A_F = \frac{356\,m^3}{m^2} \times 56\,m^2 \times \frac{ML}{1000\,m^3} = 19.95 = \underline{20\,ML}$$

A decline in UFRV indicates a drop in filter performance.

21.1.3 BACKWASH RATE

Backwash rate is the velocity of flow upwards, that is, backwash flow divided by the area of the filter. To ensure proper cleaning, backwash rates are 4 to 5 times greater than filtration rate, with 10 L/s.m^2 for 5 to 10 minutes very common.

21.1.4 BACKWASH VELOCITY

Backwash rate is occasionally expressed as rise rate in mm/s or m/h. This is a measure of the upward velocity of the water during backwashing.

Backwash rate/velocity

$$v_{BW} = \frac{Q_{BW}}{A_F} = \frac{Backwash\,flowrate}{Filter\,top\,surface}$$

21.1.5 BACKWASH VOLUME

To determine the volume of water required for backwashing, you must know both the desired backwash flow rate and the duration of backwash cycle. Once the volume of water required for backwashing has been calculated, the required depth of water in the backwash water tank can be determined. This can simply be found by dividing the volume of backwash by the surface area of the storage tank.

Volume of backwash

$$V_{BW} = Q_{BW} \times t_{BW}$$

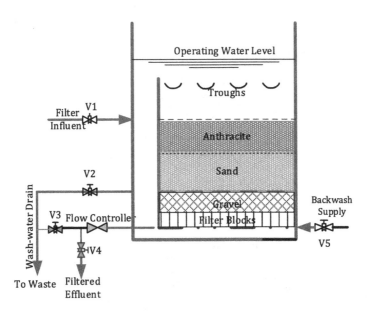

FIGURE 21.2 Backwashing of a filter.

21.1.6   BACKWASH PUMPING RATE

Backwash pumping rate can be calculated by multiplying the desired backwash flow rate by the surface area of the filter. To avoid upsetting filter media, backwash rates are kept low to begin the backwash cycle. Typically, 4.6% of water filtered is used up in backwashing.

21.1.7   PERCENT OF FILTERED WATER USED FOR BACKWASHING

In addition to UFRV and filter run, the percent of finished water used for backwashing is monitored for evaluation filter performance. To determine the percent of water used for backwashing, divide the volume of backwash water by the volume of water filtered and multiply by 100. A typical value is 4%.

Percent backwash water

$$\boxed{\frac{V_{BW}}{V_{WF}} \times 100\% = \frac{V_{BW}}{UFRV \times A_F} \times 100\%}$$

21.1.8   PERCENT MUD BALL VOLUME

Mud balls are floc particles and sand combined. If not removed in the backwash process, they will result in poor performance. To prevent mud balls, the filter media is checked periodically using a mud ball sampler (core sampler). After the filter is backwashed, the filter is drained to lower the water level 30 cm below the surface of the media. A number of samples covering the entire filter surface area are collected. The samples are separated by using sieves. The next step is to determine the volume of the mud balls. The mud balls are placed in a graduated cylinder with a known volume of water. The mud balls are immersed, and the volume of water displaced is noted by observing the rise in water level. Generally, this volume is expressed as a percent of the total sample volume.

EXAMPLE PROBLEM 21.10

A mixed media filter is 8.0 m wide and 11 m long. If the backwash pumping rate is 1.1 m³/s, what is the filter backwash rate in L/s.m² and m/h?

Given:

A_F = 11 m × 8.0 m = 88 m² Q_{BW} = 1.1 m³/s v_{BW} =?

Solution:

Backwash rate/rise rate

$$v_{BW} = \frac{Q_{BW}}{A_F} = \frac{1.1\,m^3}{s} \times \frac{1}{88\,m^2} \times \frac{1000\,L}{m^3} = 12.5\,L/s.m^2$$

$$= \frac{12.5\,L}{s.m^2} \times \frac{m^3}{1000\,L} \times \frac{3600\,s}{h} = 45.0 = \underline{45\,m/h}$$

EXAMPLE PROBLEM 21.11

Determine the backwash pumping rate in gpm for a filter surface measuring 30 ft × 20 ft. The desired backwash rate is 15 gpm/ft². Also determine the volume of backwash water if the filter is backwashed for 8.0 minutes.

Given:

$$A_F = 30 \text{ ft} \times 20 \text{ ft} = 600 \text{ ft}^2 \qquad v_{BW} = 30 \text{ gpm/ft}^2$$

Solution:

Filter backwash rate

$$Q_{BW} = v_{BW} \times A_F = \frac{15\,gal}{min.\,ft^2} \times 600.\,ft^2 = 9000\,gpm$$

Volume of water used for backwash

$$V_{BW} = Q_{BW} \times t_{BW} = \frac{9000\,gal}{min} \times 8.0\,min = \underline{72\,000\,gal}$$

EXAMPLE PROBLEM 21.12

A multimedia filter measures 12 m long and 8.0 m wide. The desired average backwash rate is 8.5 L/s.m² for a period of 10 minutes. What is the required depth of water in the backwash water tank to provide this amount of water if the diameter of the tank is 15 m?

Given:

$$A_F = 12\,m \times 8\,m = 96\,m^2 \quad v_{BW} = 8.5 \text{ L/s.m}^2 \quad D_{Stor} = 15\,m \quad t_{BW} = 10\,min$$

Solution:

Backwash water volume

$$V_{BW} = v_{BW} \times A_F \times t_{BW} = \frac{8.5\,L}{m^2.s} \times 96\,m^2 \times 10\,min \times \frac{60\,s}{min} \times \frac{m^3}{1000\,L}$$

$$= 489.6 = 490\,m^3$$

Depth of water in backwash water tank

$$d = \frac{V}{A} = 489.6\,m^3 \times \frac{4}{\pi(15m)^2} = 2.77 = \underline{2.8\,m}$$

EXAMPLE PROBLEM 21.13

A filter is backwashed every 24 hours, and the average filtration rate is 2.5 L/s·m². The filter is backwashed at a rate of 10 L/s.m² for 15 minutes. Calculate the percent of treated water used in backwashing.

Given:

$$v_F = 2.5 \text{ L/s.m}^2 \qquad v_{BW} = 10 \text{ L/s.m}^2 \qquad t_{FR} = 24 \text{ h}$$

Solution:

Depth of filtered water

$$d_{WF} = v_F \times t_{FR} = \frac{2.5\,mm}{s} \times 24\,h \times \frac{3600\,s}{h} \times \frac{m}{1000\,mm}$$

$$= 216\,m = 220\,m$$

Depth of backwash water

$$d_{BW} = v_{BW} \times t_{BW} = \frac{10\,mm}{s} \times 15\,min \times \frac{60\,s}{min} \times \frac{m}{1000\,mm}$$

$$= 9.0\,m = \frac{9.0\,m}{216\,m} \times 100\% = 4.16 = \underline{4.2\%}$$

EXAMPLE PROBLEM 21.14

A filter measuring 15 ft × 30 ft after producing 2.5 MG in a 24-h period is backwashed at the rate of 15 gal/min·ft² for 12 min. Compute the average filtration rate and percentage of treated water used in backwashing.

Given:

$$V_F = 2.5 \text{ MG} \quad A_F = 15 \text{ ft} \times 30 \text{ ft} \quad v_{BW} = 15 \text{ gal/min.ft}^2 \quad t_{FR} = 24 \text{ h } t_{BW} = 12 \text{ min}$$

Solution:

Filtration velocity

$$v_F = \frac{Q}{A_F} = \frac{2.5 \times 10^6\,gal}{24\,h} \times \frac{1}{15\,ft \times 30\,ft} \times \frac{h}{60\,min} = 3.86 = \underline{3.9\,gpm/ft^2}$$

Volume of backwash water

$$V_{BW} = v_{BW} \times t_{BW} = \frac{15\,gal}{min.\,ft^2} \times 12\,min \times 15\,ft \times 15\,ft \times 30\,ft$$

$$= \frac{81000\,gal}{2.5 \times 10^6\,gal} \times 100\% = 3.24 = \underline{3.2\%}$$

EXAMPLE PROBLEM 21.15

The water level in a clear well rises by 2.0 ft in 145 min. If the clear well has a length of 380 ft, a width of 75 ft and the plant is producing 65 MGD, what is the average discharge rate of the treated water discharge pumps in gallons per minute?

Given:

$L = 380$ ft $W = 75$ ft $Q = 65$ MGD $t = 145$ min $\Delta d = 2.0$ ft $Q_{pump} = ?$

Solution:

Change in storage

$$\Delta Q = A_s \times \frac{\Delta d}{t} = 380 \, ft \times 75 \, ft \times \frac{2.0 \, ft}{145 \, min} \times \frac{7.48 \, gal}{ft^3} = 2940 = 2900 \, gal/min$$

Pumping rate

$$Q_{pump} = Q - \Delta Q = \frac{65 \times 10^6 \, gal}{d} \times \frac{1d}{1440 \, min} - 2940 \, gal / min = 1570.0 = \underline{160 \, gpm}$$

PRACTICE PROBLEMS

PRACTICE PROBLEM 21.1

A high-rate filter has surface of 10 ft × 15 ft, and it treats a flow of 1.2 MGD. Compute the filtration rate and express it as gpm/ft^2 and ft/min. (5.6 gpm/ft^2, 0.74 ft/min)

PRACTICE PROBLEM 21.2

During a particular filter run of 72 hours, a 10 m × 10–m filter produces 66,000 m^3 of water. What is the average filtration rate in mm/s? (2.5 mm/s)

PRACTICE PROBLEM 21.3

The velocity of filtration is known to be 7.4 cm/min. Calculate the flow through the filter in L/s when the filter surface is 55 m^2. (68 L/s)

PRACTICE PROBLEM 21.4

Design five slow sand filter beds to treat a maximum flow of 12 ML/d with a filtration rate not to exceed 6.0 m^3/m^2·d. The filter box is rectangular with a length twice its width. Also assume that one unit out of five will be kept as standby. (32 m × 16 m)

PRACTICE PROBLEM 21.5

A water filtration plant has eight filters, each with an average filtration rate of 5.7 gal/min·ft^2. A flow meter indicates the plant flow is 86 ft^3/s. Assuming the filter surface is a square, find the length of each filter. (29 ft)

PRACTICE PROBLEM 21.6

On average, 12 ML of water is filtered per filter run. If the filter is 6.0 m × 6.0 m, what is the unit filter run volume in m^3/m^2? (330 m^3/m^2)

PRACTICE PROBLEM 21.7

A filter with a media surface of 30 ft × 25 ft is operated, maintaining a filtration rate of 5.0 gpm/ft^2 during a 2-d filter run. What is the unit filter run volume? (14,000 gal/ft^2)

PRACTICE PROBLEM 21.8

Determine the UFRV for a filter in ML/m^2 if the average filtration rate was 2.1 L/s.m^2 during a 45-hour filter run. (0.34 ML/m^2)

PRACTICE PROBLEM 21.9

The average filtration rate during a particular filter run was determined to be 7.8 m/h. If the filter run time was 42.5 hours, calculate UFRV. (330 m^3/m^2)

PRACTICE PROBLEM 21.10

Express the backwash rate in the previous problem in $m^3/m^2 \cdot d$ and cm/min. ($1100 \, m^3/m^2 \cdot d$, 75 cm/min)

PRACTICE PROBLEM 21.11

Calculate the backwash pumping rate in gpm for a filter measuring 30 ft × 30 ft if the desired backwash rate is 15 gpm/ft². What volume of backwash water is required if the filter is to be backwashed at this rate for 10 min? (14,000 gpm, 0.14 MG)

PRACTICE PROBLEM 21.12

A backwash pumping rate of 850 L/s is desired for a total backwash time of 7 minutes. Calculate the depth of water required in the 20-m-diameter backwash water tank to supply this amount of water. (1.1 m)

PRACTICE PROBLEM 21.13

During a filter run, the total volume of water filtered was 7.4 ML. The filter was backwashed at a flow rate of 0.90 m³/s for 7 minutes. Calculate the percent of the product water used for backwashing. (5.1%)

PRACTICE PROBLEM 21.14

A filter is operated, maintaining a filtration rate of 3.5 gal/min·ft². After running for 36 h, the filter is backwashed at the rate of 15 gal/min·ft² for 15 min. Compute the percentage of treated water used in backwashing. (3.0%)

PRACTICE PROBLEM 21.15

The level in a clear well falls 3.0 ft in 2 h and 12 min. If the clear well has a length of 245 ft, a width of 85 ft and the plant is producing 14 MGD, what is the average discharge rate of the treated water discharge pumps in gallons per minute? (13,000 gpm)

22 Chlorination

In water treatment, chlorination is primarily used for disinfection to destroy pathogens. Disinfection is usually the last process before water is pumped into a distribution system. Chlorination is achieved by using liquid chlorine or chlorine compounds, including hypochlorites and chlorine dioxide. Liquid chlorine supplied in gas cylinders is common for larger plants, and smaller plants prefer to use hypochlorites due to safety concerns. Chlorine as an oxidant is used in iron and manganese removal and the destruction of taste- and odour-causing compounds.

22.1 HYPOCHLORITE

Most treatment plants use liquid chlorine for chlorinating the water, since it is less expensive than hypochlorite. Hypochlorites are salts of hypochlorous acid (HOCl). Calcium hypochlorite, $Ca(OCl)_2$, is predominantly in dry form. High-test calcium hypochlorite is available commercially in granular powdered or tablet form, readily dissolves in water and contains about 65–70% available chlorine.

Sodium hypochlorite (NaOCl) is commercially available in liquid form (house bleach) at concentrations between 5 and 15% as available chlorine. Hypochlorite is relatively easy to handle compared to chlorine gas, which is highly toxic. Hypochlorites are commonly used in emergencies and in small water plants.

22.2 CHEMISTRY OF CHLORINATION

Chlorine gas combines with water to form hypochlorous and hydrochloric acid, which in turn can ionize to the hypochlorite ion.

$Cl_2 + H_2O \rightarrow HCl + HOCl$
$HOCl \rightarrow H^+ + OCl^-$

The ionization (breaking up into ions) depends on the pH value. Water with pH around 7.0 contains equal amounts of HOCl and OCl ions. The higher the pH level, the greater the percentage of OCl ions. Since HOCl is more powerful than OCl, the rate of disinfection depends on the pH value. The chemical reactions of hypochlorite in water are similar to those of chlorine gas.

$Ca(OCl)_2 + 2\, H_2O \rightarrow 2\, HOCl + Ca(OH)_2$
$NaOCl + H_2O \rightarrow HOCl + NaOH$

22.3 BREAKPOINT CHLORINATION

22.3.1 Stage 1

A typical chlorination curve is shown in Figure 22.1. When chlorine is added to water, it first oxidizes any reducing compounds, including iron, manganese and

DOI: 10.1201/9781003468745-22

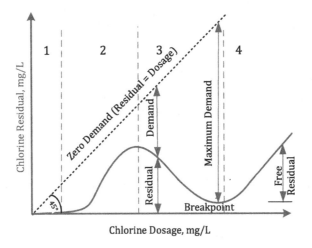

FIGURE 22.1 Breakpoint chlorination.

nitrates. At this stage no disinfection occurs, and no chlorine residual is produced (Points 1–2).

22.3.2 STAGE 2

With the addition of more chlorine, chloroorganics will form as chlorine reacts with organics and ammonia. The chloramines produce a **combined chlorine residual**—a chlorine residual combined with other substances with a reduced disinfectant power. The combination can result in chlorine taste and odour problems (Points 2–3).

22.3.3 STAGE 3

With the addition of more chlorine, the chloramines and some of the chloroorganics are destroyed (Points 3–4). Point 4 is called the **breakpoint**, as any further additions of chlorine do not combine, thus producing a **free chlorine residual**. It is common practice to go beyond the breakpoint, as the free chlorine residual is more effective. In some cases, it is preferred to have a combined residual and add ammonia to produce chloramines (combined residual). This is done to avoid the reaction of chlorine with organics like phenols which cause odour problems. Beyond the breakpoint, any chlorine added will result in an equal increase in the free chlorine residual, as indicated by the straight line with 1 to 1 slope.

22.3.4 CHLORINE DEMAND

Chlorine demand is the difference between chlorine dosage and chlorine residual and is indicated by the vertical distance between applied and residual lines. This represents the amount of chlorine reduced in chemical reactions and no longer available. *Chlorine demand is highest at the breakpoint.* For past breakpoint chlorination (free chlorine residual), this maximum demand has to be satisfied before any free residual is produced.

Dosage = demand + residual

If the increase in chlorine dosage results in similar increase in free chlorine, it is assumed that water is chlorinated past the breakpoint.

EXAMPLE PROBLEM 22.1

If the chlorine dose is 4.65 mg/L and the chlorine residual is 1.83 mg/L, what is the chlorine demand?

Solution:

Chlorine demand
Demand = dose − residual = 4.65 mg/L − 1.83 mg/L = 2.82 mg/L

EXAMPLE PROBLEM 22.2

A chlorinator is set at 10.0 kg/d when the water flow rate is 3.4 ML/d. a) What is the chlorine dosage in mg/L? b) Find the chlorine demand if the residual after 10 min contact time is 0.60 mg/L.

Given:

M = 10.0 kg/d   Q = 3.4 ML/d   C = ?

Solution:

Chlorine dosage

$$C = \frac{M}{Q} = \frac{10\ kg}{d} \times \frac{d}{3.4\ ML} \times \frac{mg/L}{kg/ML} = 2.94 = 2.9\ mg/L$$

Demand = dose − residual = 2.94 − 0.60 = 2.34 = 2.3 mg/L

EXAMPLE PROBLEM 22.3

What is the chlorine dosage if 15 MGD is treated with 290 lb/d of chlorine? If the test indicates a chlorine residual of 0.55 mg/L, what is the chlorine demand of this water?

Given:

M = 290 lb/d   Q = 15 MGD   Residual = 0.55 mg/L   Dosage, demand = ?

Solution:

Chlorine dosage

$$C = \frac{M}{Q} = \frac{290\ lb}{d} \times \frac{d}{15\ MG} \times \frac{mg/L}{8.34\ lb/MG} = 2.32 = \underline{2.3\ mg/L}$$

Chlorine demand
$$= dosage - residual = 2.32 - 0.55 = 2.34 = 1.77 = \underline{1.8\ mg/L}$$

EXAMPLE PROBLEM 22.4

In a small community water supply system, water is disinfected using powered calcium hypochlorite with 70% available chlorine. What should be the setting in kg/d on the gravity feeder to apply 2.2 mg/L of chlorine for a treating a flow of 15.4 L/s?

Given:

C = 2.2 mg/L   Q = 15.4 L/s   M = ?

Solution:

Chlorine feed rate

$$M = Q \times C = \frac{15.4\ L}{s} \times \frac{2.2\ mg}{L} \times \frac{kg}{10^6\ mg} \times \frac{3600\ s}{h} \times \frac{24\ h}{d} \times \frac{kg\ hypo}{0.7\ kg}$$

$$= 4.18 = \underline{4.2\ kg\ of\ hyp/d}$$

EXAMPLE PROBLEM 22.5

On a Monday morning, a chlorine gas cylinder weighs 84 lb. At the same time on Tuesday, after chlorination, the cylinder weighs 58 lb. Calculate the chlorine dosage applied in mg/L if the amount of flow treated for the same period is 890,000 gallons.

Given:

m = (84 − 58) = 26 lb   V = 890,000 gal   C = ?

Solution:

Chlorine dosage

$$C = \frac{m}{V} = \frac{26\ lb}{8.9 \times 10^5\ gal} \times \frac{10^6\ gal}{8.34\ lb} \times \frac{mg}{L} = 3.50 = \underline{3.5\ mg/L}$$

If the values don't compare, you are overdosing or under dosing.

EXAMPLE PROBLEM 22.6

A 2% sodium hypochlorite solution is used to disinfect well water. Experience has shown a chlorine dosage of 1.6 mg/L is necessary to maintain adequate residual in the water supply. During a 1-week time period, the flow totalizer indicated that 7856 m^3 of water was pumped. During the same period, the level of hypochlorite in a 1-m-diameter tank dropped by 72 cm. Determine if the chlorine feed rate is too high, too low or about right.

Given:

Hypochlorite tank: D = 1 m   H = 72 cm
Hypochlorite: C$_1$ = 2% = 20 g/L   V$_1$ = ?
Well water C$_2$ = 1.6 mg/L   V$_2$ = 9856 m^3

Solution:

Volume of hypochlorite pumped

$$V = \frac{\pi(1.0\ m)^2 \times 72\ cm}{4} \times \frac{m}{100\ cm} \times \frac{1000\ L}{m^3} = 565 = 570\ L$$

Chlorine dosage applied

$$C_2 = C_1 \times \frac{Q_1}{Q_2} = \frac{20\ g}{L} \times \frac{565.4\ L}{9856\ m^3} = 1.44\ g/m^3 = \underline{1.4\ mg/L}$$

Since 1.4 mg/L < 1.6 mg/L, it may be concluded that feed is less than desired.

EXAMPLE PROBLEM 22.7

A hypochlorinator is used to chlorinate the well water. To provide an adequate chlorine residual, a chlorine dosage of 1.3 mg/L is required. If the well is pumped at the rate of 300 gpm, what should be the solution feed rate in gal/h given that the solution contains 2.0% available chlorine?

Given:

Parameter	Solution Fed = 1	Treated Water = 2
Pump or flow rate	= ? (gal/h)	300 gpm
Strength/dosage	2.0%	1.3 ppm

Solution:

Feed pump rate

$$Q_1 = Q_2 \times \frac{C_2}{C_1} = \frac{300\ gal}{min} \times \frac{1.3\ lb}{10^6\ lb} \times \frac{100\%}{2.0\%} \times \frac{60\ min}{h} = 1.17 = \underline{1.2\ gal/h}$$

EXAMPLE PROBLEM 22.8

A sodium hypochlorite solution with a strength of 15% is fed to disinfect water. How many litres of the hypochlorite would be required for a dosage of 1.6 mg/L to disinfect 4500 m³/d of water?

Given:

Variable	Solution Fed = 1	Treated Water = 2
Pump or flow rate	?	4500 m³/d
Strength/dosage	150 g/L	1.6 mg/L

Solution:

Feed pump rate

$$Q_1 = Q_2 \times \frac{C_2}{C_1} = \frac{4500 \, m^3}{d} \times \frac{1.6 \, g}{m^3} \times \frac{L}{150 \, g} = 48.0 = \underline{48 \, L/d}$$

EXAMPLE PROBLEM 22.9

Estimate the desired strength (as percent chlorine) of a hypochlorite solution which is pumped by a hypochlorinator that delivers 5.0 gal/h at the desired setting. The water being treated requires a chlorine feed rate of 18 lb/d.

Given:

C = ?   Q = 5.0 gal/h   M = 18 lb/d

Solution:

Strength of hypochlorite solution

$$C = \frac{M}{Q} = \frac{18 \, lb}{d} \times \frac{h}{5.0 \, gal} \times \frac{gal}{8.34 \, lb} \times \frac{d}{24 \, h} \times 100\% = 1.80 = 1.8\%$$

How many gallons of water must be added to 50 gal of 5% hypochlorite solution to make the desired strength of 1.8%?

Given:

Variable	Liquid Hypo = 1	Dil. Solution = 2
Volume	50 gal	?
Strength	5.0%	1.8%

Solution:

Volume of diluted solution

$$V_2 = \frac{C_1}{C_2} \times V_1 = \frac{5.0\%}{1.8\%} \times 50 \; gal = 139 = 140 \; gal$$

Volume of water to be added

$$V_w = 139 \; gal - 50 \; gal = 89 \; gal = \underline{90 \; gal}$$

EXAMPLE PROBLEM 22.10

A new water main is disinfected using a 50 g/m³ chlorine dosage by applying a 2.0% hypochlorite solution.

    a) How many kg of dry hypochlorite powder containing 70% available chlorine must be dissolved in water to make 100 L of 2.0% solution?

    b) How many litres of solution should be applied to each m³ of the water entering the main to provide a concentration of 50 mg/L?

    c) How many litres of hypochlorite solution are used to fill the 300-m-long, 400-mm-diameter main?

Solution:

Kg of hypochlorite powder

$$m = 100 \; L \times \frac{20 \; g}{L} \times \frac{kg \; powder}{0.70 \; kg} = 2.86 = \underline{2.9 \; kg}$$

Volume of hypochlorite required

$$\frac{V_1}{V_2} = \frac{C_2}{C_1} = \frac{50 \; g}{m^3} \times \frac{L}{20 \; g} = 2.50 = \underline{2.5 \; L/m^3}$$

Volume of solution to be used

$$V_1 = \frac{\pi (0.40 \; m)^2 \times 300 \; m}{4} \times \frac{2.5 \; L}{m^3} = 94.2 = \underline{94 \; L}$$

EXAMPLE PROBLEM 22.11

How many gallons of sodium hypochlorite (5.25% available chlorine) are required to disinfect a well with the following parameters:

Depth of well = 315 ft   12-in. diameter well casing, L1 = 100 ft   10-in casing, $L_2 = 216$ ft
Chlorine residual = 50 mg/L   Chlorine demand = 6.0 mg/L   Depth to water, d = 84 ft

Solution:

Since the depth of the water is 84 ft, the height of water in the well is $315 - 84 = 231$ ft; that is to say, all of the smaller casing is completely filled with water, and $231 - 216 = 15$ ft of the larger casing is filled with water.

Parameter	Casing 1	Casing 2
Length, ft	100	216
Height of water, ft	15	216
Diameter, in	12	1.67%

Volume of water in the casing

$$V = V_1 + V_2 = \frac{\pi}{4}\left[1.0\ ft^2 \times 15\ ft + \left(\frac{10}{12} ft\right)^2 \times 216\ ft\right] = 129.59 = 130\ ft^3$$

Chlorine dosage $= Residual + Demand = 50 + 6 = 56\ mg/L$

Strength of hypochlorite

$$C_1 = \frac{5.25\%}{100\%} \times \frac{kg}{L} \times \frac{1000\ g}{L} = 52.5 = 53\ g/L$$

Volume of hypochlorite

$$V_1 = V_2 \times \frac{C_2}{C_1} = 129\ ft^3 \times \frac{7.48\ gal}{ft^3} \times \frac{56\ mg}{L} \times \frac{L}{52.5\ g} \times \frac{L}{1000\ mg} = 1.036 = \underline{1.0\ gal}$$

EXAMPLE PROBLEM 22.12

A chlorinator is set to feed chlorine to treated water at a dose of 30 lb/d. At this dose rate, a free chlorine residual of 0.40 mg/L is produced. When the chlorinator setting is increased to 35 lb/d, the free chlorine residual jumps to 0.63 mg/L. The water flow rate is 2.5 MGD. Is the water being tested past the breakpoint?

Given:

$C = 0.40\ mg/L$ $C_{new} = 0.62\ mg/L$ $Q = 2.5\ MGD$

$M = 30\ lb/d$ $M_{new} = 35\ kg/d$ $Q_{new} = 2.5\ MGD$

Solution:

Actual change in residual

$$\Delta C = 0.63\ mg/L - 0.40\ mg/L = 0.23\ mg/L$$

Expected change in residual

$$\Delta C = \frac{\Delta M}{Q} = \frac{(35-30)\ lb}{d} \times \frac{d}{2.5\ MG} \times \frac{mg/L}{8.34\ lb/MG} = 0.239 = \underline{0.24\ mg/L}$$

EXAMPLE PROBLEM 22.13

A chlorinator is set to feed chlorine at a dosage rate of 11 kg/d. This dose results in a chlorine residual of 0.55 mg/L when the average 24-hour flow is 5.0 ML/d. What is the chlorine demand of water? When the feeder setting is increased to 12 kg/d, determine the expected increase in residual.

Given:

Dosage rate, M = 11 kg/d   Residual = 0.55 mg/L, Q = 5.0 ML/d Demand = ?

Solution:

Chlorine dosage

$$C = \frac{M}{Q} = \frac{11\ kg}{d} \times \frac{d}{5.0\ ML} = 2.2\ kg/ML = 2.2\ mg/L$$

Chlorine demand
$$= dosage - residual = (2.2 - 0.55) = 1.75 = \underline{1.8\ mg/L}$$

New chlorine dosage

$$C_{new} = \frac{M}{Q} = \frac{12\ kg}{d} \times \frac{d}{5.0\ ML} = 2.4\ kg/ML = 2.4\ mg/L$$

Change in residual, $\Delta C = 2.4 - 2.2 = \underline{0.20\ mg/L}$

EXAMPLE PROBLEM 22.14

On a Monday morning, a chlorine gas cylinder weighs 84 lb. At the same time on Tuesday, after chlorination, the cylinder weighs 58 lb. Calculate the chlorine dosage applied in mg/L if the amount of flow treated for the same period is 890,000 gallons.

Given:

m = (84 − 58) = 26 lb   V = 890,000 gal   C = ?

Solution:

Chlorine dosage

$$C = \frac{m}{V} = \frac{26\ lb}{8.9 \times 10^5\ gal} \times \frac{mg/L}{8.34\ lb/MG} = 3.50 = \underline{3.5\ mg/L}$$

EXAMPLE PROBLEM 22.15

Water pumped from a well is disinfected by a hypochlorinator. A chlorine dosage of 1.3 mg/L is applied to maintain the desired level of chlorine residual. During a 1-week period, the flow totalizer indicated 8830 m³ of water was pumped. A 2.5% sodium hypochlorite solution is stored in a 1-m-diameter tank. Find the expected drop in the level of the hypochlorite tank.

Given:

$C_1 = 2.5\% = 25$ g/L $C_2 = 1.3$ mg/L $V_1 = ?$ $V_2 = 8830$ m³ $D = 1.0$ m

Solution:

Volume of solution pumped

$$V_1 = \frac{C_2}{C_1} \times V_2 \times \frac{1.3\ mg}{L} \times \frac{L}{25\ g} \times 8830\ m^3 \times \frac{g}{1000\ mg} \times \frac{1000\ L}{m^3} = 459 = 460\ L$$

Change in hypochlorite level in the tank

$$\Delta d = \frac{\Delta V}{A} = \frac{0.459\ m^3}{0.785(1.0\ m)^2} \times \frac{100\ cm}{m} = 58.4 = \underline{58\ cm}$$

EXAMPLE PROBLEM 22.16

How many lbs of hypochlorite (65% available chlorine) are required to disinfect 4500 ft of a 24-in water main at a chlorine dosage of 50 mg/L?

Given:

D = 24 in = 2.0 ft   L = 4500 ft   C = 50 mg/L

Solution:

Quantity of hypochlorite required

$$m = \frac{\pi}{4}(2.0\ ft)^2 \times 4500\ ft \times \frac{7.48\ gal}{ft^3} \times \frac{50\ mg}{L} \times \frac{8.34\ lb/MG}{mg/L}$$

$$= 44.096\ lb\ Cl_2 \times \frac{lb\ hypo}{0.65\ lb\ Cl_2} = 67.84 = \underline{68\ lb}$$

22.4 DISINFECTION EFFICIENCY

Inactivation of certain species of pathogens mainly depends on two things: contact time and chlorine residual. The two factors are combined in a single parameter known as CT product. Other factors influencing disinfection efficiency are the type of disinfectant, temperature pH and turbidity.

22.4.1 CT Factor

The longer the chlorine contact time, the more effective is disinfection. If the chlorine concentration is decreased, then the **contact time** (t) must be increased. The disinfecting power, often referred to as the kill, is directly related to these two factors, disinfectant residual and time of contact, known as **CT factor**.

Inactivation $= C \times T = CT$ *Factor*

C = *chlorine residual at the end of contact time mg/L*
$T = t_{10}$ = *10th percentile contact time in minutes*

Contact time, t_{10} in the previous equation, is the length of time during which no more than 10% the water passes through a given disinfection process. It is calculated using an estimate of baffling in the tank or through a tracer test. For plug flow systems, as in case of pipes flowing full, contact time can be assumed to be equal to theoretical detention time.

22.4.2 Log Removal

Log removal rather than percentage removal is more commonly used to indicate inactivation of pathogens. If the concentration of a particular pathogen is reduced by a factor of 10, that is 90% removal; it amounts to 1-log removal since the log of 10 is 1. If subscripts i and e refer to influent and effluent of a process or a combination of processes, log removal can be found as follows:

Log Removal, $LR = log\ (C_0/C_e) = -log(1 - PR/100\%)$ *or*

$PR = (1 - 10^{-LR}) \times 100\%$

For example, 2-log removal means pathogens are reduced by a factor of 2, or 99% removal. The regulating authority specifies minimum values of CT factor to achieve a given level of disinfection (1–4 log removals). For surface treatment water systems, minimum CT values must be provided for peak flow conditions to achieve 4-log removals for viruses, 3-log removals for giardia and 2-log removal for crypto. If disinfection follows conventional filtration, 2-log removal of viruses is achieved. Thus, only 2-log removal need be achieved by disinfection methods. In groundwater systems, virus removal is only 2-log since filtration through the ground layers significantly contributes to pathogen removal. The effectiveness of disinfection can also be found by finding the ratio of actual CT and Ct required. If the ratio is 1 or >1, it means removal is satisfactory. The ratio <1 indicates it fails to meet the requirements.

Satisfactory disinfection

$$CT\ Ratio = \frac{CT_{actual}}{CT_{required}} \geq 1$$

EXAMPLE PROBLEM 22.17

What is the log removal for a water treatment plant if the samples show a raw water coliform count of 120/dL and the finished water shows 1.5/dL?

Given:

C_i = 120/dL C_e = 1.5/dL Log removal = ?

Sol:

$$LR = log\left(\frac{C_i}{C_e}\right) = log\left(\frac{120}{1.5}\right) = 1.90 = \underline{1.9}$$

EXAMPLE PROBLEM 22.18

A disinfection process reduces the virus content by 3500 times. What is that as log removal?

Given:

C_i/C_e = 3500   Log removal = ?

Sol:

Log removal
$$LR = log\left(\frac{C_i}{C_e}\right) = log\,(3500) = 3.54 = \underline{3.5}$$

EXAMPLE PROBLEM 22.19

Based on the operating data of a chlorination process in a water filtration plant, 3.35 log removal of viruses is achieved. What is that expressed as percent removal?

Given:

LR = 3.35   Percent removal = ?

Sol:

Percent removal
$$\frac{C_i}{C_e} = Alog\,(LR) = Alog(3.25) = 2238$$

$$PR = \left(1 - \frac{C_e}{C_i}\right) \times 100\% = \left(1 - \frac{1}{2238}\right) \times 100\% = 99.955 = \underline{99.96\%}$$

EXAMPLE PROBLEM 22.20

A well pumping system maintains a free chlorine residual of 0.8 mg/L. Contact time is provided by a 12 m³ capacity tank. Baffling in the tank is such that effective contact time is 40% (baffling factor) of the theoretical detention time. The maximum flow through the tank is 800 L/min. Determine if the system meets the requirements for 2-log removal for viruses. The water temperature is 10°C, and the pH of water is 7.5.

Given:

$C = 0.80$ mg/L $V = 12$ m³ $Q = 800$ L/min $T = 10°C$ $pH = 7.5$ BF $= 40\% = 0.40$

Solution:

Effective contact time

$$t_{eff} = BF \times \frac{V}{Q} = 0.40 \times 12 \ m^3 \times \frac{min}{800 \ L} \times \frac{1000 \ L}{m^3} = 6.0 \ min$$

CT achieved

$$CT = \frac{0.8 \ mg}{L} \times 6.0 \ min = 4.80 = \underline{4.8 \ mg.min/L}$$

Since it is a groundwater source, 2-log virus removal is required. CT value for 2-log virus removal with free chlorine residual is 3 mg.min/L. The actual value is more than required; hence disinfection for virus removal is satisfactory.

EXAMPLE PROBLEM 22.21

A small community is served by a groundwater supply consisting of a 16-in-diameter main with a length of 4760 ft until it reaches the first consumer. The peak hourly pumping rate is 2100 gpm, and the water temperature of is 5°C. The groundwater under the direct influence (GUDI) regulation demands 3-log virus inactivation. What minimum free chlorine residual is required at the outlet of the pipeline to meet the requirement?

Given:

$C = ?$ $L = 4760$ ft $Q = 2100$ gpm $D = 16$-in $= 1.33$ ft

Solution:

Contact time

$$t = \frac{A.L}{Q} \times \frac{\pi(1.33 \ ft)^2}{4} \times \frac{4760 \ ft \times 7.48 \ gal}{ft^3} \times \frac{min}{2100 \ gal} = 23.55 = 24 \ min$$

From the table, the CT factor for 3-log virus removal at 5°C is 6.0 mg.min/L; hence the minimum free chlorine residual desired to meet the requirement is:

Minimum chlorine residual required

$$C = \frac{CT}{t} = \frac{6.0 \; mg.min}{L} \times \frac{1}{23.55 \; min} = 0.254 = \underline{0.25 \; mg/L}$$

EXAMPLE PROBLEM 22.22

A filtration system is required to provide 0.5-log giardia and 2-log virus removal by post-chlorination. The effective contact time for the clear well is evaluated is 43 min, and a free residual of 0.45 mg/L is maintained. The operating temperature is 10°C. How much removal is achieved for viruses and giardia?

Given:

At 10°C, CT = 37 mg.min/L/LR for giardia   3.0 mg.min/L/2-LR for viruses
C = 0.50 mg/L t_{10} = 43 min

Solution:

Log removal of giardia

$$LR = \frac{CT}{CT/LR} = \frac{0.45 \; mg \times 43 \; min}{L} \times \frac{L. \; LR}{37 \; mg.min} = 0.523 = \underline{0.52 \; LR}$$

Log removal of virus

$$LR = \frac{CT}{CT/LR} = \frac{0.45 \; mg \times 43 \; min}{L} \times \frac{L. \; 2LR}{3.0 \; mg.min} = 12.9 = \underline{13 \; LR}$$

EXAMPLE PROBLEM 22.23

Due to problems of elevated levels of THMs, prechlorination at a conventional water treatment plant is discontinued. Consequently, the chlorine dose was increased before the filters and the clear well, and a lithium chloride tracer study was performed. The plant requires a minimum of 1.0 log removal for giardia cysts. Given the following parameters, determine if this plant is in CT compliance.

Given:

Unit	t_{10}, min	Residual, mg/L	Temp T, °C	pH
Filter	12	0.5	15	7.0
Piping	3	0.5	15	7.0
Clear well	41	1.0	17	8.0

Solution:

Unit	t_{10}, min	C, mg/L	C × T	T, °C	pH	CT_{reqd}	IR
Filter	12	0.5	6	15	7.0	25	0.24
Piping	3	0.5	1.5	15	7.0	25	0.06
Clear well	41	1.0	41	17	8.0	32	1.28
Σ	56		48.5				1.58

The last column in the table indicates the inactivation ratio. Since the total 1.32 > 1, that means this plant meets the requirements for log removal for giardia.

Sample of Calculations

For filtration process

$$CT_{act} = 12 \times 0.5 = 6 \ mg. \ min/L$$

$$CT_{reqd} = 25 \ mg. \ min/L$$

$$IR = \frac{CT_{act}}{CT_{reqd}} = \frac{6}{25} = 0.24$$

EXAMPLE PROBLEM 22.24

The filtered water at peak hourly flow from a plant with direct filtration has a pH of 7.0 and temperature 15°C. The effective detention time in the clear well is 24 min, followed by a 1.3-km transmission line at flow velocity of 1.2 m/s before entering the distribution system. Assuming 2.0 log removal of giardia before chlorination, what chlorine residual is required at the outlet of the clear well and the pipeline? Assume no loss of residual in the pipeline.

Solution:

Travel time through transmission line

$$t = \frac{L}{v} = \frac{1.3 \ km.s}{1.2 \ m} \times \frac{1000 \ m}{km} \times \frac{min}{60 \ s} = 18.05 = 18 \ min$$

Total contact time

$$T = 24 \ min + 18 = 42 \ min$$

CT required for 1-log removal using free chlorine residual at 15°C and pH of 7.0 is 25 mg.min/L.

Minimum residual required

$$C = \frac{CT}{t} = \frac{25 \ mg.min}{L} \times \frac{1}{42 \ min} = 0.595 = 0.60 \ mg/L$$

EXAMPLE PROBLEM 22.25

Calculate the CTs for a conventional filter plant with the following characteristics to determine if it meets the criteria of 0.5-log removal for giardia cysts.

Summer

Unit	Q	V or D	Length, ft	C, mg/L	T, °C	pH
Clear well	15 MGD	10 MG	—	0.5	15	8.0
Pipe I	4900 gpm	36 in	248	0.3	15	8.0
Pipe II	220 gpm	12 in	67	0.3	15	8.0

Winter

Unit	Q	V or D	Length, ft	C, mg/L	T, °C	pH
Clear well	6 MGD	10 MG	—	0.8	10	7.5
Pipe I	2900 gpm	36 in	250	0.6	10	7.5
Pipe II	110 gpm	12 in	67	0.6	10	7.5

The clear well has no baffling (factor = 0.1). One 24-in-diameter pipeline and one 36-in-diameter pipeline feed the distribution system from the clear well. The 24-in-diameter pipeline has already been shown to meet the CT requirements. The volume of the clear well, length of pipeline until water reaches the first consumer and flow rates for summer and winter conditions are shown in the table. In addition, free chlorine residual, temperature and pH of water are indicated in the table. Water flows down the 36-in-diameter pipeline and feeds into a 12-in-diameter pipeline.

Solution:

Two separate CTs need to be calculated, one for the summer and the other for the winter. The SWTR requires a 3-log removal of giardia cysts. Plants with conventional filtration have 2.5 log credits. Thus, we need at least 0.5 log removal to be in compliance.

Summer

Unit	Q	V or D	L, ft	C, mg/L	t_{10}, min	CT_{act}	T, °C	pH	CT_{reqd}	IR
Clear well	15 MGD	10 MG	—	0.50	96	48	15	8.0	18	2.67
Pipe I	4900 gpm	36 in	250	0.30	2.7	0.81	15	8.0	18	0.05
Pipe II	220 gpm	12 in	67	0.30	1.8	0.54	15	8.0	18	0.04
Σ										2.76

Winter

Unit	Q	V or D	L, ft	C, mg/L	t_{10}, min	CT_{act}	T, °C	pH	CT_{reqd}	IR
Clear well	6 MGD	10 MG	—	0.80	86	69	10	7.5	23	3.00
Pipe I	2900 gpm	36 in	250	0.60	4.52	2.7	10	7.5	23	0.12
Pipe II	210 gpm	12 in	67	0.60	1.9	1.1	10	7.5	23	0.05
Σ										3.17

The cumulative IR both for summer and winter conditions exceeds unity; hence this plant is in compliance.

Sample of Calculations

Clear well (summer)

$$t_{10} = Factor \times \frac{V}{Q} \times 0.1 \times \frac{10\ MG.d}{15\ MG} \times \frac{1440\ min}{d} = 96\ min$$

$$CT_{act} = t_{10} \times C = 96\ min \times 0.5\ mg/L = 48\ mg.min/L$$

$$IR = \frac{CT_{act}}{CT_{reqd}} = \frac{48}{18} = 2.666 = 2.67$$

36-in pipeline (summer)

$$t_{10} = Factor \times \frac{V}{Q} = Factor \times \frac{\pi D^2 L}{4Q}$$

$$= 1 \times \frac{\pi}{4} \times \frac{(3\ ft)^2\ min}{4900\ gal} \times 250\ ft \times \frac{7.48\ gal}{ft^3} = 2.698 = 2.7\ min$$

$$CT_{act} = t_{10} \times C = 2.7\ min \times 0.30\ mg/L = 0.807 = 0.81\ mg.\ min/L$$

$$IR = \frac{CT_{act}}{CT_{reqd}} = \frac{0.807}{18} = 0.045 = 0.05$$

PRACTICE PROBLEMS

PRACTICE PROBLEM 22.1

If the chlorine residual is 0.72 mg/L and the chlorine demand is 1.3 mg/L, what is the chlorine dose? (2.0 mg/L)

PRACTICE PROBLEM 22.2

It is desired to provide a chlorine residual of 0.4 mg/L to a flow of 4500 m³/d. What should be the chlorinator setting in kg/d if the chlorine demand is 2.6 mg/L? (14 kg/d)

PRACTICE PROBLEM 22.3

What is the chlorine dosage in milligrams per litre if 40 MGD is treated with 950 lb/d of chlorine? Also, find the chlorine demand if the free chlorine residual is tested to be 0.65 mg/L. (2.9 mg/L, 2.2 mg/L)

PRACTICE PROBLEM 22.4

A flow of 2.8 MGD is disinfected using powdered hypochlorite containing 65% available chlorine. If the powder was fed at the rate of 55 lb/24 h, find the chlorine dosage applied. (1.5 mg/L)

PRACTICE PROBLEM 22.5

For proper disinfection, a water supply is to be fed at dosage rate of 2.5 ppm of chlorine. How much chlorine will be left in a 150-lb chlorine tank after a week if the average daily water flow is 0.50 MGD? (77 lb)

PRACTICE PROBLEM 22.6

A well water supply is disinfected by feeding 2.5% sodium hypochlorite solution. Over a period of 12 hours, the total volume of water pumped is 16580 m³. During the same period, the hypochlorinator level drops by 57 cm in the 0.8-m-diameter solution tank. The hypochlorinator is operated continuously over the 12-hour period, and a residual of 0.2 mg/L is maintained. Calculate the chlorine demand of the well water and the feed pump rate in mL/min. (0.23 mg/L, 400 mL/min)

PRACTICE PROBLEM 22.7

A hypochlorite solution with 3.0% available chlorine is used to disinfect water. A chlorine dose of 1.8 mg/L is desired to maintain adequate chlorine residual. If the water pumping rate is 300 gpm, what should be the solution feeder setting in gal/24 h? (26 gal/24h)

PRACTICE PROBLEM 22.8

How many mL of 15% solution should be pipetted to make 1.5 L of 1.0% solution? How many grams of available chlorine are contained in each litre of 1.0% solution? (100 mL, 10 g)

PRACTICE PROBLEM 22.9

a) A hypochlorinator is used to disinfect the water pumped from a well. A chlorine dosage rate of 10 lb/d is required for adequate disinfection of the well water. What should be the strength of solution if the feed pump delivers 3.0 gal/h? (1.7%)

b) 5 gal of 12% liquid hypochlorite are added to a 55-gal drum. How much water should be added to the drum to produce the solution of the desired strength of 1.7%? (30 gal)

PRACTICE PROBLEM 22.10

A part of a water main is repaired and has to be disinfected by applying a chlorine dosage of 50 mg/L by feeding 1.0% solution of chlorine to the water entering the pipe.

a) How many kg of dry hypochlorite powder containing 70% available chlorine must be added to water to make a 250 L of 1.0% solution? (3.57 kg)?

b) How many litres of 1.0% solution are required if the volume of the effected main is estimated to be 42,000 L? (210 L)

PRACTICE PROBLEM 22.11

How many gallons of sodium hypochlorite (5.25% available chlorine) are required to disinfect a well with the following parameters:

| Depth of well = 315 ft | 12-in. diameter well casing, L1 = 100 ft | 10-in casing, L_2 = 216 ft |
| Chlorine residual = 50 mg/L | Chlorine demand = 6.0 mg/L | Depth to water, d = 84 ft |

PRACTICE PROBLEM 22.12

A chlorinator setting is increased by 3.0 kg/d. The free chlorine residual before the increase was observed to be 0.35 mg/L. After the increase, the chlorine residual was 0.46 mg/L. Is the water being chlorinated beyond the breakpoint? The average flow rate is 2.5 ML/d. (No)

PRACTICE PROBLEM 22.13

A chlorinator is set to feed chlorine at a dosage rate of 20 kg/d when the average 24-hour flow is 6.5 ML/d. What is the expected residual if the chlorine demand found in the lab is 1.5 mg/L? Due to recent storms, water quality has been affected, as indicated by the drop in chlorine residual to 0.75 mg/L. Find the increase in chlorine demand and the new setting for the chlorinator to result in the same residual as before. (1.6 mg/L, 2.3 mg/L, 26 kg/d)

PRACTICE PROBLEM 22.14

On a Monday morning, a chlorine gas cylinder weighs 41.5 kg. At the same time on Tuesday, after chlorination, the cylinder weighs 37.3 kg. Determine the average chlorine dosage applied if the amount of flow treated for the same period is 5500 m³. (0.76 mg/L)

PRACTICE PROBLEM 22.15

Water pumped from a well is disinfected by a hypochlorinator. A chlorine dosage of 0.75 mg/L is applied to maintain the desired level of chlorine residual. During a period of 1 week, the flow totalizer indicated 6830 m³ of water was pumped. A 2.5% sodium hypochlorite solution stored in a 75-cm-diameter tank is used to inject chlorine solution. Find the expected drop in the level of the hypochlorite tank. (46 cm)

PRACTICE PROBLEM 22.16

How many lbs of HTH (70% available chlorine) are required to disinfect a 330-ft section of an 18-in water main to achieve a chlorine dosage of 75 mg/L? (2.7 lb)

PRACTICE PROBLEM 22.17

What is the log removal for a water treatment plant if the samples show a raw water coliform count of 280/dL (through extrapolation) and the finished water shows 2.0/dL? (2.2)

PRACTICE PROBLEM 22.18

A disinfection process is able to achieve 2.5-log removal of virus. How many times is the virus concentration reduced? (320×)

PRACTICE PROBLEM 22.19

In a water filtration plant, 2.30-log removal for virus inactivation is achieved by chlorination. What is that expressed as percent removal? (99.5%)

PRACTICE PROBLEM 22.20

A well pumping system maintains a free chlorine residual of 0.65 mg/L. The chlorine contact time is provided by a 2.5-m-diameter tank with a water depth of 2.0 m. Baffling in the tank is such that effective contact time is 50% (baffling factor) of the theoretical time. Maximum flow through the tank is 900 L/min. Determine if the system meets the requirement for 2-log removal for viruses. The water temperature is 10°C, and the pH of water is 7.5. (Yes, CT = 3.5 > 3.0 mg.min/L)

PRACTICE PROBLEM 22.21

Well water is supplied to a community through a 16-in-diameter pipeline 4900 ft in length. The peak hourly flow is 2200 gpm, and the water temperature is 5°C. The regulating authority requires a minimum disinfection level of achieving 3-log virus inactivation. What free chlorine residual is required in the water at the end of the pipeline? (23 min, 0.26 mg/L)

PRACTICE PROBLEM 22.22

A filtration system is required to provide 0.5-log giardia and 2-log virus removal by post-chlorination. The theoretical contact time for the clear well is 62 min at the peak flow, and the baffling factor is evaluated to be 0.75. A free chlorine residual of 0.50 mg/L is maintained. Operating temperature is 5°C. How much removal is achieved for virus and giardia? (12 LR, 0.47 LR < 0.50)

PRACTICE PROBLEM 22.23

Rework Example Problem 22.23, assuming the temperature of the water in the filter and piping is 10°C and the clear well water temperature is 15°C. (1.34)

PRACTICE PROBLEM 22.24

Rework Example Problem 22.24 assuming the disinfectant is chlorine dioxide and the temperature of the water is 10°C. (0.18 mg/L)

PRACTICE PROBLEM 22.25

Rework Example Problem 22.25 assuming 1-log removal of giardia cysts is required and the winter supply is 8 MGD. (summer: IR = 1.4, winter IR = 1.2)

23 Fluoridation

Fluoridation is like chlorination but different in many ways:

- Fluoride in water up to 1.0 mg/L is beneficial in preventing dental cavities.
- Fluoride may be present naturally in water, and its level should be checked.
- The three most commonly used fluoride compounds are sodium fluoride, sodium silico fluoride and fluosilicic acid. Table 23.1 lists some of the characteristics of these compounds.
- Only a percentage of the compound is fluoride ions.
- Much tighter control in dosage rates is required, as both underdosing and overdosing are undesirable.
- The solubility of the chemical is very important when using powder or crystal form.

23.1 AVAILABLE FLUORIDE

The available fluoride ions in a commercial chemical will depend on

1. Commercial purity
2. Percent of F ions in the compound

Available fluoride = Purity × Fluoride ion content

For example, 1 kg of 98% pure commercial NaF with a fluoride content of 45% will yield 0.98 × 0.45 = 0.44 kg of F ion (44%). To add 1 mg of F ions, you will need to add 1/0.44 = 2.27 mg of commercial compound.

TABLE 23.1
Fluoride Compounds Used in the Water Industry

Parameter	Sodium Fluoride	Sodium Silico-Fluoride	Fluosilicic Acid
Chemical formula	NaF	Na_2SiF_6	H_2SiF_6
Molar mass (g/mol)	42	188	144
Fluoride ion (%)	45	61	79
Purity (%)	90–98	98–99	22–30
Commercial form	Powder, crystal	Powder, crystal	Liquid
SG	1.1–1.4	0.9–1.2	1.22–1.26
Solubility g/L 25⁰C	4.1	7.6	Liquid

DOI: 10.1201/9781003468745-23

23.2   FEED RATES

Dry feed rate

$$M = Q \times C \times f$$

f = fraction of fluoride   *Q = water flow rate*   *C = dosage of fluoride*

liquid feed rate

$$Q_1 = Q_2 \times \frac{C_2}{C_1}$$

1 = Liquid chemical/solution   *2 = Water being fluoridated*

Concentration of liquid chemical

$$C = C_{m/m} \times \rho = C_{m/m} \times SG \times \rho_w$$

As an example, 30% commercial fluosilicic acid (79% F ions) with a SG of 1.25 will have a fluoride ion concentration of

$$C = C_{m/m} \times \rho = \frac{30.0\%}{100\%} \times \frac{0.79 \, unit \, F}{unit} \times \frac{1.25 \, kg}{L} \times \frac{1000 \, g}{kg} = 296 \, g/L$$

To apply a dosage of 1.0 mg/L of F ions, $C_2 = 1$ mg/L, the feed rate can be calculated as follows:

$$\frac{Q_1}{Q_2} = \frac{C_2}{C_1} = \times \frac{1.0 \, g}{m^3} \times \times \frac{L}{296 \, g} \times \frac{1000 \, m^3}{ML} = 3.37 = 3.4 \, L/ML$$

This is to say that adding 3.4 L of 30% commercial fluosilicic acid to 1.0 ML of water will result in fluoride dosage of 1.0 mg/L. Knowing the litres of solution required to add 1 mg/L of fluoride to 1 ML of water, dosage tables can be constructed or can be calculated by direct multiplication. Let us say the water flow is 8.8 ML/d and the fluoride level is to be raised from a natural level of 0.2 to 1.0 mg/L. To achieve a dosage of 0.80 mg/L, the silly acid feed rate is calculated as follows:

Feed pump rate

$$Q_1 = Q_2 \times \frac{C_2}{C_1} = \frac{8.8 \, ML}{d} \times \frac{3.37 \, L}{ML} \times \frac{0.80 \, mg/L}{1.0 \, mg/L} = 23.9 = 24 \, L/d$$

EXAMPLE PROBLEM 23.1

Calculate the percent fluoride content of sodium silico-fluoride (Na_2SiF_6).

Solution:

Element	Atoms	Atomic Mass	Molar Mass, g/mol
Na	2	23	$2 \times 23 = 46$
Si	1	28	$1 \times 28 = 28$
F	6	19	$6 \times 19 = 114$
Formula			$\Sigma = 188$

Fraction of fluoride

$$F^+ = \frac{114\,g}{mol} \times \frac{mol}{188\,g} \times 100\% = 60.6 = \underline{61\%}$$

EXAMPLE PROBLEM 23.2

A fluoride dosage of 1.0 mg/l is desired. How many kg of dry sodium silico fluoride (Na_2SiF_6) will be required to fluoridate water flow of 7.6 ML/d? The commercial purity of the chemical is 98%, and the fraction of fluoride ions in the compound is 61%.

Given:

Q = 7.6 ML/d   C = 1.0 mg/L of F   Purity = 98%   Fluoride ions = 61% of Na_2SiF_6

Solution:

Dosage of commercial Na_2SiF_6

$$M = Q \times C = \frac{7.6\,ML}{d} \times \frac{1.0\,kg}{m^3} \times \frac{kg\,Silico}{0.61\,kg\,F} \times \frac{Commercial}{0.98\,pure} = 12.7 = \underline{13\,kg/d}$$

EXAMPLE PROBLEM 23.3

How many kg of 98% pure NaF must be added to 400 L of water to make a 1.0% fluoride solution?

Given:

C = 1.0% as F   Purity = 98%   % F ions = 61%   Water = 400 L = 400 kg

Solution:

Mass of commercial chemical

$$m = \frac{C}{(1-C)} \times m_w = \frac{0.01}{(1-0.01)} \times 400.kg \times \frac{kg\,commercial}{0.98 \times 0.61\,kg\,F} = 6.578 = \underline{6.6\,kg}$$

EXAMPLE PROBLEM 23.4

A Na_2SiF_6 solution is prepared by dissolving 5.0 kg of 98% pure commercial salt to make 100 L of solution. Calculate the feed rate of this solution in millilitres per cubic metre of water flow to increase the fluoride content of water by 0.9 mg/L.

Given:

m = 5.0 kg (98% pure, 61% as F) $V_{Soln} = 100$ L

Solution:

Solution F-ion strength

$$C = \frac{m}{V} = \frac{5.0 \text{ kg comm.}}{100 \text{ L}} \times \frac{980 \text{ g pure}}{\text{kg comm.}} \times \frac{0.61 \text{ g } F-ion}{g \text{ chemical}} = 30.0 = 30 \text{ g/L} = \underline{3.0\%}$$

Given:

Solution = 1   Water = 2 $C_1 = 30$ g/L $C_2 = 0.9$ mg/L

Solution:

Feed pump rate

$$\frac{Q_1}{Q_2} = \frac{C_1}{C_2} = \frac{0.90 \text{ kg}}{ML} \times \frac{1000 \text{ g}}{kg} \times \frac{L}{30.0 \text{ g}} = 30.0 = \underline{30 \text{ L/ML}}$$

30 L of 3.0% F-ion solution needed to dose 1 ML of water @ 0.9 mg/L of F-ion.

EXAMPLE PROBLEM 23.5

A fluoridation installation consists of a feed pump that applies hydro-fluosilicic acid with 25% purity to a water supply directly from the shipping drum placed on a platform scale. If the desired concentration of fluoride in the water is 1.0 mg/L, what should be the solution feed rate in mL/min if the water flow rate is 80 L/s? The SG for the acid is 1.22 and fluoride content is 79%. Assume there is no natural fluoride present in the water.

Given:

Parameter	Solution Fed = 1	Treated Water = 2
Pump or flow rate	? mL/min	80 L/s
Strength/dosage	25%@ 79%, SG = 1.2	1.0 mg/L

Solution:

Strength of solution as F-ion

$$C_1 = \frac{25\%}{100\%} \times \frac{1.22 \text{ kg}}{L} \times \frac{0.79 \text{ g ion}}{g \text{ acid}} \times \frac{1000 \text{ g}}{kg} = 241 = 240 \text{ g/L}$$

Fed pump rate

$$Q_1 = Q_2 \times \frac{C_2}{C_1} = \frac{80\,L}{s} \times \frac{1\,mg}{L} \times \frac{L}{241\,g} \times \frac{g}{1000\,mg} \times \frac{60\,s}{min} \times \frac{1000\,mL}{L} = 19.9$$

$$= 20\,mL/min$$

EXAMPLE PROBLEM 23.6

A flow of 0.50 ML/d is treated with 1.0 kg/d of commercial salt of NaF. The commercial purity is 98%, and the fluoride content is 45%. Determine the fluoride dosage in mg/L.

Given:

M = 1.0 kg/d   Q = 0.50 ML/d Purity = 98%   F content = 45%

Solution:

Fluoride dosage

$$C = \frac{M}{Q} = \frac{1.0\,kg}{d} \times \frac{d}{0.50\,ML} \times \frac{0.98 \times 0.45\,kg\,F}{kg\,salt} \times \frac{mg/L}{kg/ML} = 0.882 = \underline{0.88\,mg/L}$$

EXAMPLE PROBLEM 23.7

Calculate the feed rate for hydrofluosilicic acid (H_2SiF_6) in gallons per day given the following data:

Plant Q = 43.5 MGD   F-ion dosage = 1.0 mg/L   Strength of acid = 21%
F-content = 79%   Fluoride present = 0.30 mg/L   Density of acid = 9.83 lb/gal

Given:

Parameter	Solution Fed = 1	Treated Water = 2
Pump or flow rate	? mL/min	44 MGD
Strength/dosage	21%@ 79%, SG = 1.2	1.0 mg/L

Solution:

Strength of solution as F-ion

$$C = \frac{21\%}{100\%} \times \frac{1.2\,kg}{L} \times \frac{0.79\,g\,ion}{g\,acid} \times \frac{1000\,g}{kg} = 199\,g/L$$

F-ion dosage required
Added = desired − Present = 1.0 − 0.30 = 0.70 mg/L

Feed pump rate

$$Q_1 = Q_2 \times \frac{C_2}{C_1} = \frac{44\,MG}{d} \times \frac{0.70\,mg}{L} \times \frac{L}{199.g} \times \frac{g}{1000\,mg} = 154.7 = \underline{150\,gal/d}$$

23.3 SOLUTION MIXTURES

When mixing solutions of different strengths, the volume or flow of chemicals and their strengths must be known to determine the strength of the mixture solution. Based on mass balance, the concentration of the mixture solution is given by the following expression.

Mixture concentration

$$C_{mix} = \frac{(V_1 \times C_1 \times V_2 \times C_2)}{(V_1 + V_2)}$$

This formula is also applicable if the concentrations are expressed as mass to mass provided the specific gravities of the two solutions being mixed is about the same. If the densities are very different, the factor of *specific gravity G* must be included.

Combined concentration

$$C_{mix} = \frac{(V_1 C_1 G_1 + V_2 C_2 G_2)}{(V_1 G_1 + V_2 G_2)}$$

EXAMPLE PROBLEM 23.8

A hydro-fluosilicic acid tank contains 1200 L of acid with a strength of 18.0%. A vendor delivers 7500 L of 20.0% acid to the tank. What is the resulting strength of the mixture as a percentage?

Given:

Parameter	Solution = 1	Solution = 2	Mix = 3
Concentration, %	18.0	20.0	?
Volume, L	1200	7500	8700

Solution:

Resultant concentration

$$C_{mix} = \frac{(V_1 C_1 G_1 + V_2 C_2 G_2)}{(V_1 G_1 + V_2 G_2)} = \frac{(1200\, L \times 18\% + 7500 \times 20\%)}{8700\, L} = 19.72 = \underline{19.7\%}$$

- *Fluoride doses must never be metered against suction head.*
- *Chemical purity indicates the amount of pure or active ingredient.*
- *The SG is not measured until the tank is ready to be placed in service.*

23.4   FEED RATES USING GRAPHS/CHARTS

From the mass balance equation, the chemical feed pump rate to treat a given flow rate is in an inverse proportion to the concentration. For a given fluoride chemical, the F-ion concentration C_1 is known and if desired concentration of fluoride is C_2, the ratio C_2/C_1, becomes a constant. For example, fluosilicic acid with 25% purity and SG of 1.22 is used to add 1 mg/L of F-ion.

$$\frac{Q_1}{Q_2} = \frac{C_2}{C_1} = \frac{1\,kg}{ML} \times \frac{100\%}{25\% \times 0.79} \times \frac{L}{1.22\,kg} = 4.15\,L/ML$$

Chemical feed pump rate to dose @ 1.0 mg/L

$$\boxed{Q_{acid}\,(L/d) = 4.15 \times Q_{water}\ (ML/d)}$$

In other words, 1 L of silly acid added to 0.241 ML of water will yield an F ion concentration of 1 mg/L, or 4.15 L of acid are required to treat a flow of 1 ML of water to increase the F ion concentration by 1 mg/L. This relationship can also be represented by a graph or chart by choosing the Y-axis as ML/d of water treated and X-axis as L/d of the solution and drawing a straight line through the origin with a slope of 0.241 ML/L. Each straight line represents the relationship for a chemical solution of given strength. A chemical with higher strength will yield a different line. A chemical with higher strength will be represented by a straight line with the steepest slope. Graphs for strengths of the solutions are consolidated on one chart. Some operators prefer to use graphs or charts over the formula. Nevertheless, it is strongly recommended that the values read from the chart be confirmed by calculating the feed rates using the formula. The steps for determining solution feed rates using the chart are as follows:

- Mark the flow rate on the Y-axis 1.
- Draw a horizontal line to the right until it intersects with the straight line representing the given solution strength (2).
- Drop a vertical line from (2) to the X-axis and locate point (3).
- Read the value indicated by point (3) and multiply it by the desired dosage of F ions to obtain the required feed rate.

EXAMPLE PROBLEM 23.9

At a water treatment plant, fluoride is added by feeding 23%, SG = 1.2 hydrofluosilicic acid. Work out the chemical feed pump rate to add 1.0 mg/L of fluoride to every ML of water fluoridated.

Given:

$C_1 = 23\%$ SG = 1.2 (1.2 kg/L) $C_2 = 1.0$ mg/L

Solution:

Chemical feed pump rate

$$\frac{Q_1}{Q_2} = \frac{C_2}{C_1} = \frac{1\,kg\,Fion}{ML} \times \frac{100\%}{23\%\,acid} \times \frac{kg\,acid}{0.79\,kg\,F} \times \frac{L}{1.2\,kg} = 4.58 = \underline{4.6\,L/ML}$$

EXAMPLE PROBLEM 23.10

Calculate the dosage rate for fluorosilicic acid in lb/d for treating 4.5 MGD. The desired dosage of fluoride ions is 1.20 mg/L. The purity of acid is 98%, and the fluoride content is 60.6%. The natural fluoride in the raw water is 0.37 mg/L.

Given:

C (desired) = 1.2 mg/L   C (natural) = 0.37 mg/L   Q = 4.5 MGD   M = ?

Solution:

F-ion dosage required
C = 1.2 − 0.37 = 0.83 *mg/L*

Chemical dosage rate

$$M = Q \times C = \frac{4.2\,MG}{d} \times \frac{0.83 \times 8.34\,lb}{MG} \times \frac{100\%}{0.98\%} \times \frac{lb}{0.606\,lb\,F} = 48.9 = 49\,lb/d$$

EXAMPLE PROBLEM 23.11

Calculate the feed pump rate for fluorosilicic acid in gal/h when the plant's daily flow rate is 22 MGD, acid strength is 25% and desired dosage of fluoride ions is 1.20 mg/L. Fluoride ions in acid are 79% and weigh 9.8 lb/gal. The natural fluoride in the raw water is 0.60 mg/L.

Given:

C_2 = 1.2 mg/L −0.60 mg/L (natural) = 0.60 mg/L Q_2 = 22 MGD Q_1 = ?
C_1 = 25%   F content = 79%   ρ = 9.8 lb/gal

Solution:

Strength of solution as F-ion

$$C_1 = \frac{25\%}{100\%} \times \frac{9.8\,lb}{gal} \times \frac{0.79\,lb\,ion}{lb\,acid} = 1.93\,lb/gal$$

F-ion dosage required
C_2 = *desired − Natural* = 1.2 − 0.60 = 0.60 *mg/L*

Feed pump rate

$$Q_1 = Q_2 \times \frac{C_2}{C_1} = \frac{22\ MG}{d} \times \frac{0.60 \times 8.34\ lb}{MG} \times \frac{gal}{1.93\ lb} \times \frac{d}{24\ h}$$

$$= 2.37 = \underline{2.4\ gal/h}$$

EXAMPLE PROBLEM 23.12

Calculate the feed rate for fluorosilicic acid in mL/min given the following data:

Flow rate = 45 ML/d Solution strength = 20%   Fluoride (F) desired = 1.20 mg/L
Fluoride content = 79% SG = 1.18   Natural fluoride = 0.50 mg/L

Given:

C_2 = 1.2 mg/L – 0.50 mg/L (natural) = 0.70 mg/L   Q = 45 ML/d
C_1 = 20% F-content = 79% SG = 1.18 Q_1 = ? mL/min

Solution:

Strength of solution as F-ion

$$C_1 = \frac{20\%}{100\%} \times \frac{1.18\ kg}{L} \times \frac{0.79\ g\ ion}{g\ acid} = 1.864 = 1.86\ kg/L$$

Feed pump rate

$$Q_1 = Q_2 \times \frac{C_2}{C_1} = \frac{45\ ML}{d} \times \frac{0.70\ kg}{ML} \times \frac{L}{1.86\ kg} \times \frac{d}{1440\ min} \times \frac{1000\ mL}{L}$$

$$= 11.7 = \underline{12\ mL/min}$$

Alarms are important to signal and prevent the loss of feed and overfeeding.

PRACTICE PROBLEMS

PRACTICE PROBLEM 23.1

Calculate the percent fluoride ions in fluosilicic acid. (79%)

PRACTICE PROBLEM 23.2

Calculate kg/d of commercial sodium silicofluoride to be fed if the natural level of fluoride in the water is 0.2 mg/L and the flow to be treated is 56 L/s. The level of fluoride desired is 1.0 mg/L. (6.5 kg/d)

PRACTICE PROBLEM 23.3

A total of 3.5 kg of 98% pure NaF is dissolved in 200 L of water. What is the mg/L fluoride concentration of the solution? (0.76%)

PRACTICE PROBLEM 23.4

A liquid feeder applies a 4.0% sodium fluoride solution to a water supply, increasing the fluoride concentration from the natural level of 0.4 mg/L to 1.0 mg/L. The commercial NaF powder contains 45% fluoride by mass. What is the dosage of 4.0% NaF solution in mL per cubic meter of water? (33 mL/m^3)

PRACTICE PROBLEM 23.5

A flow of 24 ML/d is to be treated with a 20% solution of hydro-fluosilicic acid with 79% F content and SG of 1.2. What should be the acid feed rate of the acid in mL/min to increase the fluoride content of water from 0.2 to 1.0 mg/L? (70 mL/min)

PRACTICE PROBLEM 23.6

A dry feeder recorded a weight loss of 13 kg of Na_2SiF_6 in the treatment of 7500 m^3 of water. The commercial powder is 98% pure, and the fluoride content in the pure compound is 61%. Calculate the dosage of fluoride ions added to treat the water. (1.0 mg/L)

PRACTICE PROBLEM 23.7

How many litres of water must be mixed with 400 L of 20% fluosilicic acid to result in a 150 g/L fluoride solution? The SG of 20% acid is 1.2. (106 L)

PRACTICE PROBLEM 23.8

In a well water system, water contains natural fluoride on the order of 0.25 mg/L. For the hydrofluosilicic described in the example problem, work out the ratio of L/ML to elevate the fluoride level to 1.0 mg/L. (3.4 L/ML)

PRACTICE PROBLEM 23.9

At a water treatment plant, fluoride is added by feeding 22% hydrofluosilicic acid of SG = 1.2. Find the feed pump rate setting to apply a dosage of 1.2 mg/L of fluoride to every ML of water fluoridated. (5.8 L/ML)

PRACTICE PROBLEM 23.10

A fluoride dose of 1.0 mg/L is used to treat a flow of 7.5 MGD. How many pounds per day of sodium silicofluoride with a commercial purity of 98% and a fluoride ion content of 61% are needed? The water being treated contains 0.15 mg/L fluoride. (89 lb/d)

PRACTICE PROBLEM 23.11

Calculate the feed rate for fluorosilicic acid in gallons per day given the following data: Daily flow = 8.3 MGD treated with 20.0% solution of H_2SiF_6, desired F-ion = 1.20 mg/L, fluoride ion content = 79%. Fluoride in raw water is 0.22 mg/L. H_2SiF_6 weighs 9.8 lb/gal. (44 gal/d)

PRACTICE PROBLEM 23.12

Calculate the feed rate for fluorosilicic acid in mL/min given the following data: Flow rate = 25 ML/d, solution strength = 25%, fluoride (F) desired = 1.0 mg/L, fluoride content = 79%, SG = 1.21, natural fluoride = 0.20 mg/L. (58 mL/min)

24 Water Softening

Hardness is due to **bivalent** ions present in water. Calcium and magnesium are usually the only cations present in significant quantities. Therefore, hardness is considered a general expression of the total concentration of calcium and magnesium ions in terms of common compound calcium carbonate ($CaCO_3$). Thus, to estimate hardness, the concentration of these constituents is expressed in equivalents or milliequivalents per litre. To do this, it is essential to know the equivalent mass of the different constituents.

24.1 EQUIVALENT MASS

Equivalent mass is the atomic mass or molecular mass divided by the net positive valence. Atomic mass of an element can be read from the periodic table. For example, the equivalent mass of $CaCO_3$ is:

Equivalent mass of CaCO₃

$$\frac{100\ g}{mol} \times \frac{mol}{2\ equivalents} = 50\ g/eq = 50\ mg/meq$$

24.2 HARDNESS AS CACO₃

Since hardness can be due to various ions like calcium magnesium, it is defined as the equivalent of calcium bicarbonate. To calculate the total hardness of given water, the concentrations of calcium and magnesium are expressed in meq/L and then added. The sum is multiplied by the equivalent mass of $CaCO_3$ to express as the equivalent of $CaCO_3$.

Total hardness

$$TH = \frac{(Ca+Mg)meq}{L} \times \frac{50\ mg\ CaCO_3}{meq} = mg/L\ CaCO_3$$

EXAMPLE PROBLEM 24.1

Water from a well sample has 72 mg/L of magnesium (Mg) as $CaCO_3$ and 112 mg/L of calcium (Ca) as $CaCO_3$. What is the total hardness of the sample as $CaCO_3$?

Solution:

Since both Ca and Mg are expressed as equivalent $CaCO_3$, no conversion is required.

DOI: 10.1201/9781003468745-24

Total hardness
$$TH = CaH + MgH = 72 \ mg/L + 112 \ mg/L = 184 = 180 \ mg/L$$

EXAMPLE PROBLEM 24.2

What is the calcium hardness as $CaCO_3$ if the water sample has a calcium content of 84 mg/L?

Solution:

Calcium hardness
$$CaH = \frac{84 \ mg}{L} \times \frac{meq}{20 \ mg} \times \frac{50 \ meq}{meq} = 210.0 = 210 \ mg/L$$

EXAMPLE PROBLEM 24.3

The magnesium content of a water sample is 8.4 mg/L. What is the magnesium hardness expressed as milligrams per litre $CaCO_3$?

Magnesium hardness
$$MgH = \frac{8.4 \ mg}{L} \times \frac{meq}{12 \ mg} \times \frac{50 \ meq}{meq} = 35.0 = 35 \ mg/L$$

EXAMPLE PROBLEM 24.4

The total hardness of well water is known to be 210 mg/L. If the magnesium content is 18 mg/L, what is the calcium content?

Solution:

Total hardness
$$TH = \frac{210 \ mg}{L} \times \frac{meq}{50 \ mg} = 4.2 \ meq/L$$

Magnesium hardness
$$MgH = \frac{18 \ mg}{L} \times \frac{meq}{12 \ mg} = 1.5 \ meq/L$$

Calcium content
$$Ca = \frac{(4.2 - 1.5) \ meq}{L} \times \frac{20 \ meq}{meq} = \underline{54 \ meq/L}$$

EXAMPLE PROBLEM 24.5

A sample of water has a calcium content of 40 mg/L and magnesium content of 15 mg/L. What is the total hardness of water expressed as $CaCO_3$?

Solution:

Calcium hardness

$$CaH = \frac{40\ mg}{L} \times \frac{meq}{20\ mg} = 2.00 = 2.0\ meq/L$$

Magnesium hardness

$$MgH = \frac{15\ mg}{L} \times \frac{meq}{12\ mg} = 1.25 = 1.3\ meq/L$$

Total hardness

$$TH = \frac{(2.0-1.25)\ meq}{L} \times \frac{50\ meq}{meq} = 162 = \underline{160\ mg/L}$$

24.3 SOFTENING

Hardness in the range of 80–120 mg/L is desirable. Excessive hardness cause scaling and add a salty taste to water. In such cases, water is softened. Removal of hardness in water is called softening. The lime and soda ash process is a common process used for softening; Figure 24.1 shows the process scheme of lime soda ash softening.

24.4 CARBONATE AND NON-CARBONATE HARDNESS

Total hardness is equal to the sum of Ca and Mg ions, which may be associated with anions like carbonates (HCO_3) or sulphates (SO_4). Depending on the type of anion, the hardness is further classified as **carbonate hardness (temporary)** or **non-carbonate hardness (permanent)**. Carbonate hardness is due to bicarbonate alkalinity, and non-carbonate hardness is hardness in excess of bicarbonate alkalinity. Total alkalinity is due to carbonates; non-carbonates; hydroxides; and occasionally borates,

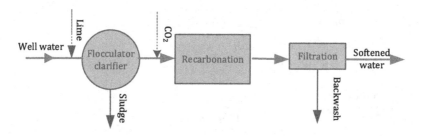

FIGURE 24.1   Single-stage lime soda ash process.

silicates, and phosphates. In the majority of natural waters, it is mainly bicarbonates. For this reason, the terms alkalinity and bicarbonate are used interchangeably.

As hardness and alkalinity can be due to different ions, both are expressed as equivalent of $CaCO_3$

$TH > ALK$	$CH = ALK$	$NCH = TH - ALK$
$TH < ALK$	$NCH = 0$	$CH = TH$

$TH = Total\ hardness$ $ALK = Alkalinity$
$CH = Carbonate\ hardness$ $NCH = Noncarbonate\ hardness$

24.5 HYPOTHETICAL COMBINATIONS

In a sample of water, the sum of the meq/L of the cations must equal the sum of the anions. In a perfect evaluation, they would be the same, since water in equilibrium is electrically balanced. After expressing the concentration of major ions in meq/L, the next step is to find which ion is combined with another ion. To accomplish this, the major cations arranged in the order of Ca, Mg, Na and K and the major anions including HCO_3, SO_4 and Cl are tabulated. This allows us to make hypothetical combinations. These combinations are helpful in determining the amount of carbonate hardness and non-carbonate hardness.

EXAMPLE PROBLEM 24.6

The alkalinity and total hardness of a water sample are found to be 92 mg/l and 115 mg/L. What are the carbonate hardness and non-carbonate hardness of the water?

Given:

TH = 115 mg/L   Alk = CH = 92 mg/L   NCH = ?

Solution:

Non-carbonate hardness
$NCH = TH - Alk = (115 - 92)\ mg/L = 23.0 = 23\ mg/L$

Carbonate hardness
$CH = Alk = \underline{92\ mg/L}$

EXAMPLE PROBLEM 24.7

Given the following water analysis, make hypothetical combinations and determine total hardness and alkalinity. Concentrations of various ions in meq/L are given in the following table.

Given:

Ca⁺⁺	Mg⁺⁺	Na⁺	K⁺	HCO₃⁻	SO₄⁻	Cl⁻
2.0	2.0	1.0	1.0	?	1.0	?

Solution:

Hypothetical combinations

	Ca⁺⁺	Mg⁺⁺	Na⁺	K⁺	Σ
HCO₃⁻	2.0	1.0	0.0	0.0	3.0
SO₄⁻	0.0	1.0	0.0	0.0	1.0
Cl⁻	0.0	0.0	1.0	1.0	2.0
Σ	2.0	2.0	1.0	1.0	6.0

Total hardness

$$TH = \frac{(2.0+2.0)\,meq}{L} \times \frac{50\,meq}{meq} = 200.0 = 200\ mg/L$$

Carbonate hardness

$$CH = \frac{3.0\ meq}{L} \times \frac{50.0\ mg}{meq} = 150.0 = 150\ mg/L$$

Non-carbonate hardness
$$NCH = TH - Alk = (200 - 150)\,mg/L = 50.0 = \underline{50\ mg/L}$$

EXAMPLE PROBLEM 24.8

Express the concentrations in meq/L and list the hypothetical combinations for the following groundwater data.

Given:

Ca = 175 mg/L   Mg = 40 mg/L   Na = 14 mg/L   K = 4 mg/L
Alk = 200 mg/L   SO₄ = 29 mg/L   Cl = 14 mg/L

Solution:

	Concentration						
	Cations				Anions		
Units	Ca	Mg	Na	K	HCO₃	SO₄	Cl
mg/L	70	9.5	14	4	200	29	14
mg/meq	20	12	23	39	50	48	35.5
meq/L	3.5	0.79	0.61	0.10	4.0	0.60	0.39

$$\Sigma\ Cations = Ca + Mg + Na + K = 3.5 + 0.79 + 0.61 + 0.10 = 5.0\ meq/L$$

$$\Sigma\ Anions = HCO_3 + SO_4 + Cl = (4.0 + 0.60 + 0.39)\ meq/L = 4.99 = 5.0\ meq/L$$

Total cations = Total anions; thus, analysis is okay.

	HCO_3	SO_4	Cl	Σ
Ca	3.5	—	—	3.50
Mg	0.5	0.29	—	0.79
Na	—	0.31	0.30	0.61
K	—	—	0.10	0.10
Σ	4.0	0.60	0.40	5.0

Carbonate hardness

$$CH = \frac{4.0\ meq}{L} \times \frac{50.0\ mg}{meq} = 200.0 = 200\ mg/L\ as\ CaCO_3$$

Non-carbonate hardness

$$NCH = \frac{0.29\ meq}{L} \times \frac{50.0\ mg}{meq} = 14.5 = 14\ mg/L\ as\ CaCO_3$$

Total hardness

$$TH = NCH + CH = (200.0 + 14.5)\ mg/L = 214.5 = 210\ mg/L \quad or$$

$$TH = CaH + MgH = (3.50 + 0.79)\frac{meq}{L} \times \frac{50.0\ mg}{meq} = 214.5 = 210\ mg/L$$

24.6 CHEMISTRY OF LIME SODA ASH SOFTENING

The chemical reactions in precipitation softening are:

$$CO_2 + Ca(OH)_2 = CaCO_3 + H_2O.$$

$$Ca(HCO_3)_2 + Ca(OH)_2 = 2CaCO_3 + 2H_2O$$

$$Mg(HCO_3)_2 + Ca(OH)_2 = MgCO_3 + 2H_2O$$

$$MgCO_3 + Ca(OH)_2 = Mg(OH)_2 + CaCO_3$$

$$MgSO_4 + Ca(OH)_2 = CaSO_4 + Mg(OH)_2$$

$$CaSO_4 + Na_2CO_3 = CaCO_3 + Na_2SO_4$$

Lime added to water first reacts with any free carbon dioxide. Next, the lime reacts with bicarbonate alkalinity. $MgCO_3$ being soluble, the alkalinity (bicarbonate) associated with magnesium reacts again with lime to precipitate as $Mg(OH)_2$. Non-carbonate hardness requires the addition of soda ash for precipitation. Precipitation

TABLE 24.1
Chemical Requirements

Type of Hardness	Constituent	Lime	Soda Ash
Carbonate hardness, CH	Ca	1	none
	Mg	2	none
Non-carbonate hardness, NCH	Ca	none	1
	Mg	1	1
Carbon dioxide	CO_2	1	none

of the magnesium ions demands a higher pH and the presence of the excess lime in the amount of about 35 mg/L CaO = 1.25 meq/L.

One equivalent of each requires 1 meq of the lime or soda ash. The practical limits of precipitation softening are 30 mg/L of $CaCO_3$ (0.6 meq/L) and 10 mg/L of $Mg(OH)_2$ as $CaCO_3$ (0.2 meq/L). Chemical requirements for lime soda ash softening are shown in Table 24.1.

24.6.1 SELECTIVE CALCIUM CARBONATE REMOVAL

Selective calcium removal may be used to soften a water low in magnesium hardness, less than 40 mg/L as $CaCO_3$. Enough lime is added to precipitate calcium carbonate without adding any excess.

24.6.2 SPLIT TREATMENT SOFTENING

In split treatment, a portion of the water is treated with excess lime. After settling, the bypassed water is mixed with treated water to lower the pH and further remove hardness.

24.7 EXCESS LIME TREATMENT

As discussed before, excess lime in addition to stoichiometric requirements is needed to precipitate magnesium carbonate hardness.

Excess lime softening

$$Lime\,dosage = CO_2 + Alk + Mg + Excess$$

Selective Ca hardness removal is conducted when Mg hardness is less than 40 mg/L.

All terms in the previous equation are expressed as meq/L. Lime can be expressed as CaO (28 mg/meq) or $Ca(OH)_2$ (37 mg/meq). The purity of commercial lime ranges from 85% to 92%. Soda ash is used to remove noncarbonate hardness from water.

Excess lime softening

$$\boxed{Soda\ ash\ dosage\ =\ NCH\ =\ TH\ -\ Alk}$$

24.8 SELECTIVE CALCIUM REMOVAL

In selective calcium removal, no excess lime is required and only calcium hardness is considered for determining lime and soda ash dosage.

Selective calcium removal

$$\boxed{Lime\ dosage = CO_2 + Ca - CH}$$

Selective calcium removal

$$\boxed{Soda\ ash\ dosage = Ca - NCH}$$

24.9 RECARBONATION

Recarbonation is used to stabilize lime-treated water, reducing its scale-forming potential. Carbon dioxide neutralizes excess lime, precipitating it as calcium carbonate. Further recarbonation converts carbonate to bicarbonate. Recarbonation converts the excess lime to $CaCO_3$ precipitate. Further carbon dioxide converts CO_3 to HCO_3. Hardness in the finished water remains equal to practical limits of removal of $0.6 + 0.2 = 0.8$ meq/L = 40 mg/L.

$$Ca(OH)_2 + CO_2 = CaCO_3 + H_2O$$
$$CaCO_3 + CO_2 + H_2O = Ca(HCO_3)_2$$

Recarbonation

$$\boxed{CO_2\ Dosage = Excess\ lime + Residual\ Hardness}$$

EXAMPLE PROBLEM 24.9

Groundwater is softened by lime precipitation. What dosage of lime with a purity of 78% CaO is required to combine with 70 mg/L of calcium (selective calcium removal)?

Solution:

Lime dosage

$$C = \frac{70\ mg}{L} \times \frac{meq}{20\ mg} \times \frac{28\ mg\ CaO}{meq} \times \frac{comm.}{0.78\ pure} = 125.6 = 130\ mg/L$$

EXAMPLE PROBLEM 24.10

How much soda ash is required (lb/d) to remove 40 mg/L of non-carbonate hardness from a flow of 6.0 MGD?

Given:

$Q = 6.0$ MGD   NCH $= 40$ mg/L

Solution:

Non-carbonate hardness

$$NCH = \frac{40 \ mg}{L} \times \frac{meq}{50 \ mg} = 0.800 = 0.80 \ meq/L$$

Soda-ash feed rate

$$M_{soda} \ \frac{0.80 \ mg}{L} \times \frac{53 \ mg}{meq} \times \frac{8.34 \ lb/MG}{mg/L} \times \frac{6.0 \ MG}{d} = 2121 = \underline{2100 \ lb/d}$$

EXAMPLE PROBLEM 24.11

For the chemical analysis shown in the following, list the hypothetical combinations and calculate Ca hardness and Mg hardness.

Ca	Mg	Na	K	HCO$_3$	SO$_4$	Cl
2.0	0.6	0.2	0.1	2.2	0.5	0.2

Solution:

Hypothetical combinations chart

	HCO$_3$	SO$_4$	Cl	Σ
Ca	2.0			2.0
Mg	0.2	0.4		0.6
Na			0.2	0.2
K		0.1		0.1
Σ	2.2	0.5	0.2	2.9

Calcium carbonate hardness (temporary hardness)

$$CaCH = \frac{2.0 \ meq}{L} \times \frac{50.0 \ mg}{meq} = 100.0 = 100 \ mg/L$$

Magnesium carbonate hardness (temporary hardness)

$$MgCH = \frac{0.20\ meq}{L} \times \frac{50.0\ mg}{meq} = 10.0 = \underline{10\ mg/L}$$

Total carbonate hardness
$$TCH = (100.0 + 10.0)\ mg/L = 110.0 = \underline{110\ mg/L}$$

Magnesium non-carbonate hardness

$$MgNCH = \frac{0.40\ meq}{L} \times \frac{50.0\ mg}{meq} = 20.0 = \underline{20\ mg/L}$$

$$NaCl = 0.2\ meq/L, \quad K_2SO_4 = 0.1\ meq/L$$

EXAMPLE PROBLEM 24.12

For the water discussed in Example Problem 24.11, calculate the chemical require-ments for selective removal of calcium hardness.

Solution:

The only hypothetical combination involving Ca is 2.0 meq/L of $Ca(HCO_3)_2$; there-fore, no soda ash is needed.

Lime dosage
$$C_{lime} = \frac{2.0\ meq}{L} \times \frac{37\ mg\ Ca(OH)_2}{meq} = 74.0 = \underline{74\ mg/L}$$

Total hardness of finished water
$$TH = \frac{(0.60 + 0.60)meq}{L} \times \frac{50.0\ mg}{meq} = 60.0 = 60\ mg/L$$

Carbonate hardness (alkalinity) of finished water
$$CH = \frac{(0.60 + 0.20)meq}{L} \times \frac{50.0\ mg}{meq} = 40.0 = 40\ mg/L$$

The difference of 20 mg/L is non-carbonate hardness due to $MgSO_4$.

EXAMPLE PROBLEM 24.13

For the data of Example Problem 24.11, calculate the chemical requirements (85% lime as CaO, 95% soda ash) for removal of hardness using excess lime treatment. Assume a CO_2 amount of 11 mg/L.

Solution:

Concentration of carbon dioxide

$$CO_2 = \frac{11 \ mg}{L} \times \frac{meq}{22 \ mg} = 0.50 \ meq/L$$

Lime dosage

$$= CO_2 + Alk + Mg + Excess$$

$$= (0.50 + 2.2 + 0.6 + 1.25)\frac{meq}{L} \times \frac{28 \ mg}{meq} \times \frac{1}{0.85} = \underline{150 \ mg/L}$$

Soda ash dosage

$$= NCH = \frac{0.4 \ meq}{L} \times \frac{53 \ mg}{meq} \times \frac{1}{0.95} = 22.3 = \underline{22 \ mg/L}$$

EXAMPLE PROBLEM 24.14

The water indicated by the following analysis is to be softened by excess lime treatment. Make hypothetical combinations and find the total hardness of the water.

Component	CO_2	Ca	Mg	Na	Alk	SO_4	Cl
meq/L	0.40	2.0	1.21	0.60	2.7	0.60	0.51

Solution:

	HCO_3	SO_4	Cl	Σ
Ca	2.0			2.0
Mg	0.7	0.51		1.21
Na		0.09	0.51	0.60
Σ	2.7	0.60	0.51	3.81

The hypothetical combinations are:

$Ca(HCO_3)_2 = 2.0$ meq/L $Mg(HCO_3)_2 = 0.7$ mg/L $MgSO_4 = 0.51$ meq/L

$Na_2SO_4 = 0.09$ meq/L $NaCl = 0.51$ meq/L

Total hardness

$$TH = Ca + Mg = \frac{(2+1.21)meq}{L} \times \frac{50.0 \ mg}{meq} = 161 = 160 \ mg/L$$

EXAMPLE PROBLEM 24.15

For the data of Example Problem 24.14, do the following:

a) Calculate the softening chemicals required, expressing lime dosage as CaO and soda ash as Na_2CO_3.
b) Give the chemical characteristics of the softened water before and after recarbonation.

Solution:

Component	Conc. meq/L	Lime meq/L	Soda ash meq/L
CO_2	0.4	0.4	—
Alk	2.7	2.7	—
$Mg(HCO_3)_2$	0.7	0.7	—
$MgSO_4$	0.51	0.51	0.51
Excess		1.25	—
		$\Sigma = 5.56$	$\Sigma = 0.51$

Lime dosage

$$Lime = \frac{5.56 \ meq}{L} \times \frac{28 \ mg \ CaO}{meq} = 156 = \underline{160 \ mg/L}$$

Soda ash dosage

$$Soda \ ash = \frac{0.51 \ meq}{L} \times \frac{53 \ mg}{meq} = 27.03 = \underline{27 \ mg/L}$$

Finished water

1. Excess lime equal to 35 mg/L as CaO is added to increase the pH high enough to precipitate Mg $(OH)_2$ remains in the treated water.
2. The 0.6 meq/L of Ca (30 mg/L as $CaCO_3$) and 0.2 meq/L of Mg (10 mg/L as $CaCO_3$) are the practical limits of hardness reduction.
3. 0.60 meq/L of Na is already present, and the amount increased by soda ash addition is 0.51 meq/L. The total sodium in the finished water is 0.6 + 0.51 = 1.11 meq/L.
4. The SO_4 and Cl content remain unchanged.

24.10 ION EXCHANGE SOFTENING

In the ion exchange process, ions of Ca and Mg in solution are replaced by sodium ions attached to insoluble resin zeolites. After the resin is exhausted, it is regenerated using brine solution. In ion exchange softening, hardness is most commonly expressed in terms of grains per gallon. A grain is a mass unit equivalent to 65 mg.

24.10.1   REMOVAL CAPACITY

Removal capacity is the total amount of hardness which can be removed by a unit volume of resin. Given the removal capacity, the exchange capacity of a softener can be calculated as follows:

Removal capacity of the resin

$$\boxed{Exchange\ Capacity\ Removal\ Capcity \times Volume\ of\ the\ media}$$

Removal capacity can be expressed as g or kg/m^3. The removal capacity of resins ranges from 15 to 100 kg/m^3.

24.10.2   WATER TREATMENT CAPACITY

To calculate the volume of water that can be softened before the resin must be regenerated, the exchange capacity and the hardness of water must be known.

$$\boxed{Treatment\ Capacity = \frac{Exchange\ Capacity}{Hardness\ Capacity}}$$

This will allow you to calculate the volume of water that can be softened with one cycle of operation. This can also be expressed as hours or days of operation if the water flow rate is known.

24.10.3   BRINE REQUIREMENT

Exhausted resin is regenerated by pumping a concentrated brine solution (10–15 % NaCl) onto the resin. The salt dosage required ranges from 3–4 kg/kg of hardness. The feed rate of solution is typically 40 L/m^2.min.

EXAMPLE PROBLEM 24.16

Estimate the exchange capacity in kg of hardness for an ion exchanger with diameter of 3.0 m and media depth of 1.4 m. The resin has a removal capacity of 14 kg/m^3.

Given:

$$D = 3.0\ m \quad d = 1.4\ m \quad Exchange = 1.4\ kg/m^3$$

Solution:

Capacity of the ion exchanger

$$V = \frac{\pi(3.0\ m)^2}{4} \times 1.4\ m = 9.89 = 9.9\ m^3$$

Hardness removal capacity

$$= \frac{14 \ kg}{m^3} \times 9.89 \ m^3 = 138 = 140 \ kg \ of \ hardness$$

EXAMPLE PROBLEM 24.17

Determine the ion exchange softener operating time in hours. The exchange capacity of the softener is 230 lb of hardness, and the flow rate is 150 gpm. The raw water contains 64 ppm of hardness

Given:

Capacity = 230 lb   hardness of water = 64 ppm   Q = 150 gpm

Solution:

Hours of operation

$$t = \frac{V}{Q} = 230 \ lb \times \frac{10^6}{64} \times \frac{gal}{8.34 \ lb} \times \frac{min}{150 \ gal} \times \frac{h}{60 \ min} \times 47.9 = \underline{48 \ h}$$

EXAMPLE PROBLEM 24.18

An ion exchange softener has an exchange capacity of 270 kg of hardness. a) How many litres of water with a hardness of 190 mg/L can be treated? b) How many hours will this unit operate without regenerating when treating an average flow of 20 L/s?

Given:

Removal capacity = 270 kg   hardness of water = 190 mg/L   Q = 2.0 L/s

Solution:

Volume of water that can be treated

$$V = 270 \ kg \times \frac{m^3}{190 \ g} \times \frac{1000 \ g}{kg} = 1421 \ m^3$$

Hours of operation

$$t = \frac{V}{Q} = 1421 \ m^3 \times \frac{L}{190 \ mg} \times \frac{s}{2.0 \ L} \times \frac{h}{3600 \ s} \times \frac{1000 \ L}{m^3} = 197 = \underline{200 \ h}$$

EXAMPLE PROBLEM 24.19

An ion exchange softener has 48.0 ft³ of cationic resin with a capacity of 1.1 lb/ft³. If the softener becomes exhausted after softening 98,000 gal, what is the hardness of the water being treated?

Given:

$V = 48.0 \text{ ft}^3$ $Capacity = 1.1 \ lb/ft^3$ $Q = 98000 \ gal$ $Hardness \ of \ water = ?$

Solution:

Exchange capacity

$$EC = 48.0 \ ft^3 \times \frac{1.1 \ lb}{ft^3} = 53 \ lb$$

Hardness of raw water

$$\frac{EC}{Q} = \frac{53 \ lb}{98000 \ gal} \times \frac{gal}{8.34 \ lb} = 6.49 \times 10^{-5} = \underline{65 \ ppm}$$

EXAMPLE PROBLEM 24.20

An ion exchange softener will remove 80 kg of hardness before the resin becomes exhausted. If 3 kg of salt are required per kg of hardness, how many kg of salt are needed? If a 15% salt solution is used to regenerate the unit, how many litres of brine are required? The SG of the 15% brine solution is 1.1.

Given:

$Hardness = 80 \ kg, \ Salt \ requirement = 3 \ kg/kg \ hardness$ $C = 15\%$ $SG = 1.1$

Solution:

Quantity of brine required

$$Salt = 80 \ kg \times \frac{3.0 \ kg \ salt}{kg \ hardness} = 240.0 = 240 \ kg$$

Volume of brine required

$$V = \frac{m}{C} = 240 \ kg \times \frac{100\%}{15\%} \times \frac{L}{1.1 \ kg} = 1454 = \underline{1500 \ L}$$

PRACTICE PROBLEMS

PRACTICE PROBLEM 24.1

Water from a well sample has 62 mg/L of magnesium (Mg) as $CaCO_3$ and 98 mg/L of calcium (Ca) as $CaCO_3$. What is the total hardness of the sample as $CaCO_3$? (160 mg/L)

PRACTICE PROBLEM 24.2

What is the calcium hardness as $CaCO_3$ if the water sample has a calcium content of 99 mg/L? (250 mg/L)

PRACTICE PROBLEM 24.3

The magnesium content of well water is 14 mg/L. What is the maximum content of Ca allowed if the total hardness is not to exceed 150 mg/L as $CaCO_3$? (37 mg/L)

PRACTICE PROBLEM 24.4

The calcium content of a water sample is 17 mg/L. What is the calcium hardness as milligrams per litre $CaCO_3$? (43 mg/L)

PRACTICE PROBLEM 24.5

The total hardness of given water is 135 mg/L as $CaCO_3$. What is the alkalinity of this water if the noncarbonate hardness component is 25 mg/L as $CaCO_3$? (110 mg/L)

PRACTICE PROBLEM 24.6

Determine the total hardness of water as $CaCO_3$ which was analysed to have a calcium and magnesium content of 30 mg/L and 8 mg/L, respectively. (110 mg/L as $CaCO_3$)

PRACTICE PROBLEM 24.7

Given the following water analysis, determine the unknowns. Alkalinity and non-carbonate hardness are respectively 3.0 and 1.0 meq/L. ($Mg^{++} = 2$, $Na = 1$, $HCO_3 = 3$ meq/L)

Ca^{++}	Mg^{++}	Na^+	K^+	$HCO_3^{"}$	$SO_4^{"}$	Cl^-
2.0	?	?	1.0	?	2.0	1

PRACTICE PROBLEM 24.8

The results of a water analysis are calcium 29.0 mg/L, magnesium 16.4 mg/L, sodium 23.0 mg/L, potassium 17.5 mg/L, bicarbonate 117 mg/L, sulphate 36.0 mg/L and chloride 24.0 mg/L. Express all concentrations in meq/L, list hypothetical combinations and calculate total hardness and non-carbonate hardness as mg/L $CaCO_3$. (141 mg/L, 23 mg/L)

PRACTICE PROBLEM 24.9

$Ca(OH)_2$ (lime slurry) reacts with calcium bicarbonate $(Ca(HCO_3)_2)$ in solution to precipitate $CaCO_3$. Calculate the amount of 78% CaO required reacting with 185 mg/L of calcium hardness as $CaCO_3$. (133 mg/L)

PRACTICE PROBLEM 24.10

Based on the chemical analysis of a water sample, following hypothetical combinations are found: $(Ca(HCO_3)_2 = 3.6, CaCl_2 = 0.40, MgCl_2 = 2.4, MgSO_4 = 1.0, CaSO_4 = 1.0, NaCl = 1.1$ all in meq/L). Calculate the dosage rate for 95% pure soda ash to treat 4.0 MGD. (8900 lb/d)

PRACTICE PROBLEM 24.11

The meq/L concentrations of various components in a water sample are reported as follows:

Ca	Mg	Na	K	HCO_3	SO_4	Cl
3.5	0.8	0.6	0.1	4.0	0.6	0.4

Calculate the Ca hardness, Mg hardness, carbonate hardness and non-carbonate hardness as mg/L of $CaCO_3$. (175, 40, 200, 15 mg/L)

PRACTICE PROBLEM 24.12

For the data of Practice Problem 24.11, do the following:

a) Calculate the lime dosage in mg/L of $Ca(OH)_2$ required for selective removal of calcium hardness. (130 mg/L)
b) Calculate the hardness and alkalinity of the softened water. (70 mg/L, 55 mg/L)

PRACTICE PROBLEM 24.13

For data of Practice Problem 24.11, calculate chemical requirements (85% lime as CaO, 95% soda ash) for removal of hardness using excess lime treatment. Assume a CO_2 amount of 11 mg/L. (220 mg/L, 17 mg/L)

PRACTICE PROBLEM 24.14

The analysis of water (all components measured as meq/L) is as follows:

Ca	Mg	Na	K	HCO_3	SO_4	Cl
3.7	1.0	1.0	0.5	4.0	1.2	1.0

List the hypothetical combinations and calculate the total hardness. (4.7 meq/L = 235 mg/L)

PRACTICE PROBLEM 24.15

Calculate the chemical doses of lime as CaO and soda ash as Na_2CO_3 for excess lime softening. Assume 1.25 meq/L of excess lime. (175 mg/L as CaO, 37 mg/L as Na_2CO_3)

PRACTICE PROBLEM 24.16

An ion exchange water softener has a diameter of 3.0 m. It is filled with a resin to a depth of 1.8 m. If the removal capacity of the resin is 21 kg/m^3, what is the total exchange capacity of the softener? (270 kg)

PRACTICE PROBLEM 24.17

Determine the ion exchange softener operating time in hours given the following: Exchange capacity of the softener is 310 lb. Flow rate is 140 gpm. Raw water contains 45 ppm of hardness. (98 h)

PRACTICE PROBLEM 24.18

The removal capacity of an ion exchange softening unit is 35 kg/m^3. The softener contains a total of 6.5 m^3 of resin. How many million litres of water with a hardness of 210 mg/L can be treated? (1.1 ML)

PRACTICE PROBLEM 24.19

An ion exchange softener has 250 ft^3 of cationic resin with a capacity of 0.12 lb/gal. If the softener becomes exhausted after softening 600,000 gal, what is the hardness of the water being treated? (45 ppm)

PRACTICE PROBLEM 24.20

A total of 185 kg of salt will be required to regenerate an ion exchange softener. If the solution is to be 12% (SG = 1.09) brine solution, how many kL of brine will be required? (1.4 kL)

25 Iron and Manganese Removal

Iron and manganese hardness may have no direct adverse effect on health, but high concentration of these elements is cause for concern. Bivalent forms of iron (Fe^{++}) and manganese (Mn^{++}) are soluble and may exist in well waters or anaerobic reservoirs. When exposed to air (oxidized), these elements slowly transform to insoluble reduced forms (Fe^{+++} and Mn^{+++}), which impart a brownish colour to water.

25.1 REMEDIAL ACTION

Preventive measures may sometimes be used with reasonable success. The addition of **polyphosphate** may bind iron and keep it in soluble form. This kind of treatment is more successful when water contains manganese up to 0.3 mg/L and the iron content is less than 0.1 mg/L. The chlorine dose for phosphate treatment should be sufficient to produce a free chlorine residue of about 0.25 mg/L. The proper phosphate dose is determined by jar testing (refer to Section 10.4). The polyphosphate dosage is the lowest dosage that allows noticeable discolouration for a period of 4 days.

The chlorine should never be fed before the addition of polyphosphate, as it will oxidize iron and manganese to produce insoluble precipitates.

25.2 AERATION, FILTRATION

Plain aeration followed by filtration is the simplest form of treatment. This treatment will be successful when manganese content is very low. Insufficient removal of manganese can cause serious problems with post-chlorination. The oxidation of manganese will cause plugging of the chlorine injection mechanism as well as deposition in the distribution system.

25.3 AERATION, OXIDATION, FILTRATION

This process scheme is more commonly used for removing iron and manganese from well waters without softening treatment. In addition to oxidation by plain aeration, chemical oxidation by using free chlorination residual or potassium permanganate ($KMnO_4$) is practised to ensure complete removal. Potassium permanganate is a dark purple crystal or powder available at about 98% purity. The theoretical requirements of $KMnO_4$ for oxidation of iron and manganese are as follows:

Stoichiometric requirement $KMnO_4$

$$\frac{kMnO_4}{F^{++}} = 0.95 \qquad \frac{kMnO_4}{Mn^{++}} = 1.92$$

DOI: 10.1201/9781003468745-25

When using chlorine, for each unit of ferrous iron (Fe^{++}), the dosages of chlorine and potassium permanganate are 0.64 and 0.60, respectively.

25.3.1 POTASSIUM PERMANGANATE DOSAGE

To calculate the dosage of $KMnO_4$, you need to know the concentration of iron and manganese being treated at the location in the process where the $KMnO_4$ is added.

Dosage KMnO$_4$

$$\boxed{0.6 \times F^{++} + 2.0 \times Mn^{++}}$$

EXAMPLE PROBLEM 25.1

Calculate the potassium permanganate (KMnO4) dosage if, after aeration, 0.25 mg/L iron (Fe) and 0.28 mg/L of manganese (Mn) remain.

Given:

$$Fe = 0.25 \text{ mg/L} \qquad Mn = 0.28 \text{ mg/L} \qquad KMnO_4 = ?$$

Solution:

KMnO$_4$ dosage for oxidation

$$= 0.95 \times F^{++} + 1.92 \times Mn^{++} = 0.95 \times 0.25 \ mg/L + 1.92 \times 0.28 \ mg/L$$
$$= 0.775 = \underline{0.78 \ mg/L}$$

EXAMPLE PROBLEM 25.2

A stock solution of polyphosphate is prepared by dissolving 1.0 g of polyphosphate and making up 1.0 L of solution by adding distilled water.

  a) If 10 mL of this solution are added to 1.0 L of sample, what is the polyphosphate dose in mg/L?
  b) How many litres of this stock solution should be added to each kL of water to result in a dose of 4.5 mg/L?

Given:

Parameter	Solution Fed = 1	Water Sample = 2
Volume	10 mL	1.0 L
Strength	1.0 g/L	?

Solution:

a) Dosage of polyphosphate in water

$$C_2 = C_1 \times \frac{V_1}{V_2} = \frac{1000 \ mg}{L} \times \frac{10 \ mL}{L} \times \frac{L}{1000 \ mL} = \underline{10 \ mg/L}$$

Given:

$1 =$ Solution $2 =$ Water treated
$V_1 = ?$ $C_1 = 1000$ mg/L
$V_2 = 1.0$ kL $C_2 = 4.5$ mg/L

Solution:

b) Chemical solution feed rate

$$V_1 = \frac{C_2 V_2}{C_1} = \frac{4.5\ mg}{L} \times \frac{L}{1000\ mg} \times kL \times \frac{1000\ L}{kL} = \underline{4.5\ L}$$

EXAMPLE PROBLEM 25.3

Polyphosphate is fed to sequester iron and manganese. What is the required dosage rate of polyphosphate to treat a flow of 1.2 MGD to dose the water at 4.0 mg/L.? If the polyphosphate solution strength is 2.5%, what will be the daily use of the solution in gallons?

Given:

Polyphosphate solution $= 1$ Treated water $= 2$
$Q_1 = ?$ $Q_2 = 1.2$ MGD
$C_1 = 2.5\% = 2.5$ pph $C_2 = 4.0$ ppm

Solution

Dosage rate

$$M = \frac{1.2\ MG}{d} \times \frac{4.0\ mg}{L} \times \frac{8.34\ lb/MG}{mg/L} = 40.0 = 40\ lb/d$$

Feed pump rate

$$Q_1 = \frac{40\ lb}{d} \times \frac{100}{2.5} \times \frac{gal}{8.34\ lb} = 1.10 = 192 = \underline{190\ gal/d}$$

EXAMPLE PROBLEM 25.4

Determine the chemical pump rate of polyphosphate solution in L/d to dose a water flow of 2.6 ML/d at the rate of 3.5 mg/L. The strength of polyphosphate solution is 2.0%.

Given:

Parameter	Feed Solution = 1	Treated Water = 2
Q	?	2.6 ML/d
C	2.0%=20 g/L	3.5 mg/L

Solution:

Solution feed pumping rate

$$Q_1 = Q_2 \times \frac{C_2}{C_1} = \frac{2.6\ ML}{d} \times \frac{3.5\ kg}{ML} \times \frac{L}{20.0\ g} \times \frac{1000}{kg} = 455 = \underline{460\ L/d}$$

EXAMPLE PROBLEM 25.5

A chemical feed pump setting is 250 mL/min. Average daily water flow is 2.5 ML/d. Determine the strength of polyphosphate solution to provide a dosage of 3.5 mg/L.

Given:

Parameter	Feed Solution = 1	Treated Water = 2
Q	250 mL/min	2.5 ML/d
C	?	3.5 mg/L

Solution:

Strength of feeding solution

$$C_1 = \frac{3.5 \ kg}{ML} \times \frac{2.5 \ ML}{d} \times \frac{1000 \ g}{kg} \times \frac{min}{250 \ mL} \times \frac{d}{1440 \ min} \times \frac{1000 \ mL}{L} = 24.3 \ g/L = \underline{2.4\%}$$

EXAMPLE PROBLEM 25.6

A cylindrical reactor is 4.2 m in diameter and 1.1 m deep. What maximum flow in ML/d can be allowed to provide a minimum reaction time (detention time) of 20 min?

Given:

$$D = 4.2 \ m \qquad H = 1.1 \ m \qquad HDT = 20 \ min \qquad Q = ?$$

Solution:

Maximum daily flow

$$Q = \frac{V}{t} = \frac{\pi (4.2 \ m)^2}{4} \times \frac{1.1 \ m}{20 \ min} \times \frac{1440 \ min}{d} \times \frac{ML}{1000 \ m^3} = 1.09 = \underline{1.1 \ ML/d}$$

EXAMPLE PROBLEM 25.7

A 25-ft-square reactor is 3.5 ft deep. If water flow through the reactor is 1.0 MGD, what is the theoretical detention time?

Given:

$$L = W = 5.0 \ m \qquad H = 3.5 \ ft \qquad Q = 1.0 \times 10^6 \ gal/d \qquad HDT = ?$$

Solution:

Detention time

$$t = \frac{V}{Q} = \frac{(25 \ ft)^2 \times 3.5 \ ft.d}{1.0 \times 10^6 \ gal} \times \frac{1440 \ min}{d} \times \frac{7.48 \ gal}{ft^3} = 23.56 = \underline{24 \ min}$$

EXAMPLE PROBLEM 25.8

What dosage of $KMnO_4$ is theoretically needed to oxidize 3.5 mg/L of iron (Fe) and 0.80 mg/L of manganese (Mn)?

Given:

$$Fe = 3.5 \text{ mg/L} \qquad Mn = 0.80 \text{ mg/L} \qquad KMnO_4 = ?$$

Solution:

$KMnO_4$ *dosage for oxidation*

$$= 0.95 \times F^{++} + 1.92 \times Mn^{++}$$

$$= 0.95 \times \frac{3.5 \text{ mg}}{L} + 1.92 \times \frac{0.8 \text{ mg}}{L} = 4.86 = \underline{4.9 \text{ mg/L}}$$

EXAMPLE PROBLEM 25.9

What dosage of $KMnO_4$ is needed to treat city well water with 3.5 mg/L iron before aeration and 0.2 mg/L after aeration? The manganese content of 1.2 mg/L remains the same before and after aeration.

Given:

$$Fe = 0.2 \text{ mg/L (after aeration)} \qquad Mn = 1.2 \text{ mg/L} \quad KMnO_4 = ?$$

Solution:

$KMnO_4$ *dosage for oxidation after aeration*

$$= 0.95 \times F^{++} + 1.92 \times Mn^{++}$$

$$= 0.95 \times 0.2 \text{ mg/L} + 1.92 \times 1.2 \text{ mg/L} = 2.49 = \underline{2.5 \text{ mg/L}}$$

EXAMPLE PROBLEM 25.10

A 2.5% solution of potassium permanganate is fed into a manganese zeolite pressure filter. The desired dosage is 2.0 ppm. What should the feed pump rate be in gal/h if the water is being pumped at a rate of 230 gpm?

Given:

$$KMnO_4 \text{ Solution} = 1 \qquad \text{Well water} = 2$$

$$Q_1 = ? \text{ gal/h} \qquad\qquad Q_2 = 230 \text{ gpm}$$

$$C_1 = 2.5\% = 2.5 \text{ pph} \qquad C_2 = 2.0 \text{ ppm}$$

Solution

Solution feed pumping rate

$$Q_1 = \frac{C_2}{C_1} \times Q_2 = \frac{2.0}{10^6} \times \frac{100}{2.5} \times \frac{230 \text{ gal}}{min} \times \frac{60 \text{ min}}{h} = 1.10 = 1.1 \text{ gal/h}$$

Plain aeration is too slow to be effective for waters with high manganese.
Chlorine should never be fed ahead of polyphosphate.

PRACTICE PROBLEMS

PRACTICE PROBLEM 25.1

Calculate the potassium permanganate ($KMnO_4$) dosage if, after aeration, 0.15 mg/L iron (Fe) and 0.21 mg/L of manganese (Mn) remain. (0.55 mg/L

PRACTICE PROBLEM 25.2

0.675 grams of polyphosphate are weighed out. What volume of solution should be made by adding water such that the solution concentration is 0.1% or 1 g/L? How many mL of this solution should be added to 1.5 L of water sample to dose water at the rate of 4.0 mg/L? Also calculate how many litres of this solution will be required to dose 1 m^3 (kL) of water at a rate of 2.5 mg/L. (675 mL, 6 mL, 2.5 L)

PRACTICE PROBLEM 25.3

Polyphosphate is fed to sequester iron and manganese. What is the required dosage rate of polyphosphate to treat a flow of 0.6 MGD with a dosage of 3.5 mg/L? If the polyphosphate solution strength is 2.0%, what will be the daily use of the solution in gallons? (18 lb/d, 110 gpd)

PRACTICE PROBLEM 25.4

A water supply of 1.7 ML/d containing 0.2 mg/L of manganese is treated with polyphosphate with a dose of 4.0 mg/L to prevent oxidation. Determine the feed pump rate of the phosphate solution if the strength of the polyphosphate solution is 2.5%. (190 mL/min)

PRACTICE PROBLEM 25.5

Rework Example Problem 25.5 assuming the feed pump setting is changed to 350 mL/min and the desired dosage is 2.8 mg/L. (1.4%)

PRACTICE PROBLEM 25.6

A reaction basin 4.0 m in diameter and 1.1 m deep treats a flow of 0.85 ML/d. What is the average detention time? (23 min)

PRACTICE PROBLEM 25.7

A 16 ft × 12 ft reaction basin with water depth of 4.0 ft treats a flow of 250 gpm. What is the average detention time? (23 min)

PRACTICE PROBLEM 25.8

A 25-ft-square reactor is 3.5 ft deep. If water flow through the reactor is 1.0 MGD, what is the theoretical detention time?

PRACTICE PROBLEM 25.9

What dosage of $KMnO_4$ is required to oxidize 5.5 mg/L of iron and 1.2 mg/L of manganese? (7.5 mg/L)

PRACTICE PROBLEM 25.10

By aeration, the iron content of city well water is reduced to 0.3 mg/L. If the manganese concentration after aeration is 1.0 mg/l, calculate the $KMnO_4$ dosage in mg/L. (2.2 mg/L)

PRACTICE PROBLEM 25.11

A well is being pumped at the rate of 340 gpm. How many gallons of 5% $KMnO_4$ solution should be fed per hour to provide a dosage of 2.5 mg/L? (1.0 gal/h)

26 Water Stabilization

Corrosive water can change problems both in water supply and treatment plants. Before the water is pumped into the distribution system, it should be chemically treated so that water is saturated or slightly supersaturated with calcium carbonate.

26.1 CARBONATE EQUILIBRIUM

One way to prevent corrosion is the formation of a thin protective layer of calcium carbonate on the walls of the pipe. The formation of the film will depend on control of the following chemical reaction:

$$CaCO_3 + CO_2 + H_2O + Ca(HCO_3)_2$$

For stability of a given water, the pH must be compatible with carbonate-carbon dioxide equilibrium at the prevailing temperature. It is recommended that 5–10 mg/L of $CaCO_3$ be kept in water above the saturation value. The minimum calcium and alkalinity levels should be in the range of 40–70 mg/L as $CaCO_3$ depending on concentration of other ions.

26.2 LANGELIER INDEX

Any corrosivity index is a way of determining the tendency of water to form scale or to corrode. The Langelier saturation index is the most common and is applicable in the pH range of 6.5 to 9.5. The Langelier index (LI) is defined as follows:

Langelier index
$$LI = pH - pH_S$$

$pH = Actual\ pH \qquad pH_s = pH\ at\ which\ water\ just\ saturated$

pH at saturation
$$pH_S = A + B - log(Ca \times Alk)$$

The values of A and B can be read from Table 26.1. "Alk" is alkalinity, and Ca^{++} is the calcium hardness, both expressed in mg/L as $CaCO_3$. This calculation is accurate enough up to a pH_s value of 9.3. A positive (>1) LI indicates that the water is supersaturated with $CaCO_3$ and will tend to form scale. If the actual pH is less than the pH_S, (LI < 1), the water is corrosive. If the pH and pH_S are equal, (LI = 0), the water is stable.

DOI: 10.1201/9781003468745-26

TABLE 26.1
Values of A and B

T°C	A	TDS, mg/L	B
0	2.34	0	9.63
5	2.27	50	9.72
10	2.20	100	9.75
15	2.12	200	9.80
20	2.05	400	9.86
25	1.98	800	9.94
30	1.91	1000	10.04

26.3   AGGRESSIVE INDEX

The aggressive index is equal to pH plus the logarithm of the product of alkalinity and calcium hardness as $CaCO_3$.

Aggressive index

$$AI = pH + log(Ca \times Alk)$$

An aggressive index of 12 corresponds to an LI of zero.

26.4   DRIVING FORCE INDEX

The relationship between the driving force index and Langelier index is as follows:

Driving force index

$$DFI = 10^{LI}$$

This relationship is fairly accurate for LI ranging from −5 to +1.

EXAMPLE PROBLEM 26.1

The test results of distribution water give a pH of 7.2 and a pHs of 7.4. What is the Langelier index, and what does this tell you about the distribution system in the area where the water was collected?

TABLE 26.2
Comparison of Various Indices

Corrosivity	LI	AI	DFI
High	< −2.0	< 10.0	< 0.01
Moderate	−2.0 < LI < 0.0	10 < LI < 12	0.01 < DFI < 1.0
Non-aggressive	> 0.0	> 12	> 1.0
Stable	0.0	12	1.0

Given:

$pH = 7.2 \qquad pH_s = 7.4$

Solution:

Langelier index

$LI = pH - pH_s = 7.2 - 7.4 = \underline{-0.20}$

Since the LI is negative, water in question is bit corrosive.

EXAMPLE PROBLEM 26.2

Water from a distribution system has the following characteristics: pH is 7.9, temperature is 10°C, TDS is 300 mg/L and alkalinity is 150 mg/L. Find the Langelier index.

Given:

$Ca = 100.\, mg/L \qquad Alk = 150\, mg/L \qquad TDS = 300\, mg/L \qquad pH = 7.9\, at\, 10°C$

Solution:

Constant A, from Table 26.1 $\qquad T = 10°C, A = 2.20$
Constant B, from Table 26.1 $\qquad TDS = 300\, mg/L,\ B = 9.83$

pH at saturation
$pH_S = A + B - log(Ca \times Alk) = 2.2 + 9.83 - log(100 \times 150) = 7.85$

Langelier index
$LI = pH - pH_S = 7.9 - 7.85 = 0.046 = \underline{0.05}$

EXAMPLE PROBLEM 26.3

Calculate the Langelier index of water at 10°C with a total dissolved solids concentration of 210 mg/L, alkalinity and calcium content of 110 and 40 mg/L as $CaCO_3$, respectively.

Given:

$Ca = 40.\, mg/L \qquad Alk = 110\, mg/L \qquad TDS = 210\, mg/L \qquad pH = 8.4\, at\, 10°C$

Solution:

Constant A, from Table 26.1 A = 2.20
Constant B, from Table 26.1 TDS $= 210\, mg/L,\ B = 9.80$

pH at saturation
$pH_S = A + B - log(Ca \times Alk) = 2.2 + 9.8 - log(40 \times 110) = 8.35 = 8.4$

Langelier index
$LI = pH - pH_S = 8.4 - 8.35 = 0.05 = \underline{0.1}$

EXAMPLE PROBLEM 26.4

Calculate the Langelier index of water with the following characteristics.

Given:

$Ca = 65$ mg/L Alkalinity $= 172$ mg/L $T = 15°C, A = 2.12$
$TDS = 365$ mg/L, B $= 9.85$ pH $= 8.4$

Solution:

Calcium hardness

$$CaH = \frac{65\ mg}{L} \times \frac{meq}{20\ mg} \times \frac{50\ mg\ CaCO_3}{meq} = 162.5 = 160\ mg\ /\ L$$

pH at saturation
$$pH_S = A + B - \log(Ca \times Alk) = 2.12 + 9.8 - \log(162.5 \times 172) = 7.52 = 7.5$$

Langelier index
$$LI = pH - pH_S = 8.4 - 7.52 = 0.88 = \underline{0.9}$$

EXAMPLE PROBLEM 26.5

Calculate the Langelier index of water at 10°C with a TDS of 100 mg/L and alkalinity of 80 mg/L as $CaCO_3$. Also determine the corresponding value of the AI.

Given:

$TDS = 100$ mg / L $T = 10°C$ Alk $= 80$ mg / L   Ca $= 40$ mg / L as $CaCO_3$

Solution:

A at 10°C $= 2.2$ B at 100 mg/L $= 9.75$

pH at saturation
$$pH_S = A + B - \log(Ca \times Alk) = 2.2 + 9.75 - \log(40 \times 80) = 8.44$$

Langelier index
$$LI = pH - pH_S = 8.4 - 8.44 = \underline{-0.04}$$

Aggressive index
$$AI = pH + \log(Ca \times Alk) = 8.4 + \log(40 \times 80) = 11.9 = \underline{12}$$

EXAMPLE PROBLEM 26.6

For the data of Example Problem 26.5 (LI = –0.04), calculate the driving force index (DFI).

Solution:

Driving force index
$$DFI = 10^{LI} = 10^{-0.04} = 0.912 = \underline{0.91}$$

Aggressive index
$AI = LI + 12 = -0.04 + 1211.96 = \underline{12}$

All indices indicate that this water is marginally corrosive.

26.5   MARBLE TEST

This test is used to determine the degree to which a given water is saturated with calcium carbonate ($CaCO_3$). The test is based on the principle that water in contact with powdered $CaCO_3$ (calcite) will approach saturation. During the test, the temperature of water should not be allowed to rise, and water being tested should not be exposed to atmospheric carbon dioxide. After saturating the sample with $CaCO_3$, the change in pH is recorded. This change in pH closely represents the Langelier index. If this value is less than 0.2, this indicates the water is very near saturation and hence relatively stable.

Langelier index
$$\boxed{LI = initial\ pH - finial\ pH}$$

After recording the final pH, the sample is filtered, and hardness and alkalinity tests are performed on the filtrate (water that passed through the filter). The $CaCO_3$ precipitation potential can be calculated using the following equation:

Precipitation potential = Initial hardness – Final hardness

EXAMPLE PROBLEM 26.7

Results from a marble test performed on a water sample are in the following table. Calculate calcium carbonate precipitation potential and LI.

Given:

Temp. °C	pH	Hardness, mg/L
11.0	8.50	55
11.4	8.95	58

Solution:

Langelier index
$LI = pH - PH_S = 8.50 - 8.95 = -0.45 = \underline{-0.45}$

TABLE 26.3
Precipitation Potential

$CaCO_3$ precipitation potential

Positive	Supersaturated	Scale forming
Negative	Unsaturated	Corrosive
Zero	Equilibrium	Stable

PRACTICE PROBLEMS

PRACTICE PROBLEM 26.1

The test results of distribution water give a pH of 7.8 and a pHs of 7.6. What is the Langelier index, and what does this tell you about the distribution system in the area where the water was collected? (0.2, scaling)

PRACTICE PROBLEM 26.2

Find the pHs and the LI for water with the following characteristics: pH is 7.6, temperature is 10°C, TDS is 200 mg/L, alkalinity is 200 mg/L and calcium hardness is 50 mg/L. (8.0, −0.4)

PRACTICE PROBLEM 26.3

The chemical characteristics of a water supply at 15°C are as follows: Calcium and alkalinity are 105 and 94 mg/L as $CaCO_3$, respectively, and TDS = 180 mg/L; pH = 7.90. Calculate the Langelier index. (−0.03)

PRACTICE PROBLEM 26.4

A water sample has the following characteristics:

Alkalinity $= 80$ mg/L $CaCO_3$ Ca $= 72$ mg/L   TDS $= 400$ mg/L,
$T = 12°C$, $pH = 7.80$

Calculate the Langelier saturation index. (−0.07)

PRACTICE PROBLEM 26.5

Finished water has the following characteristics:

TDS $= 300$ mg/L   T $= 10$ pH $= 7.5$ Alk $= 60$ mg/L Ca $= 30$ mg/ as $CaCO_3$

a) Based on LI, determine whether the water is corrosive or not. (corrosive)
b) Express the water stability as AI. (10.8)

PRACTICE PROBLEM 26.6

For a sample of water, the DFI is determined to be 1.65. What is the corresponding value of the LI? (0.22)

PRACTICE PROBLEM 26.7

For the data of Example Problem 26.7, calculate the calcium carbonate precipitation potential. (−3.0 mg/L)

27 Laboratory Tests

27.1 DISTRIBUTION SAMPLING

The need to collect a representative sample can't be overemphasized. However, it is not easy to secure a representative sample from the distribution system. Local conditions at the tap and the connection of the tap to the main are two important considerations in selecting the sampling point. An ideal sampling point would have a short direct connection with the main made of corrosion-resistant material. In many water systems, special sample taps are not available. The sample therefore must be collected from a customer faucet.

27.1.1 FLUSHING TIME

To ensure that the sample collected is representative of the water supply, the service line should be flushed adequately. Flushing is considered adequate when the water in the service line has been replaced twice while maintaining a flow of about 2 L/min.

27.1.2 ESTIMATING FLOW FROM A FAUCET

Flow from a faucet can be measured by recording the time taken to fill a container of known capacity. Flushing time can then be calculated by determining the volume of the water held in the line at a given time.

Flushing time (two replacements)

$$t = \frac{V}{Q} = \frac{\pi D^2 \times L}{2Q} = \frac{1.57 D^2 \times L}{Q}$$

EXAMPLE PROBLEM 27.1

The flow from a faucet filled up a 5-L container in 2 minutes and 10 seconds. What is the flow rate in L/min?

Given:

$V = 5\,L$ $t = 2$ min. and 10 s $Q = ?$

Solution:

Time period

$$t = 2\ min + 10\ s \times \frac{min}{60\ s} = 2.17 = 2.2\ min$$

DOI: 10.1201/9781003468745-27

Flow rate

$$Q = \frac{V}{t} = \frac{5.0\ L}{2.17\ L/min} = 2.30 = \underline{2.3\ min}$$

Flow exceeding 2 L/min may disturb deposits and make the sample unrepresentative.

EXAMPLE PROBLEM 27.2

How long should a 19-mm–inside diameter line 27 m long be flushed (replace two volumes) if the faucet is opened to maintain a flow of 2.2 L/min?

Given:

D = 19 mm = 0.019 m   L = 27 m   Q = 2.2 L/min

Solution:

Time period for which to run the faucet

$$t = \frac{\pi D^2 \times L}{2Q} = \frac{\pi (0.019\ m)^2 \times 27\ m}{2} \times \frac{min}{2.2\ L} \times \frac{1000\ L}{m^3} = 6.95 = \underline{7.0\ min}$$

EXAMPLE PROBLEM 27.3

A sampling tap is connected to water main via an 85-ft-long and ½-in-diameter pipeline How long it must be run at the rate of 0.5 gpm before collecting a sample of water? Assume the water sample is collected when two volumes of water are replaced.

Given:

D = 0.5 in   L = 85 ft   Q = 1.0 gpm

Solution:

Time period for which to run the tap

$$t = \frac{\pi D^2 L}{2Q} = \frac{\pi (0.5/12\ ft)^2 \times 85\ ft}{2} \times \frac{min}{0.5\ gal} \times \frac{7.48\ gal}{ft^3} = 3.46 = \underline{3.5\ min}$$

27.2   CHEMICAL SOLUTIONS

27.2.1   CONCENTRATION

1. *Mass concentration* is mass of solute in a given volume of solution. Units of mg/L (g/m³) are common for dilute solutions and g/L for highly concentrated solutions.

2. *Mass to mass concentration* is preferred for very concentrated solutions (liquid chemicals) as mass of solute in 100 mass units of the solutions.

Instead of the volume, it is the mass of the solution that is used here. It can be expressed as a fraction or percentage because the units of both the numerator and denominator are the same. A 1% solution represents a solution containing 10 g (1 dag) of solute in every 1 kg of solution. To express it as mass concentration, you need to know the density or specific gravity of the solution. The density of the liquid gives the mass of one-unit volume of the liquid solution. As an example, a 48.5% liquid alum solution with a specific gravity of 1.32 has a mass concentration of 640 g/mL.

$$C = C_{m/m} \times \rho = \frac{48.5\%}{100\%} \times \frac{1.32\ g}{mL} = 640.2 = 640\ g/mL$$

Adding 1 mL of liquid alum to 1 L of water will result in an alum dosage of 640 mg/L.

3. The concentrations of *standard solutions* are generally expressed as normality. One normal solution is a 1-gram equivalent of solute in every litre of solution, that is, 1.0 eq/L. For example, the concentration of a standard 0.02 NaOH solution is 20 meq/L, and in terms of mass concentration in mg/L, it is:

$$0.02N\ NaOH = \frac{20\ meq}{L} \times \frac{40\ mg}{meq} = 800\ mg/L$$

The larger the number in front of N, the stronger the solution. For example, 1.0 N H_2SO_4 is 40 times more concentrated than a 0.025 N H_2SO_4.

4. Similar to normality, the *molarity* of a solution is the concentration in moles rather than equivalents.

$$0.02M\ H_2SO_4 = \frac{20\ mmol}{L} \times \frac{98\ mg}{meq} = 1960 = 2000\ mg/L$$

1.0 M of H_2SO_4 is twice as concentrated as 1.0 N solution, as there are two equivalents in every mole of H_2SO_4.

5. Another method of specifying the concentration of solution uses the *a + b system*. This means that "a" volume of concentrated reagent is diluted with "b" volumes of distilled water to form the required solution. For example, 1 + HCl means 1 volume of concentrated HCl is diluted with 1 volume of distilled water.

27.2.2 STANDARD SOLUTIONS

A standard solution is a solution whose exact concentration is known. Many times, pre-prepared standard solutions can be ordered from chemical supply companies. Laboratory procedures also indicate the amount of chemical and distilled water that must be mixed to produce a solution with the desired concentration. Once a standard has been established, it can be used to standardize other laboratory solutions.

27.2.2.1 Standardization

Standardizing a solution means determining and adjusting its concentration accurately. It is the process of using one solution of known concentration to determine the concentration of another solution. This will entail titration, as will be discussed in forthcoming sections.

27.2.2.2 Preparation

When preparing standard solutions or reagents, it is required to weigh the exact mass of the chemical and dilute it with 1 litre of distilled water. Occasionally an approximate quantity is weighed. Then a proportionate volume of solution is made of the chemical weighed exactly so as to maintain the same concentration.

EXAMPLE PROBLEM 27.4

Determine the molarity of a 36% solution of HCl. Assume the density of liquid solution is 1.2 kg/L.

Given:

$C\% = 36\%$ $\rho = 1.2$ kg/L

Solution:

Molar concentration
$$C = \frac{36\%}{100\%} \times \frac{1.2\ kg}{L} \times \frac{1000\ g}{kg} \times \frac{mol}{(1+35.5)g} = 11.8 = \underline{12\ mol/L}$$

EXAMPLE PROBLEM 27.5

What is the normality of a 1.0 L solution containing 5.0 mg of H_3PO_4?

Solution:

Molar mass
$$H_3PO_4 = (3 \times 1 + 31 + 4 \times 16) = 98\ mg/mmol$$

Normality or normal concentration
$$N = \frac{5.0\ mg}{L} \times \frac{mmol}{98\ mg} \times \frac{3\ meq}{mmol} = 0.153 = \underline{0.15\ meq/L}$$

EXAMPLE PROBLEM 27.6

The lab procedures indicate weighing out 7.6992 g of a chemical for making 1 L of a solution. If you weigh out 7.5353 g, how much volume of the solution should be prepared to achieve the same concentration?

Given:

Parameter	Desired = 1	Actual = 2
Mass, g	7.6992	7.5353
Volume, L	1.00	?

Solution:

Final volume of solution

$$V_2 = V_1 \times \frac{m_1}{m_2} = 1.0\,L \times \frac{7.5353\,g}{7.6992\,g} \times \frac{1000\,mL}{L} = 978.7 = \underline{979\,mL}$$

EXAMPLE PROBLEM 27.7

Standard methods indicate that a 0.0192-N solution of silver nitrate ($AgNO_3$) is to be prepared using 3.27 g of the chemical in 1 L of distilled water. If 3.14 g of the chemical are actually weighed, how many mL of distilled water should be used in making the solution?

Given:

Parameter	Desired = 1	Actual = 2
m, g	3.27	3.14
V, L	1.00	?

Solution:

$$V_2 = \frac{m_1\,V_1}{m_2} = \frac{3.14\,g}{3.27\,g} \times 1.00\,L \times \frac{1000\,mL}{L} = 960.2 = \underline{960\,mL}$$

3.14 g of the chemical must be diluted to 960 mL of the solution.

27.3 ANALYSIS BY TITRATION

A titration involves the measured addition of a standardized solution to a measured volume of the sample to determine the concentration of the constituents in the water sample. A solution with known concentration is the titrant. The titrant is added to the other solution or sample, as the case may be, until an "end point" is reached. The end point (reaction goes to completion) is indicated by a colour change or pH change.

27.3.1 NORMALITY AND TITRATION

Equivalent mass is the combining mass of a chemical, thus, volume of titrant required to reach the end point is inversely proportional to the normality.

Volume and normality

$$\boxed{V_1 \times N_1 = V_2 \times N_2}$$

2 = Sample (unknown)   1 = Standard solution

Sample normality

$$\boxed{N_2 = N_1 \times \frac{V_1}{V_2}}$$

In a given titration, a known volume (V_1) of the sample is titrated with a standard solution (N_2) known), and the volume of the titrant (V_2) used to reach the end point is measured. The unknown concentration in the sample can thus be determined using the previous equation. Concentration is more commonly expressed as mass concentration, mg/L or g/m³. When standardizing solution, the concentration is more often expressed in terms of normality. The normality can always be expressed as mass concentration.

27.3.2 ALKALINITY

Alkalinity of water is a measure of its capacity to neutralize acids. Alkalinity is measured by titrating a given sample volume with sulphuric acid of known normality, usually 0.02 N. For high-alkaline samples, the first step is titrating to a pH of 8.3. During this phase of titration, all carbonates are converted to bicarbonates. This is called phenolphthalein alkalinity, as the end point can be indicated by using phenolphthalein indicator. Phenolphthalein turns pink to red at a pH of 8.3. The second phase, or first in case of a water with an initial pH of less than 8.3, is titrating to indicate pH of 4.5. This will represent the total alkalinity. If a pH of a sample is less than 8.3, alkalinity is all in the form of bicarbonate. Samples containing both carbonates and bicarbonates have a pH greater than 8.3, and phenolphthalein alkalinity represents one-half of the carbonate alkalinity. If "P" represents phenolphthalein alkalinity and "T" represents the total, then following relationships define various forms of alkalinity. Knowing the volumes of the sample used and titrant to reach the end point, alkalinity concentration can be calculated. If the pH of the sample is greater than 8.3, you will record two readings. The first reading represents P-alkalinity. The second reading of the total volume of titrant consumed can be used to calculate total alkalinity.

TABLE 27.1

Alkalinity Relationships

	Hydroxide	Carbonate	Bicarbonate
P = 0	0	0	T
T > 2 P	0	2P	T – 2P
T = 2 P	0	2P or T	0
T < 2 P	2P – T	2(T – P)	0
T = P	T	0	0

EXAMPLE PROBLEM 27.8

A 50.0-mL sample of water with an initial pH of 7.7 is titrated to pH 4.5 using 9.2 mL of 0.020 N sulphuric acids. Calculate the total alkalinity.

Given:

Parameter	Titrant = 1	Sample = 2
N, meq/L	20.0	?
V, mL	9.2	50.0

Solution:

As pH < 8.5, P = 0, alkalinity is only present as bicarbonates.

Alkalinity of the sample

$$N_2 = N_1 \times \frac{V_1}{V_2} = \frac{20.0\ meq}{L} \times \frac{9.20\ mL}{50.0\ mL} \times \frac{50.0\ mg}{meq} = 184 = \underline{180\ mg/L}$$

EXAMPLE PROBLEM 27.9

A 100.0-mL water sample is titrated for alkalinity by using 0.02 N sulphuric acid as a titrant. To reach the pH 8.3, 1.5 mL of titrant was used. Calculate the phenolphthalein alkalinity.

Given:

Parameter	Titrant = 1	Sample = 2
N, meq/L	20.0	?
V, mL	1.5	100.0

Phenolphthalein alkalinity

$$N_2 = N_1 \times \frac{V_1}{V_2} = \frac{20.0\ meq}{L} \times \frac{1.50\ mL}{50.\ mL} \times \frac{50.0\ mg}{meq} = 15.0 = \underline{15\ mg/L}$$

This indicates that some of the alkalinity is due to carbonates.

27.3.3 HARDNESS DETERMINATION

Hardness is primarily due to the presence of calcium and magnesium salts in water. Iron, aluminium, manganese and a few other substances also produce hardness but very often too small to be considered. Thus, only total hardness and magnesium hardness are determined. The following table gives the ranges for various degrees of hardness.

TABLE 27.2
Classification of Hardness

Hardness, mg/L $CaCO_3$	Classification
0–50	Soft
50–150	Moderately hard
150	Very hard

Hardness is measured by doing a titration. The titrant used is diamine tetra acetic acid (EDTA), usually with a normality of 0.02 N. Erichrome black T solution or Minver powder pillows are used as indicators to determine the end point. As for any titration, knowing the volume of the titrant V_1 used to reach the end point for a given sample portion (aliquot) V_2 hardness N_2 can be calculated, as discussed before. If the aliquot used is 50 mL, then each mL of titrant used will indicate a hardness of 20 mg/L as $CaCO_3$. This also means if 100 mL of sample is used, then each mL of titrant will represent 10 mg/L of hardness as $CaCO_3$.

Hardness calculation based on mL of titrant used

$$\frac{N_2}{V_1} = \frac{N_1}{V_2} = \frac{20\ meq}{L} \times \frac{1}{50.0\ mL} \times \frac{50.0\ mg\ CaCO_3}{meq} = 20\ mg/L/mL\ of\ titrant$$

EXAMPLE PROBLEM 27.10

12.5 mL of 0.02-N EDTA were used to reach the end point when titrating 50 mL of a well water sample. What is the hardness of the well water?

Given:

Parameter	Titrant = 1	Sample = 2
N, meq/L	20	?
V, mL	12.5	50

Solution:

Hardness expressed as $CaCO_3$

$$N_2 = N_1 \times \frac{V_1}{V_2} = \frac{20\ meq}{L} \times \frac{12.5\ mL}{50\ mL} \times \frac{50\ mg\ CaCO_3}{meq} = 250.0 = \underline{250\ mg/L}$$

EXAMPLE PROBLEM 27.11

A 25-mL water sample is titrated with standard 0.014-N mercury nitrate. 8.4 mL of the titrant are used to reach the end point from yellow to the first purplish tinge. Calculate the chloride content of water.

Given:

Parameter	Titrant = 1	Sample = 2
N, meq/L	14.0	?
V, mL	8.4	25.

Solution:

Chloride concentration

$$N_2 = \frac{N_1 V_1}{V_2} = \frac{14.0\ meq}{L} \times \frac{8.4\ mL}{25.\ mL} \times \frac{35.5\ mg}{meq} = 166 = \underline{170\ mg/L}$$

27.4 JAR TESTING

Jar testing is conducted to determine the effectiveness of chemical treatment. Many of the chemicals we add to water can be evaluated on a bench scale by the use of a jar test. A jar testing apparatus is shown in Figure 27.1. The most important of these chemicals are those used for coagulation and flocculation, such as alum and polymers. By performing a jar test, the plant operator can determine the optimum dosage of chemicals when raw water quality changes or when new coagulants or polymers are being considered for use on a plant scale.

27.4.1 STOCK SOLUTIONS

Stock solutions of coagulants, coagulant aids and other chemicals should be prepared at concentrations such that quantities suitable for use in coagulation tests can be measured accurately and conveniently.

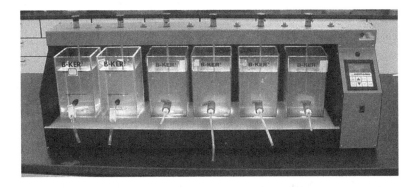

FIGURE 27.1 Jar testing apparatus.

TABLE 27.3
Working Solution Dosages

Range of Dosage	Solution Strength			1 mL of Solution in 1 L of Water
mg/L	%	g/L*	mg/L	mg/L
1–10	0.1	1.0	1000	1.0
10–50	1.0	10.0	10000	10.0
50–500	10.0	100.0	100000	100.0

* This column indicates the grams of pure chemical that should be used to prepare 1 litre of solution.

27.4.2 LIQUID CHEMICALS

For dry chemicals, the solution preparation is straightforward. With concentrated liquid solutions such as liquid alum, a dilution step has to be done. Any dilution step has to consider the specific gravity of the solution being diluted.

EXAMPLE PROBLEM 27.12

How many mL of liquid alum are required to make 1.0 L of a 1.0% solution for jar testing if the liquid alum is 48% by mass having SG of 1.34? How many times should the 1.0% solution be diluted in order to achieve strength of 0.05%?

Given:

Parameter	Liquid Alum = 1	Working Solution = 2
C	48%	1.0% = 10 g/L
V	?	1.0 L
SG	1.34 (1.34 g/mL)	1.0

Solution:

Strength of liquid alum

$$C_1 = C_{m/m} \times \rho = \frac{48\%}{100\%} \times \frac{1.34\ g}{mL} = 0.648\ g/mL$$

Volume of liquid alum

$$V_1 = \frac{C_2}{C_1} \times V_2 = \frac{10\ g}{L} \times \frac{mL}{0.648\ g} \times 1.0\ L = 15\ mL$$

$$\frac{V_2}{V_1} = \frac{C_1}{C_2} = \frac{1.0\%}{0.05\%} \times 20.0 = \underline{\underline{20 \times}}$$

EXAMPLE PROBLEM 27.13

A sample of river water is studied to determine optimum alum dosage using jar test analysis. The ranges of alum dosage selected for the six trials are 5, 10, 15, 20, 25 and 30 mg/L. How many mL of 1.0% alum stock solution should be added to each jar containing 1 L of river water?

Given:

Variable	Working Solution (1)	River Water (2)
V	?	1.0 L
C	1.0% = 10. g/L	5.0 mg/L

Solution:

Volume of stock solution required

$$V_1 = \frac{C_2}{C_1} \times V_2 = \frac{5.0\ mg}{L} \times \frac{L}{10.g} \times 1.0\ L \times \frac{1000\ mL}{1.0\ L} \times \frac{g}{1000\ mg}$$
$$= 0.500 = \underline{0.50\ mL}$$

A volume of 0.5 mL of 1.0% stock solution added will result in an alum dosage of 5.0 mg/L. Hence, the mL of working solution needed to achieve concentration of 10, 15, 20, 25 and 30 mg/L, respectively, are 1.0, 1.5, 2.0, 2.5 and 3.0 mL.

27.5 CHLORINE RESIDUAL

The methods of measuring chlorine residual chlorine fall into four categories:

1. DPD colorimetric method
2. DPD titrimetric method
3. Amperometric titration method
4. Drop dilution technique

Selection of the most practical and appropriate procedure in any particular instance generally depends on the characteristics of the water being examined.

27.5.1 DPD METHOD

In the colorimetric method, the analyst employs chemical solutions (reagents) which produce specific colours that indicate the presence, and the intensity of the colour indicates the concentration. In measuring the chlorine residual, the reagent used is DPD (N-dimethyl-p-phenylene-diamine). The intensity of the colour can be measured by photometers or comparing with standards.

27.5.2   TITRATION METHOD

Using DPD as an indicator and pH buffers, 100 mL of sample are titrated with standard ferrous ammonium sulphate (FAS) with a normality of 0.00282 N. There is a reason for choosing this normality:

V_1 = 100 mL (sample) C_1 = ? V_2 = 1 mL (titrant) C_2 = 0.00282 N

Concentration of chlorine per mL of titrant

$$\boxed{\frac{C_2}{V_1} = \frac{C_1}{V_2} = \frac{2.82 \; meq}{L} \times \frac{1}{100 \; mL} \times \frac{35.5 \; mg \; chlorine}{meq} = 1.0 \; mg/L/mL \; titrant}$$

Thus, each mL of the titrant used for 100 mL of the sample aliquot represents a chlorine residual of 1 mg/L.

27.5.3   AMPEROMETRIC TITRATION METHOD

This is the most accurate method but calls for greater skills. As the name indicates, the end point of titration is indicated by a micrometer. As more and more titrant (0.00564 N phenyl arsine oxide) is added to the sample, the concentration of chlorine decreases, causing the current to diminish. The end point is reached when titrant has been added to the extent that all the chlorine has been neutralized and no further reduction in current is possible. You might have noticed that the normality of the standard PAO solution used as titrant is 0.00564 N, twice that of FAS used in the DPD titrimetric method. Choosing a sample aliquot of 200 mL, each mL of titrant used to reach the end point will be equal to 1 mg/L of chlorine residual. This method is commonly used to study breakpoint chlorination. Each mL of titrant used is equivalent to 2 mg/L of chlorine in a 100-mL sample.

27.5.4   DROP DILUTION TECHNIQUE

When disinfecting clear wells, distribution reservoirs or mains, very high chlorine residuals need to be measured. A drop-dilution technique may be used to estimate the chlorine residual. To use this method, 10 mL distilled water and 0.5 mL of DPD solution are added to the sample tube. The water to be tested is then added to the sample tube one drop at a time until a colour is produced. The chlorine residual in the sample as a result of the colour produced is determined by reading on the meter (field kit). Remember, this reading represents the chlorine reading in the diluted sample. It must be modified by the dilution factor to estimate the chlorine residual in the sample.

Chlorine residual = Residual reading × Dilution factor, DF

$$\boxed{DF = \frac{10 \; mL}{\text{\# of drops}} \times \frac{20 \; drops}{mL} = \frac{200}{\text{\# drops}}}$$

EXAMPLE PROBLEM 27.14

The chlorine residual read by using the drop dilution technique is 0.30 mg/L. If 3 drops of water sample were used to produce the colour, what is the actual chlorine residual of the sample?

Given:

Diluted sample = 3 drops in 10 mL of water
Chlorine residual = 0.30 mg/L in the diluted sample

Solution:

Estimated chlorine residual

$$= reading \times DF = \frac{0.30\ mg}{L} \times \frac{200}{3} = \underline{20\ mg/L}$$

27.6   CHLORINE DEMAND

The chlorine demand of water is the difference between the amount of chlorine applied and the chlorine residual at the end of specific contact period. The chlorine demand test can be used by the plant operator to determine the best chlorine dosage to achieve specific chlorination objectives. Chlorine demand is measured by dosing a series of water samples with selected amounts of chlorine. After the desired contact time, the chlorine residual in each sample is determined and plotted against the applied chlorine dosage to plot the breakpoint chlorination curve.

27.6.1   PREPARING CHLORINE SOLUTION

You need to prepare a stock chlorine solution from household bleach, which is typically 5% (50 g/L as available chlorine). Dilute it 100 times with *chlorine demand free water* to reduce the concentration to about 500 mg/L and standardize to determine its exact concentration. Take 25 mL of this solution and titrate with 0.025 N sodium thiosulphate until the yellow iodine colour almost disappears. Add 1–2 mL starch indicator solution and continue to titrate until the blue colour disappears.

27.6.2   STRENGTH OF DOSING SOLUTION

The actual strength of the dosing solution can be found by a simple titration with sodium thiosulphate.

Strength of the dosing solution

$$C_2 = C_1 \times \frac{V_1}{V_2} = \frac{25\ meq}{L} \times \frac{35.5\ mg}{meq} \times \frac{V_1}{V_2} = \frac{886\ mg}{L} \times \frac{V_1}{V_2}$$

$$2 = Dosing\ solution\ (unknown) \qquad 1 = Titrant\ (Na_2S_2O_3)$$

If 25 mL of the solution took 12.8 mL of the titrant to reach the end point, the strength of the solution is:

$$= 886 \times \frac{V_1}{V_2} = \frac{886 \ mg}{L} \times \frac{12.8 \ mL}{25 \ mL} = 453.6 = 454 \ mg/L$$

27.6.3 CHLORINE DOSING OF SAMPLES

Knowing the strength of the dosing solution, calculate the mL of the dosing solution required to each sample to apply the selected chlorine dosage. For example: Sample volume V_1 = 500 mL, C_2 = 454 mg/L, C_1/V_2 = ?

$$\frac{C_1}{V_2} = \frac{C_2}{V_1} = \frac{454 \ mg}{L} \times \frac{1}{500 \ mL} = 0.908 = 0.91 \ mg/L/mL \ of \ dosed \ solution$$

Each mL of dosing solution added to 500 mL of water would result in chlorine dosage of 0.91 mg/L. Similarly, if the volume of water sample is 100 mL, 2.2 mL of 455 mg/L chlorine solution added to water results in a chlorine dosage of 1 mg/L.

EXAMPLE PROBLEM 27.15

A stock solution of chlorine was prepared by taking 10 mL of house bleach and making up to 1.0 L by adding chlorine demand free water. 25 mL of this solution was used to standardize. It took 13.5 mL of 0.025 N thiosulphate titrant to reach the end point. What is the strength of the chlorine dosing solution?

Given:

Parameter	Titrant = 1	Dosing Solution = 2
V, mL	13.5	25.0
C	25 meq/L	?

Solution:

Concentration of dosing solution

$$C_2 = C_1 \times \frac{V_1}{V_2} = \frac{25 \ meq}{L} \times \frac{35.5 \ mg}{meq} \times \frac{13.5 \ mL}{25.0 \ mL} = 478.5 = \underline{479 \ mg/L}$$

EXAMPLE PROBLEM 27.16

How many mL of dosing solution with 48 mg/L as available chlorine should be added to 50 mL of sample to result in a chlorine dosage of 0.50 mg/L?

Given:

Parameter	Water Sample = 1	Dose Solution = 2
V, mL	50	?
C, mg/L	0.50	48

Solution:

Volume of dosing solution

$$V_2 = \frac{C_1 V_1}{C_2} = \frac{0.50 \ mg/L}{48 \ mg/L} \times 50.0 \ mL = 0.520 = \underline{0.52 \ mL}$$

EXAMPLE PROBLEM 27.17

The results of a chlorine demand test on raw water after a contact time of 20 min are shown in the following table. Calculate the chlorine demand and indicate the breakpoint dosage.

Sample #	1	2	3	4	5	6
Dosage, mg/L	0.5	1.0	1.5	2.0	2.5	3.0
Residual, mg/L	0.26	0.72	1.08	0.60	0.75	1.25
Demand, mg/L	0.24	0.28	0.42	1.40	1.75	1.75

Chlorine demand is calculated by subtracting the residual from chlorine dosage and is shown in the last row of the table above. As chlorine demand is the same in samples 5 and 6, chlorine dosage corresponding to sample number 5 indicates the break point dosage, that is, 2.5 mg/L.

EXAMPLE PROBLEM 27.18

In a water sample, different chlorine dosages were applied, and chlorine residuals were tested after 12 min of contact time. For the data shown in the table, plot the breakpoint chlorination curve and find breakpoint dosage and residual and maximum chlorine demand.

Dosage, mg/L	0.20	0.4	0.60	0.80	1.0	1.2	1.4	1.6	1.8	2.0
Residual, mg/L	0.05	0.15	0.30	0.45	0.30	0.20	0.35	0.62	0.78	1.0

Solution:

The chlorine demand for each dosage is calculated as shown in the following table, and the plot is shown in Figure 27.2. From the plot: Breakpoint dosage = 1.2 mg/L, Max demand = 1.0 mg/L.

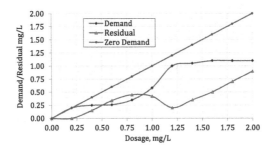

FIGURE 27.2 Breakpoint chlorination (Example Problem 27.18).

Dosage, mg/L	0.20	0.40	0.60	0.80	1.00	1.20	1.40	1.60	1.80	2.00
Residual, mg/L	0.05	0.15	0.30	0.45	0.30	0.20	0.35	0.62	0.78	1.00
Demand, mg/L	0.15	0.25	0.30	0.35	0.70	1.00	1.05	0.98	1.02	1.00

27.7 SPECTROPHOTOMETRY

Lab tests for determination of contents of iron, manganese, phosphorus and fluoride are based on the colour intensity achieved when a specific colour-developing reagent is added to the sample. The spectrophotometer is an instrument generally used to measure the colour intensity of a chemical solution. In spectrophotometry, the colour intensity is measured in two ways:

– Percent transmittance (T), arithmetic scale (1 to 100).
– Absorbency (A), a logarithmic scale, (0.0 to 2.0).

$$A = 2 - log\ T$$

A sample with a low colour intensity will have high percent transmittance but a low absorbency. The main usefulness of absorbency lies in the fact that it is a logarithmic function rather than a linear function. As Beer's law states, the concentration is directly proportional to the absorbency over a given range of concentrations. This means when absorbency is plotted versus the concentration, it will yield a straight line.

27.7.1 CALIBRATION CURVE

The calibration curve is used to determine the concentration of the parameter in question (iron, manganese, etc.). Using a series of standards and a blank, the transmittance or absorbency is read on the meter after developing the colour by using the reagents described in the standard procedure. From the data obtained, a linear equation between concentration and absorbance is fitted. Once you have established this relationship, you can determine the unknown concentration in a given sample by plugging the value of absorbency into the established equation.

EXAMPLE PROBLEM 27.19

Results from a series of tests for fluoride are as follows:

Standard mg/L, y	0.0	0.5	0.75	1.0	1.25
Absorbency, x	0.73	0.63	0.56	0.50	0.44

Develop the calibration equation. Using the equation, determine the fluoride concentration in a well water sample for which the absorbency is read to be 0.52.

Solution:

#	y	x	x^2	xy
1	0.0	0.73	0.5329	0.0
2	0.50	0.63	0.3969	0.315
3	0.75	0.56	0.3136	0.420
4	1.0	0.50	0.25	0.500
5	1.25	0.44	0.1936	0.550
Σ	3.5	2.86	1.687	1.785

Slope of the fitted line

$$b = \frac{n\Sigma xy - \Sigma x\Sigma y}{n\Sigma x^2 - (\Sigma x)^2} = \frac{5\times1.785 - 2.86\times3.5}{5\times1.687 - (2.86)^2} = -4.248 = \underline{-4.25}$$

Y-intercept of the line

$$a = \frac{\Sigma y - b\Sigma x}{n} = \frac{3.5 - (-4.248)\times2.86}{5} = 3.129 = \underline{3.13}$$

Calibration equation

$$C = 3.13 - 4.25A$$

Predicted value

$$C = 3.13 - 4.25A = 3.13 - 4.25\times0.52 = 0.920 = 0.92 \ mg/L$$

The reliability of the prediction depends upon the closeness of plotted points on the fitted line.

PRACTICE PROBLEMS

PRACTICE PROBLEM 27.1

A 10-L container is filled up to the 5-L mark in 95 s. Calculate the flow rate in L/min. (3.2 L/min)

PRACTICE PROBLEM 27.2

Maintaining a flow rate of 0.5 gpm, how long will it take to flush a 100-ft length of ½-in-diameter line such that the water in the line is replaced twice? (4.0 min)

PRACTICE PROBLEM 27.3

A sampling tap is connected to a water main via a 32-m-long and 20-mm-diameter pipeline How long it must be run at the rate of 5.0 L/min before collecting a sample of water? (2.0 min)

PRACTICE PROBLEM 27.4

Calculate the molarity of a solution prepared by dissolving 49 g of sulphuric acid and made up to 1 L. (0.5 M)

PRACTICE PROBLEM 27.5

Calculate the normality of a 1.0 L solution containing 6.5 mg of $Ca(OH)_2$. (1.76×10^{-4} N)

PRACTICE PROBLEM 27.6

In the previous problem, if 7.3465 g of the chemical are weighed, what volume of solution should be made? (954 mL)

PRACTICE PROBLEM 27.7

If 3.05 g of $AgNO_3$ are actually weighed, what volume of distilled water should be measured to make the solution? (933 mL)

PRACTICE PROBLEM 27.8

A 100-mL water sample was titrated with 0.02 N sulphuric acid. 12.5 mL of the titrant was used to reach pH of 4.5. Calculate the total alkalinity. (125 mg/L)

PRACTICE PROBLEM 27.9

In titration for alkalinity, 0.6 mL of 0.02 N sulphuric acid is used to reach phenol-phthalein end point (pH = 8.3) and 11.4 mL to reach the bromocresol green methyl red end point (pH = 4.5) for a 100-mL water sample. Calculate the carbonate and bi-carbonate alkalinity present. (12 mg/L, 102 mg/L as $CaCO_3$)

PRACTICE PROBLEM 27.10

A well water sample was studied for hardness. Due to the suspected high degree of hardness, the sample was diluted by a factor of two by adding distilled water. 6.8 mL of 0.02 N EDTA were used to reach the end point when 50 mL of the diluted sample were taken. What is the hardness of the water? (270 mg/L)

PRACTICE PROBLEM 27.11

For a 25-mL aliquot of the water sample, each mL of the 0.014 N titrant used to reach the end point represents how many mg/L of chloride in the sample? (20 mg/L)

PRACTICE PROBLEM 27.12

You are required to prepare 100 mL of 1.0% ferric chloride solution. The liquid concentrate is 42% by mass, and the SG is 1.42. What volume of concentrate should be measured? (1.7 mL)

PRACTICE PROBLEM 27.13

For the river water, another jar test is planned to study the effectiveness of a coagulant aid at a dosage of 0.5 mg/L along with alum. How many mL of 0.05% polymer solution would you need to add to dose water at this rate? (1 mL)

PRACTICE PROBLEM 27.14

Two drops of a sample were added to 10 mL of distilled water, and an indicator produced a chlorine residual of 0.3 mg/L. Calculate the actual chorine residual in the sample. (30 mg/L)

PRACTICE PROBLEM 27.15

A chlorine dosing solution was prepared by diluting household bleach. An aliquot of 50 mL of this solution took 22.5 mL of 0.025N thiosulphate titrant to reach the end point. What is the strength of the chlorine solution? (400 mg/L)

PRACTICE PROBLEM 27.16

How many mL of the 500-mg/L dosing solution should be added to 1 L of the sample to achieve a chlorine dosage of 1.0 mg/L? (2.0 mL)

PRACTICE PROBLEM 27.17

The results of a chlorine demand test on a raw water sample are as follows:

Dosage, mg/L	0.25	0.50	0.75	1.00	1.25	1.50	1.75	2.00
Residual, mg/L	0.24	0.41	0.55	0.53	0.25	0.50	0.75	1.00

What is the breakpoint dosage, and what is the chlorine demand at a dosage of 1.50 mg/L? (1.25 mg/L, 1.00 mg/L)

PRACTICE PROBLEM 27.18

For the data of Example Problem 27.8, calculate the chlorine feeder setting in kg/d for treating a flow of 12 ML/d and for achieving a residual of 0.65 mg/L. (20 kg/d)

PRACTICE PROBLEM 27.19

Six standards of iron were tested, and results are shown as follows. Fit a calibration equation of the form $y = a + bx$ and determine the concentration of iron in a well water sample, the absorbency of which was read to be 35%. (b = 1.52, a = 0.00, 0.53 mg/L)

Standard (mg/L)	0.0	0.1	0.25	0.5	0.75	1.0
Absorbency	0.000	0.066	0.16	0.328	0.495	0.658

28 Basic Water Distribution

The basics of hydraulics as applied to water distribution are discussed. It is strongly recommended to review the topic of hydraulics before completing this unit. It cannot be overemphasized that knowledge of hydraulics is a prerequisite for understanding the operation of water distribution systems.

28.1 PRESSURE AND FORCE

Pressure is force per unit area. For the same force applied, pressure is inversely proportional to area. Similarly, for a given area, pressure will vary directly with the force applied.

28.2 PRESSURE AS A LIQUID COLUMN

Hydrostatic pressure created by a water column is directly proportional to the height of the water. This height represents the water energy due to pressure and is also known as pressure head.

Pressure and height

$$p = \gamma \times h \quad or \quad h = \frac{p}{\gamma}$$

Since the weight density of water is 9.8 kN/m^3, it means 9.8 kPa or roughly 10 kPa of pressure is created due to 1.0 m of water column. In USC units, the water density of water is 62.4 lb/ft^3, or 0.433 psi/ft.

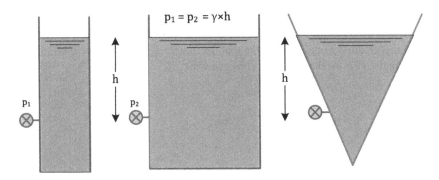

FIGURE 28.1 Pressure and height of liquid.

28.3 CONTINUITY EQUATION

Based on the principle of mass conservation, under steady state conditions, the flow that enters the pipe is the same flow that exits the pipe. This implies that any change in the flow area will result in proportional change in flow velocity at the point of interest. At a given water section, flow velocity will increase due to narrowing of the section or reduction in pipe size.

Continuity equation

$$\boxed{Q_1 \times A_1 = Q_2 \times A_2}$$

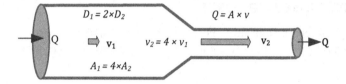

FIGURE 28.2 Flow velocity and pipe diameter.

In a circular pipe of diameter D, the flow velocity for a given flow rate can be found as follows:

Flow velocity and flow area

$$\boxed{v = \frac{Q}{A} = \frac{4Q}{\pi D^2} = \frac{1.27Q}{D^2}}$$

Velocity and diameter

$$\boxed{\frac{v_2}{v_1} = \frac{A_1}{A_2} = \left(\frac{D_1}{D_2}\right)^2}$$

EXAMPLE PROBLEM 28.1

The sight glass readings indicate that the liquid level in a chemical feed tank dropped by 6.0 inches during one shift. Given that the diameter of the tank is 7.0 ft, find how much chemical is used over a period of one shift.

Given:

D = 7.0 ft   h = 6.0 in = 0.50 ft   V = ?

Solution:

Volume of chemical used

$$V = \frac{\pi D^2 h}{4} = \frac{\pi (7.0 \ ft)^2 \times 0.5 \ ft}{4} \times \frac{7.48 \ gal}{ft^3} = 143 = \underline{140 \ gal}$$

EXAMPLE PROBLEM 28.2

After routine maintenance and cleaning, a 25-m-diameter reservoir is filled at the rate of 80 L/s. If the overflow is at a height of 12 m, how many hours will it take to fill the reservoir?

Given:

$D = 12$ m $Q = 80$ L/s $t = ?$

Solution:

Capacity of the reservoir

$$V = \frac{\pi D^2 h}{4} = \frac{\pi (25\ m)^2 \times 12\ m}{4} \times \frac{ML}{1000\ m^3} = 5.89 = 5.9\ ML$$

Time required to fill

$$t = \frac{V}{Q} \times \frac{5.89\ ML.s}{80\ L} \times \frac{h}{3600\ s} \times \frac{10^6\ L}{ML} = 20.45 = \underline{20\ h}$$

EXAMPLE PROBLEM 28.3

A flow gauge reads 3.5 gpm. How many gallons of water will flow in 2.5 hours?

Given:

$Q = 3.5$ gal/min $t = 2.5$ h $V = ?$

Solution:

Volume of flow

$$V = Q \times t = \frac{3.5\ gal}{min} \times 2.5\ h \times \frac{60\ min}{h} = 525 = \underline{530\ gal}$$

EXAMPLE PROBLEM 28.4

A chemical solution tank is 120 cm in diameter and 150 cm deep. How long will it take to fill this tank at the rate of 12 L/min?

Given:

$D = 120$ cm $= 1.2$ m $d = 150$ cm $= 1.5$ m $Q = 12$ L/min $t = ?$

Solution:

Capacity of the chemical tank

$$V = \frac{\pi D^2 h}{4} = \frac{\pi (1.2m)^2 \times 1.5\ m}{4} \times \frac{1000\ L}{m^3} = 1696 = 1700\ L$$

Time required to fill

$$t = \frac{V}{Q} = \frac{1696 \, L.min}{12 \, L} \times \frac{h}{60 \, min} = 2.35 = \underline{2.4 \, h}$$

EXAMPLE PROBLEM 28.5

Over a period of 48 hours, the level in a hypochlorite solution feed tank drops by 8.0 inches. If the diameter of the tank is 3.5 ft, what is the solution feed rate in gal/h?

Given:

$D = 3.5 \, ft \qquad d = 8 \, in = 0.667 \, ft \qquad t = 48 \, h \qquad Q = ?$

Solution:

Solution feed pump rate

$$Q = \frac{V}{t} = \frac{\pi D^2 h}{4t} = \frac{\pi (3.5 \, ft)^2 \times 8 \, in}{4 \times 48 \, h} \times \frac{ft}{12 \, in} \times \frac{7.48 \, gal}{ft^3} = 0.99 = \underline{1.0 \, gal/h}$$

EXAMPLE PROBLEM 28.6

A 1.5-m-diameter solution tank is filled to a height of 1.0 m. If the chemical is fed at the rate of 0.50 L/min, how many hours will it take to empty half of the tank?

Given:

$D = 1.5 \, m \qquad d = 1.0 \, m/2 = 0.50 \, m \qquad Q = 0.50 \, L/min \qquad t = ?$

Solution:

Half capacity of the chemical tank

$$V = \frac{\pi D^2 d}{4} = \frac{\pi (1.5 \, m)^2 \times 0.5 \, m}{4} \times \frac{1000 \, L}{m^3} = 883 = 880 \, L$$

Time required to empty half the tank

$$t = \frac{V}{Q} = \frac{883 \, L.min}{0.50 \, L} \times \frac{h}{60 \, min} = 29.45 = \underline{29 \, h}$$

EXAMPLE PROBLEM 28.7

The level in a storage tank rises 4 ft and 4 in in 5.0 h. If the tank has a diameter of 225 ft and the plant is producing 24 MGD, what is the average discharge rate of the treated water discharge pumps in gallons per minute?

Given:

$t = 5.0 \, h \qquad D = 225 \, ft \qquad \Delta h = 4 \, ft \, 4 \, in = 4.33 \, ft \qquad Q_o = ? \qquad Q_i = 24 \, MG/d$

Solution:

According to the principle of continuity, the rate of inflow minus the rate of outflow must equal the rate of the change in storage. The rate of the outflow can be expressed as $Q_o = Q_i - \Delta Q_s$

Inflow rate

$$Q_i = \frac{24 \times 10^6 \ gal}{d} \times \frac{d}{1440 \ min} = 16600 \ gpm$$

Change in storage

$$\Delta Q_S = \frac{\pi D^2}{4} \times \frac{\Delta h}{t} = \frac{\pi}{4} \times (225 \ ft)^2 \times \frac{4.33 \ ft}{5.0 \ h} \times \frac{7.48 \ gal}{ft^3} \times \frac{h}{60 \ min}$$

$$= 4292 \ gpm = 4300 \ gpm$$

Outflow rate

$$Q_o = Q_i = \Delta Q_s = 16666.6 - 4292.6 = 12374 = 12000 \ gpm$$

EXAMPLE PROBLEM 28.8

The flushing velocity recommended to clean a water main is 1.5 m/s. To flush a 300-mm-diameter main, what discharge rate should be maintained at the hydrant?

Given:

D = 300 mm = 0.30 m $\qquad$ Q = ? $\qquad$ v = 1.5 m/s

Solution:

Hydrant discharge rate

$$Q = A \, v = \frac{\pi (0.3 \ m)^2}{4} \times \frac{1.5 \ m}{s} \times \frac{1000 \ L}{m^3} = 106.0 = 110 \ L/s$$

EXAMPLE PROBLEM 28.9

Your supervisor has asked you to conduct a maintenance flushing procedure on a 150-mm-diameter main. To achieve the desired results, the line needs to be flushed maintaining a flow velocity of 1.5 m/s. Find the discharge rate in L/s.

Given:

D = 150 mm = 0.150 m $\qquad$ v = 1.5 m/s $\qquad$ Q = ?

Solution:

Hydrant discharge rate

$$Q = A \, v = \frac{\pi (0.15 \ m)^2}{4} \times \frac{1.5 \ m}{s} \times \frac{1000 \ L}{m^3} = 26.49 = 26 \ L/s$$

EXAMPLE PROBLEM 28.10

A pressure gauge on a fire hydrant reads 51 m of pressure head. What is the water pressure in kPa?

Given:

$h = 51$ m $p = ?$ $\gamma = 9.81$ kPa/m

Solution:

Hydrostatic pressure

$$p = \gamma h = \frac{9.81 \; kPa}{m} \times 51 \; m = 500.3 = \underline{500 \; kPa}$$

EXAMPLE PROBLEM 28.11

The hydraulic grade for a remote point in the upper pressure zone of a given water distribution system is read from the hydraulic profile for peak hourly flow If the hydraulic and elevation head are 345.0 ft and 255.0 ft, respectively, what minimum pressure can be expected in the region during the peak flow hour?

Given:

$h = 345.0$ ft $p = ?$ $\gamma = 0.433$ psi/ft $Z = 255.0$ ft

Solution:

Hydrostatic pressure

$$p = \gamma h = \frac{0.433 \; psi}{ft} \times (345.0 - 255.0) \; ft = 38.97 = \underline{39 \; psi}$$

EXAMPLE PROBLEM 28.12

What will be the pressure at the bottom of a 10-m-diameter reservoir when the water level reaches a height of 9.0 m?

Given:

$h = 9.0$ m $p = ?$ $\gamma = 9.81$ kPa/m

Solution:

Hydrostatic pressure

$$p = \gamma h = \frac{9.81 \; kPa}{m} \times 9.0 \; m = 88.2 = \underline{88 \; kPa}$$

28.4   JET VELOCITY

Under flowing conditions, energy due to gravity or pressure as indicated by the hydraulic head is converted to velocity head. Assuming no head loss, the velocity of the jet is given by the following relationship:

Water jet flow velocity

$$\boxed{v_j = \sqrt{2gh} = \sqrt{2gp/\gamma}}$$

Depending on the design and type of opening, the actual velocity of the jet would be less than the ideal flow velocity as given by the previous relationship. The reduction is indicated by the coefficient, which is typically 80–90%. Knowing the actual flow velocity, the water discharge rate, as in the case of a flowing hydrant, can be worked out. In hydrant testing, we need to know the maximum flow capacity that can be expected when the residual pressure drops to 150 kPa. This test is run by measuring the velocity head by putting the pitot head meter in front of the free-flowing jet, and velocity pressure is read. Discharge rate is calculated by multiplying the actual flow velocity by the area of the hydrant nozzle.

EXAMPLE PROBLEM 28.13

The velocity pressure of a water jet is read using a pitot gauge. For a velocity pressure reading of 3.5 psi, calculate the flow velocity of the jet.

Given:

p = 3.5 psi   γ = 2.31 ft/psi   v = ?

Solution:

Jet flow velocity

$$v = \sqrt{\frac{2gp}{\gamma}} = \sqrt{\frac{2 \times 32.2\ ft}{s^2} \times \frac{3.5\ psi \times 2.31\ ft}{psi}} = 22.8 = \underline{23\ ft/s}$$

EXAMPLE PROBLEM 28.14

The velocity pressure of a water jet is read 32 kPa when a pitot gauge is held in front of a free-flowing jet of water out of a fire hydrant. Assuming the actual flow velocity is 85% of the theoretical flow velocity, what the discharge rate from a 60-mm-diameter nozzle?

Given:

p = 32 kPa   C = 85% = 0.85   Q = ?

Solution:

Hydrant discharge rate

$$Q = A \times C \times v = \frac{\pi D^2}{4} \times C \times \sqrt{\frac{2gp}{\gamma}} = \frac{\pi (0.06\ m)^2}{4} \times 0.85 \sqrt{\frac{2 \times 9.81\ m \times 32\ kPa}{s^2} \times \frac{m}{9.81\ kPa}}$$

$$= 1.92 \times 10^{-2} m^3/s = \underline{19\ L/s}$$

EXAMPLE PROBLEM 28.15

After maintenance and repairs, a 12-m-diameter surface reservoir is to be disinfected by dosing it with a chlorine concentration of 100 mg/L. It is thought that a dosage of 100 mg/L should be able to maintain a residual of 50 mg/L during the disinfection period. What volume of 15% sodium hypochlorite solution should be added to fill the tank to a depth of 3.0 m?

Given:

D = 12 m   d = 3.0 m C_2 = 100 mg/L C_1 = 15% = 150 g/L V_1 = ?

Solution:

Volume of water dosed

$$V_2 = \frac{\pi D^2 h}{4} \times \frac{\pi (12m)^2 \times 3.0\ m}{4} = 339 = 340\ m^3$$

Volume of hypochlorite needed

$$V_1 = 339\ m^3 \times \frac{1000\ L}{m^3} \times \frac{0.10\ g/L}{150\ g/L} = 226 = \underline{230\ L}$$

EXAMPLE PROBLEM 28.16

A surface reservoir contains 8.0 ML of water. If you mix 500 L of 5.0% bleach with the water in the reservoir, what is the chlorine dosage in mg/L?

Given:

Variable	Bleach = 1	Water = 2
Volume	500 L	8.0 ML
Conc.	5.0%	?

Solution:

Chlorine dosage

$$C_2 = \frac{50\ g}{L} \times \frac{500\ L}{8.0\ ML} \times \frac{ML}{1000\ m^3} = 3.12 = \underline{3.1\ g/m^3 (mg/L)}$$

EXAMPLE PROBLEM 28.17

Work out the volume of 4.5% bleach required to disinfect a 400-m-diameter well casing and well screen applying a chlorine dosage of 100 mg/L. The depth of water in the well is 25 m.

Given:

Variable	Bleach = 1	Well Water = 2
Volume	?	400 mm, 25 m deep
Conc.	4.5% = 45 g/L	100 mg/L

Solution:

Volume of water standing in well

$$V_2 = \frac{\pi D^2 h}{4} \times \frac{\pi (0.40\ m)^2 \times 25\ m}{4} \times \frac{1000\ L}{m^3} = 3141.5 = 3100\ L$$

Volume of bleach required to disinfect

$$V_1 = V_2 \times \frac{C_2}{C_1} = 3141.5\ L \times \frac{0.10\ g/L}{45\ g/L} = 6.98 = \underline{7.0\ L}$$

EXAMPLE PROBLEM 28.18

At the Queen Well, the plant site is a 500-gal-capacity hypochlorite tank that is filled every 10 d. The well chlorination system is set to apply a chlorine dosage of 1.5 mg/L. The strength of the liquid chemical is 12.5%, and the density is 1.2 kg/L. What is the water pumping rate at this station?

Given:

Variable	Hypo = 1	Well Water = 2
Rate	200 L/d	?
Conc.	12.5%	1.5 mg/L
Density	1.2 kg/L	1.0 kg/L

Solution:

Mass concentration

$$C_1 = \frac{12.5\%}{100\%} \times \frac{8.34\ lb \times 1.2}{gal} = 1.250 = 1.25\ lb/gal$$

Water pumping rate

$$Q_2 = Q_1 \times \frac{C_1}{C_2} = \frac{500\ gal}{10\ d} \times \frac{1.25\ lb}{gal} \times \frac{L}{1.5\ mg} \times \frac{mg/L}{8.34\ lb/MG} = 4.99 = \underline{5.0\ MGD}$$

EXAMPLE PROBLEM 28.19

At the Queen Well plant site, there is a 2.0-kL-capacity hypochlorite tank used to feed chlorine. The well chlorination system is set to apply a chlorine dosage of 2.3 mg/L. The strength of the hypochlorite is 140 g/L as chlorine. If the well is pumped @ 12 ML/d, find after how many days the tank needs to be filled or changed.

Given:

Variable	Hypo = 1	Well Water = 2
Rate	?	12 ML/d
Conc.	140 kg/kL	2.3 mg/L

Solution:

Feed pump rate

$$Q_1 = Q_2 \times \frac{C_2}{C_1} = \frac{12\ ML}{d} \times \frac{2.3\ kg}{ML} \times \frac{kL}{140\ kg} = 0.197 = \underline{0.20\ kL/d}$$

Time period after which tank needs to be refilled

$$t = \frac{V}{Q} = 2.0\ kL \times \frac{d}{0.197\ kL} = 10.1 = \underline{10\ d}$$

EXAMPLE PROBLEM 28.20

Water is pumped from a well into a 12-in water main at the rate of 1250 gpm. The length of the water main before the first consumer is 1.1 mile. What is the theoretical detention time achieved in the water main?

Given:

D = 12 in = 1.0 ft   L = 1.1 mile   Q = 1250 gpm t_d = ?

Solution:

Detention time achieved

$$t_d = \frac{\pi D^2}{4} \times \frac{L}{Q} = \frac{\pi (1.0\ ft)^2}{4} \times \frac{1.1\ mile.min}{1250\ gal} \times \frac{5280\ ft}{mile} \times \frac{7.48\ gal}{ft^3} = 27.29 = \underline{27\ min}$$

PRACTICE PROBLEMS

PRACTICE PROBLEM 28.1

During flushing of a 2000-ft length of an 8-in-diameter water main, you need to replace two water volumes. How much water you should discharge at the flowing hydrant? (10,000 gal)

PRACTICE PROBLEM 28.2

A 300-m-long section of a 200-mm-diameter water main is to be flushed by discharging the hydrant at the rate of 20 L/s. How long does the hydrant need to be kept open to replace two water volumes? (8 min)

PRACTICE PROBLEM 28.3

During a fire flow test lasting for 23 min, the flowing hydrant discharged water at the rate of 19 L/s. How many kL of water are discharged? (26 kL)

PRACTICE PROBLEM 28.4

How long will it take to fill a 1.2-m-diameter tank to a depth of 1.2 m when filled at the rate of 8.0 L/min? (2 h and 50 min)

PRACTICE PROBLEM 28.5

A 1.2-m-diameter tank is filled to a height of 2.5 m. Find how many days it will take to drop the liquid level in the tank by 1.0 m, maintaining a constant feed rate of 400 mL/min. (2.0 d)

PRACTICE PROBLEM 28.6

A 1200-mm-diameter solution tank is used to feed hypochlorite solution at a well pumping station. Over a period of an 8-h shift, the liquid level dropped by 15 cm. Express the chemical feed rate in mL/min. (350 mL/min)

PRACTICE PROBLEM 28.7

In a period of 5.0 h, the water level in a storage tank rises by 4.1 ft. If the tank has a diameter of 250 ft and the plant is producing 21 MGD, what is the average discharge rate of the treated water discharge pumps in gallons per minute? (9600 gpm)

PRACTICE PROBLEM 28.8

What flushing velocity is achieved when a 200-mm-diameter main is flushed by maintaining a flow of 50 L/s at the downstream flowing hydrant? (1.6 m/s)

PRACTICE PROBLEM 28.9

To achieve proper cleaning, it is recommended to flush the line, causing a flow veloc-ity of 1.5 m/s. To flush a 300-mm-diameter main, what discharge rate at the flowing hydrant or hydrants should be achieved? (110 L/s)

PRACTICE PROBLEM 28.10

A pressure gauge attached to a fire hydrant reads 350.0 kPa. What is the hydraulic grade at this location given that the elevation is 251.4 m? (287.1 m)

PRACTICE PROBLEM 28.11

The hydraulic grade at a fire hydrant is expected to be 135.0 m during the average day flow conditions. What hydrostatic pressure can be expected at this hydrant during average day flow conditions? The elevation is of this location is 81.4 m. (526 kPa)

PRACTICE PROBLEM 28.12

In the hill area of the town, water pressure in the water main is on the order of 350 kPa. If a fire hydrant breaks open, how high will the water go? (36 m)

PRACTICE PROBLEM 28.13

If the water jet from a fire hose nozzle reaches a height of 12 ft, what is the velocity of the jet? (28 ft/s)

PRACTICE PROBLEM 28.14

The velocity pressure of a water jet issuing from a 60-mm hydrant nozzle is read as 28 kPa. Assuming the actual flow velocity is 80% of the theoretical flow velocity, what the discharge rate of the flowing hydrant? (17 L/s)

PRACTICE PROBLEM 28.15

A reservoir needs to be disinfected before being placed online. The rectangular tank is 1.8 m deep, 3.0 m wide and measures 6.0 m. An initial dose of 100 mg/L is expected to maintain the desired residual of 50 mg/L during the disinfec-tion period of 24 hours. How much bleach with 5.0% available chlorine will be needed? (65 L)

PRACTICE PROBLEM 28.16

If you fill a 180-L capacity drum that already has 60 L of 12% sodium hypo-chlorite solution in it with water, what will be the strength of the diluted solu-tion? (4.0%)

PRACTICE PROBLEM 28.17

How many litres of 5.0% sodium hypochlorite will be needed to disinfect a well with a 500-mm-diameter casing and well screen? The well is 110 m deep, and there are 35 m of water in it. It is recommended to provide an initial dose of 100 mg/L of chlorine. (14 L)

PRACTICE PROBLEM 28.18

The chlorine feed pump at the King well pumping station is fed at the rate of 8.0 gal/24 h to provide a chlorine dosage of 0.80 mg/L. If the well pumping rate is 1.2 MGD, what is the strength of the hypochlorite solution being fed? (12%)

PRACTICE PROBLEM 28.19

How long will a 1.0-kL-capacity chemical tank filled with 12%, SG = 1.2 hypochlorite last when the well pumping rate is 9.3 ML/d and well water is dosed @ 2.2 mg/L? (7.0 d)

PRACTICE PROBLEM 28.20

Water is pumped from a well into a 8-in water main at the rate of 650 gpm. What is the minimum length of water main required before it reaches the first consumer to achieve a minimum detention time of 30 min? (7400 ft)

29 Advanced Water Distribution

The main purpose of this chapter is to illustrate the key hydraulic concepts related to water distribution system operation and maintenance. This includes the use of the energy equation and continuity equation and working out the pumping head and power.

Water distribution is an important part of any water system. The water flow takes place under pressure conditions which may be due to an elevated reservoir or directly pumped into the water main by the high lift pumps. The key equation in the following examples is as follows:

Head added

$$h_a = \frac{p_2}{\gamma} - \frac{p_1}{\gamma}$$

29.1 HEAD LOSSES IN A WATER MAIN

The energy equation can also be used to estimate the head losses or pressure drop for given flow conditions. Under normal operating conditions, the head loss rate as indicated by the hydraulic grade line is typically 0.1–0.2%. Head losses at this rate amount to 10–20 kPa for every km length of the water main. During emergencies, as in the case of firefighting, flows in the water main may be as much as two to three times the daily average. This will cause significant head losses. It is for this reason

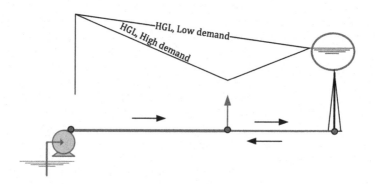

FIGURE 29.1 A simplified water distribution plant.

DOI: 10.1201/9781003468745-29

that during such events, pressure can drop below the normal 250–400 kPa (35–60 psi) down to as much as 150 kPa (20 psi). However, under no conditions should pressure be allowed to drop below 150 kPa. This may happen during a main break, opening too many hydrants on the same water main or excessive withdrawals for firefighting. This causes serious water quality problems due to *back siphonage*, so great caution must be used when operating hydrants.

Head loss in a water line

$$h_l = Z_2 - Z_1 + \frac{p_2}{\gamma} - \frac{p_1}{\gamma}$$

As shown in the above equation, head losses are found from the hydraulic grade for which the pipe is designed. Pressure at two points, for example, two hydrants, can be read for a given flow condition, and the actual head loss as found by the equation can be compared to the design head loss.

EXAMPLE PROBLEM 29.1

A hydraulic gradient test is performed in the field. Two hydrants 750 m apart were chosen to observe the hydraulic head. For the maximum flow conditions, pressure readings at hydrants 1 and 2 were observed to be 477 kPa and 495 kPa. From the map for the area, the elevations of hydrants 1 and 2 are 112.4 m and 111.9 m, respectively. Find the hydraulic head at each of the hydrants and hence the head loss in the pipeline connecting the two hydrants.

Given:

Variable	Hydrant 1	Hydrant 2
Pressure, kPa	477	495
Elevation, m	112.4	111.9
Head loss, h_L	?	
Length, m	750	

Solution:

Head loss

$$h_l = (Z_2 - Z_1) + \frac{(p_2 - p_1)}{\gamma}$$

$$= (112.4 - 111.9)\,m + (477 - 495)\,kPa \times \frac{m}{9.81\,kPa} = -1.33 = \underline{-1.3\,m}$$

Since head loss cannot be negative, it means the flow is in the opposite direction.

EXAMPLE PROBLEM 29.2

The hydrostatic pressure at the hydrant located at the corner of Pine and Pim streets is read to be 440 kPa. Given that the elevation of the hydrant is 112.4 m, find the hydraulic grade at the hydrant.

Given:

p = 440 kPa   γ = 9.81 kPa/m   Z = 112.4 m   h = ?

Solution:

Hydraulic grade or head

$$h = \frac{p}{\gamma} + Z = \frac{840 \ kPa.m}{9.81 \ kPa} + 112.4 = 198.0 = \underline{198 \ m}$$

29.2   HEAD LOSS DUE TO FRICTION

Head loss due to friction can also be found using either the flow equation, Darcy–Weisbach or Hazen–Williams, as discussed in the unit on hydraulics. Though the Darcy–Weisbach flow equation is more accurate, the Hazen–Williams flow equation is easy to use. Whereas friction is indicated by the friction factor f in the Darcy–Weisbach flow equation, the roughness of the pipe is indicated by the roughness coefficient C. Values of C range from 50 to 150, with higher values indicating smoother pipes. Typical C values for water mains are in the range of 80–130 depending on age, degree of encrustation and tuberculation. Since this is an empirical relationship, the constant in the formula will change if the diameter of the pipe is given in units other than m. The Darcy–Weisbach equation is a theoretical equation and hence can be used for any consistent units. The parameter friction slope is a dimensionless parameter and indicates head loss due to friction per unit length of the flow pipe. If, instead of head loss, pressure loss is given, make sure to convert it to head units.

Friction slope/gradient

$$\boxed{S_f = \frac{h_f}{L} = \frac{\Delta h}{L} = \frac{\Delta p}{\gamma} \times \frac{1}{L}}$$

Darcy–Weisbach equation (Consistent Units)

$$\boxed{h_f = \frac{f}{1.23 \ g} \times L = \frac{Q^2}{D^5}, \quad Q = \sqrt{\frac{1.23 \ g}{f} \times \frac{h_f}{L} \times D^5}, \quad D = \sqrt[5]{\frac{f}{1.23 \ g} \times \frac{1}{S_f} \times Q^2}}$$

29.2.1   FLOW CAPACITY Q

The flow-carrying capacity of a given pipe size for a maximum permissible head loss or friction slope can be calculated using the flow equation. Note that in the

Hazen–Williams equation, the exponent of the parameter diameter D is 2.63 to indicate that an increase in pipe size results in a much greater increase in flow capacity.

Hazen–Williams equation (SI)

$$Q = 0.278C \times D^{2.63} \times S_f^{0.54} \ SI$$

Hazen–Williams equation (D = in, Q = gpm)

$$Q = 0.281C \times D^{2.63} \times S_f^{0.54}$$

Darcy–Weisbach equation (SI)

$$Q = 3.5x \sqrt{\frac{S_f}{f}} \times D^5$$

EXAMPLE PROBLEM 29.3

Estimate the flow capacity of a 400-mm-diameter main for an allowable friction slope of 0.15%. Assume the Hazen–Williams coefficient of friction is 110.

Given:

D = 400 m = 0.4 m S_f = 0.15% = 0.0015   C = 110

Solution:

Flow-carrying capacity

$$Q = 0.278C \times D^{2.63} \times S_f^{0.54} = 0.278 \times 110 \times 0.40^{2.63} \times 0.0015^{0.54}$$

$$= 8.20 \times 10^{-2} = 8.2 \times 10^{-2} \, m^3 / s = \underline{82L/s}$$

EXAMPLE PROBLEM 29.4

Estimate the flow-carrying capacity of a 12-in water main for an allowable friction loss of 10 ft ft/mile length of pipe. Assume the Hazen–Williams roughness coefficient to be 100.

Given:

D = 12 in S_f = 10 ft/mile   C = 100

Solution:

Friction slope

$$S_f = \frac{10 \, ft}{mile} \times \frac{mile}{5280 \, ft} = 0.001893 \, ft/ft$$

Flow-carrying capacity

$$Q = 0.281C \times D^{2.63} \times S_f^{0.54} = 0.281 \times 100 \times 12^{2.63} \times 0.00189^{0.54} = 655 = \underline{660 \, gpm}$$

EXAMPLE PROBLEM 29.5

Estimate the flow capacity of a 400-mm-diameter main for an allowable friction slope of 0.15%. Assume the Darcy friction factor f = 0.02.

Given:

D = 400 m = 0.4 m $\qquad$ S_f = 0.15% = 0.0015 $\qquad$ f = 0.020

Solution:

Flow-carrying capacity

$$Q = 3.5x\sqrt{\frac{S_f}{f}} \times D^5 = 3.5x\sqrt{\frac{0.0015}{0.020}} \times 0.4^5$$

$$= 9.70 \times 10^{-2} = 9.7 \times 10^{-2} \, m^3/s = \underline{97 \, L/s}$$

EXAMPLE PROBLEM 29.6

Estimate the flow-carrying capacity of a 12-in water main for an allowable friction loss of 10 ft ft/mile length of pipe. Assume the Darcy friction factor = 0.022.

Given:

D = 12 in = 1.0 ft $\qquad$ S_f = 10 ft/mile $\qquad$ f = 0.022

Solution:

Friction slope

$$S_f = \frac{10 \, ft}{mile} \times \frac{mile}{5280 \, ft} = 0.001893 \, ft/ft$$

Flow-carrying capacity

$$Q = 6.3x\sqrt{\frac{S_f}{f}} \times D^5 = 6.3x\sqrt{\frac{0.001893}{0.022}} \times 1^5$$

$$= \frac{1.85 \, ft^3}{s} \times \frac{7.48 \, gal}{ft^3} \times \frac{60 \, s}{min} = 829 = \underline{830 \, gpm}$$

29.2.2 FRICTION SLOPE

For a given pipe size (D) and material (C), head loss due to friction can be found for various flow (Q) conditions. For the maximum flow, the expected pressure drop can be found.

Friction slope (Hazen–Williams)

$$S_f = 10.7 \times \left(\frac{Q}{C}\right)^{1.85} \times \frac{1}{D^{4.87}} - \left(D \, in \, m, Q \, in \, m^3/s\right)$$

$$S_f = 10.4 \times \left(\frac{Q}{C}\right)^{1.85} \times \frac{1}{D^{4.87}} - \left(D = in, Q = gpm\right)$$

EXAMPLE PROBLEM 29.7

Water from a well is pumped through a 12-in-diameter and 950-ft-long transmission main. Calculate the head loss for a flow of 850 gpm assuming C = 100 (rough) and C = 130 (smooth).

Given:

D = 12-in h_f = ?   L = 950 ft   Q = 850 gpm   C= 100, 130

Solution:

Head loss due to friction

$$h_f = 10.4 \times \left(\frac{Q}{C}\right)^{1.85} \times \frac{L}{D^{4.87}} = 10.4 \times \left(\frac{850}{100}\right)^{1.85} \times \frac{L}{D^{4.87}} = 2.87 = \underline{2.9 \, ft}$$

Head loss due to friction

$$h_f = 10.4 \times \left(\frac{Q}{C}\right)^{1.85} \times \frac{L}{D^{4.87}} = 10.4 \times \left(\frac{850}{130}\right)^{1.85} \times \frac{950 \, ft}{12^{4.87}} = 1.769 = \underline{1.7 \, ft}$$

EXAMPLE PROBLEM 29.8

A 600-mm-diameter and 12-km-long water main carries water from the pumping station to the load centre. Calculate the hydraulic gradient/frictional slope and hence the head loss in the water main with a coefficient C of 110 when carrying a flow of 200 L/s.

Given:

C = 110   Q = 200 L/s = 0.20 m³/s   D = 600 mm = 0.60 m S_f, h_f = ?

Solution:

Friction slope

$$S_f = 10.7 \times \left(\frac{Q}{C}\right)^{1.85} \times \frac{1}{D^{4.87}} = 10.7 \times \left(\frac{0.20}{110}\right)^{1.85} \times \frac{1}{0.60^{4.87}} = 1.09 \times 10^{-3} = 0.11\%$$

Head loss

$$h_f = S_f \times L = 1.09 \times 10^{-3} \times 122\,km \times \frac{1000\,m}{km} = 13.16 = \underline{13\,m}$$

29.2.3 PIPE ROUGHNESS, COEFFICIENT C

When testing the roughness of the pipe in a given part of a water system, a hydrant is opened to cause a certain flow through the section of the pipe being studied. Once the flow becomes steady, flow reading, and pressure drop are observed. Knowing the flow and the pressure drop, roughness coefficient of the pipe material is computed.

Roughness coefficient

$$C = \frac{Q}{0.278D^{2.63} \times S_f^{0.54}}$$

EXAMPLE PROBLEM 29.9

A field test was performed on a section of a newly installed main. A drop in pressure of 22 kPa is observed across a 400-m length of 250-mm main line when a flow of 61 L/s is carried. What is roughness coefficient C?

Given:

$\Delta p = 22$ kPa $Q = 61$ L/s $L = 400$ m $D = 250$ mm $= 0.25$ m $C = ?$

Solution:

Friction slope

$$S_f = \frac{h_f}{L} = \frac{22\,kPa}{400\,m} \times \frac{m}{9.81\,kPa} = 5.60 \times 10^{-3}\,m/m$$

Roughness coefficient

$$C = \frac{Q}{0.278D^{2.63} \times S_f^{0.54}} = \frac{0.061}{0.275 \times 0.25^{2.63} \times 0.0056^{0.54}} = 138 = \underline{140}$$

EXAMPLE PROBLEM 29.10

A flow test was conducted on an old existing main to determine the roughness coefficient C. A pressure head drop of 12 ft is observed across an 1800-ft length of the 8-in main line when a flow of 350 gpm is carried. What is the friction factor C?

Given:

$h_f = 12$ ft $Q = 250$ gpm $L = 1800$ ft $D = 8$-in $C = ?$

Solution:

Friction slope

$$S_f = \frac{h_f}{L} = \frac{12\,ft}{1800\,ft} = 6.667 \times 10^{-3}\,ft/ft$$

Roughness coefficient

$$C = \frac{Q}{0.281 D^{2.63} S_f^{0.54}} = \frac{350}{0.281 \times 8^{2.63} \times (0.00667)^{0.54}} = 78.5 = \underline{79}$$

29.2.4 Pipe Size D

Friction formulas can be applied to select pipe size to meet the peak demand without exceeding the desired friction slope. The Hazen–Williams formula rearranged to find diameter becomes:

Pipe diameter

$$D^{2.63} = \frac{Q}{0.278 C \times S_f^{0.54}}$$

EXAMPLE PROBLEM 29.11

A new subdivision is being planned. The peak water demand is worked out to be 50 L/s. Select the size of the main (C = 110) without exceeding the friction loss of 0.25%.

Given:

$S_f = 0.12\% = 0.0025$ $Q = 15$ L/s $= 0.015$ m³/s $C = 110$ $D = ?$

Solution:

Dimeter of the water main

$$D^{2.63} = \frac{Q}{0.278 C \times S_f^{0.54}} = \frac{1}{0.278 \times 110} = \frac{0.050}{0.0012^{0.54}} = 0.0416$$

$$D = 0.409^{1/2.63} = 0.298\,m = \underline{300\,mm}$$

29.3 MINOR LOSSES

Minor losses in water systems are due to changes in flow velocity and flow path. Hence, accessories, including valves, reducers, bends and elbows, cause head losses. Such losses are called minor since in water supply systems, they are usually less than head loss due to friction in pipes. However, that may not be always true. For example, in a pumping system, suction pipe length is usually small, and sometimes head loss due to foot valves or elbow tees may exceed head loss due to friction. Minor losses can be estimated with the same flow equation as for minor losses by adding length hydraulically equivalent to each of the accessories. A second method is to use the head loss coefficient. Head loss for expansion or contraction can be found by multiplying the velocity head by an appropriate coefficient. Tables for equivalent length and head loss coefficient are available in any standard book on hydraulics.

EXAMPLE PROBLEM 29.12

A 16-in main carries a flow of 1500 gpm. At a junction, a conical reducer is placed to reduce the diameter to 8 in. If the pressure before the junction is 40 psi, what will be the water pressure after the junction, assuming a 1.5-ft head loss due to contraction?

Given:

$p_1 = 40$ psi $Q = 1500$ gal/min $D_1 = 16$ in $D_2 = 8$ in

Solution:

Flow velocity before and after contraction

$$v_1 = \frac{4Q}{\pi D^2} = \frac{4}{\pi} \times \frac{500\,gal}{min} \times \frac{min}{60} \times \frac{ft^3}{7.48\,gal} \times \left(\frac{12}{16\,ft}\right)^2 = 2.39\,ft/s$$

$$v_2 = v_1 \times \left(\frac{D_2}{D_1}\right)^2 = \frac{1.76\,ft}{s} \times \left(\frac{16}{8}\right) = 9.57\,ft/s$$

Applying energy equation

$$p_2 = p_1 - \gamma h_1 - \frac{\gamma\left(v_2^2 - v_1^2\right)}{2g} = p_1 - \gamma\left[h_1 + \frac{\left(v_2^2 - v_1^2\right)}{2g}\right]$$

$$= 40\,psi - \frac{0.433\,psi}{ft}\left[1.5 + \frac{\left(7.76^2 - 1.91^2\right)}{2 \times 32.2}\right]ft = 38.97 = \underline{39\,psi}$$

EXAMPLE PROBLEM 29.13

A 450-mm-diameter water main carries a flow of 150 L/s. At a junction, the pipe diameter is reduced to 200 mm. If the pressure before the junction is 320 kPa, what will be the water pressure after the junction, assuming head loss due to contraction is half of the velocity head?

Given:

$p_1 = 320$ kPa $Q = 150$ L/s $= 0.15$ m³/s $D_1 = 450$ mm $D_2 = 200$ mm

Solution:

Flow velocity before and after contraction

$$v_1 = \frac{4Q}{\pi D^2} = \frac{4}{\pi} \times \frac{0.15 \ m^3}{s} \times \left(\frac{1}{0.45}\right)^2 = 0.943 \ m/s$$

$$v_2 = v_1 \times \left(\frac{D_1}{D_2}\right)^2 = \frac{0.94 \ m}{s} \times \left(\frac{450 \ mm}{200 \ mm}\right)^2 = 4.77 \ m/s$$

Applying energy equation

$$p_2 = p_1 - \gamma \times h_1 + \frac{\gamma\left(v_1^2 - v_2^2\right)}{2g} = p_1 - \gamma\left(\frac{0.5v_2^2}{2g} - \frac{\left(v_1^2 - v_2^2\right)}{2g}\right) = p_1 - \gamma\left(\frac{1.5v_2^2 - v_1^2}{2g}\right)$$

$$= 320 \ kPa - \frac{9.81 \ kPa}{m} \times \frac{(1.5 \times 4.77^2 - 0.943^2)m}{2 \times 9.81} = 303 = 300 \ kPa$$

The drop in pressure is both due to head loss and increase in flow velocity.

EXAMPLE PROBLEM 29.14

Determine the head loss in a 400-mm water main (f = 0.022) carrying a flow of 100 L/s. The length of the water main in question is 250 m, and for losses due to fittings, assume an equivalent length of 1%.

Given:

D = 400 mm = 0.40 m   Q = 100 L/s = 0.1 m³/s   L+ L$_e$ = 250 + 25 = 275 m
f = 0.022   h$_l$ = ?

Solution:

Major and minor head loss

$$h_l = \frac{f(L+L_e)}{1.23 \ g} \times \frac{Q^2}{D^5} = \frac{0.022}{1.23} \times \frac{s^2}{9.81 \ m} \times 275 \ m \times \left(\frac{0.10 \ m^3}{s}\right) \times \frac{1}{(0.40 \ m)^5}$$

$$= 0.489 = 0.48 \ m$$

EXAMPLE PROBLEM 29.15

What head loss can be expected when an 8-in water main (f = 0.02) is carrying a flow of 500 gpm? The length of the water main in question is 840 ft, and assume an equivalent length of 50 ft to account for minor losses.

Given:

D = 8-in = 0.667 ft   Q = 450 gpm   L + L$_e$ = 840 ft + 50 ft = 890 ft   h$_l$ = ?

Solution:

Water flow rate

$$Q = \frac{500 \, gal}{min} \times \frac{min}{60s} \times \frac{ft^3}{7.48 \, gal} = 1.114 = 1.1 \, ft^3/s$$

Head loss

$$h_l = \frac{f(L+L_e)}{1.23 \, g} \times \frac{Q^2}{D^5} = \frac{0.02}{1.23} \times \frac{s^2}{32.2 \, ft} \times 890 \, ft \times \left(\frac{1.114 \, ft^3}{s}\right)^2 \times \frac{1}{(0.667 \, ft)^5}$$

$$= 4.211 = 4.2 \, ft$$

EXAMPLE PROBLEM 29.16

A 450-mm-diameter water main is an old concrete pressure pipe with a friction factor f = 0.026. Calculate the pressure drop and friction slope between two hydrants 300 m apart when the pipe is carrying water at the rate of 500 L/s. Assume that hydrants 1 and 2 are at the same elevation.

Given:

D = 450 mm $D_1 = D_2$ $v_1 = v_2$, $Z_1 = Z_2 = ?$ f = 0.026   L = 300 m   Q = 500 L/s

Solution:

Manipulation of general energy equation

$$h_l = \frac{v_1^2}{2g} - \frac{v_2^2}{2g} + h_a + Z_1 - Z_2 + \frac{p_1}{\gamma} - \frac{p_2}{\gamma}$$

$$= \frac{p_1}{\gamma} - \frac{p_2}{\gamma} \quad or \quad p_1 - p_2 = \gamma h_l$$

Flow velocity

$$v = \frac{Q}{A} = \frac{4Q}{\pi D^2} = \frac{4}{\pi} \times \frac{500 \, L}{s} \times \frac{m^3}{1000 \, L} \times \left(\frac{1}{0.45 \, m}\right)^2 = 3.14 \, m/s$$

Head loss due to friction

$$h_f = \frac{fLv^2}{2gD} = 0.026 \times \frac{300 \, m}{0.45 \, m} \times \left(\frac{3.14 \, m}{s}\right)^2 \times \frac{s^2}{19.62 \, m} = 8.71 \, m$$

$$p_1 - p_2 = \gamma h_f = \frac{9.81\ kPa}{m} \times 8.71\ m = 85.45 = 85\ kPa$$

Friction slope

$$S_f = \frac{8.71\ m}{300\ m} \times 100\% = 2.90 = 2.9\%$$

EXAMPLE PROBLEM 29.17

As shown in Figure 29.2, water at the filtration plant is pumped into a 350-mm-diameter water main (C = 110) at a discharge pressure of 350 kPa. The 1.5-km-long water main connects to the load centre, where the residual pressure during the peak demand period is known to be 170 kPa. The elevations of the pump and the load centre, respectively, are 102.0 m and 105.0 m.

1. Calculate the percent hydraulic gradient in the water main section A to B.
2. Applying the Hazen–Williams equation, find how much demand in L/s is served by the clear well at the water plant.

Given:

Parameter	A	B	C
Elevation, Z, m	102.0	105.0	110.0
Pressure, kPa	350	170	294

Parameter	AB	CB
Length, km	1.5	0.90
Diameter, mm	350	250
Coefficient, C	110	110

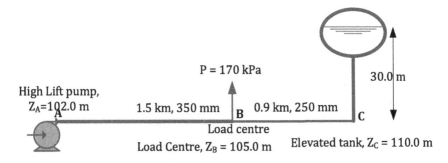

FIGURE 29.2 Water supply system (Example Problem 29.9).

Solution:

Hydraulic head

$$h_A = \frac{p_A}{\gamma} + Z_A = \frac{350\,kPa.m}{9.81\,kPa} + 102.0\,m = 137.67\,m$$

$$h_B = \frac{p_B}{\gamma} + Z_B = \frac{170\,kPa.m}{9.81\,kPa} + 105.0\,m = 122.32\,m$$

Friction slope

$$S_f = \frac{h_f}{L} = \frac{(h_A - h_B)}{AB} = \frac{(137.67 - 122.23)\,m}{1500\,m} = 0.0102 = 1.0\%$$

Flow in pipeline AB

$$Q_{AB} = 0.278C \times D^{2.63} \times S_f^{0.54} = 0.278 \times 110 \times 0.35^{2.63} \times 0.0102^{0.54}$$

$$= 0.1628\,m^3/s = \underline{160\,L/s}$$

29.4   PUMPING HEAD VERSUS SYSTEM HEAD

The terms suction and discharge refer to the inlet side and outlet side of the pump. Heads measured on the suction side of a pump are called suction heads, and heads measured on the discharge side are called discharge heads. The term **system head** refers to the head required to pump water into a distribution system at a certain flow rate. Since flow velocity will be higher at higher flows, system head will increase exponentially with an increase in Q (Figure 29.3). Pumping head is the head added by the pump to water. It is used to overcome head losses and lifting of water. As shown in Figure 29.3, system head is equal to fixed head (lift) plus head losses (variable head).

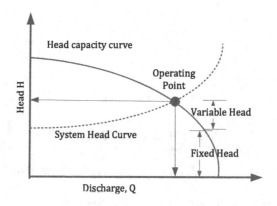

FIGURE 29.3   Pump operating point.

System head

$$h_{sys} = (Z_2 - Z_1) + h_1 = h_{fixed} + h_{variable}$$

By superimposing the pump curve on the system curve, the pump **operating point** is the point where the two curves meet. Any changes in the system into which water is pumped would affect pump operation. For example, if the system head increases due to an increase in lift, head losses or both, pumping head increases and hence results in a low discharge rate. Under static (no pumping) conditions, head losses are zero, as friction will come into play only when pumping starts.

EXAMPLE PROBLEM 29.18

Work out the pump power for the following data: H = 52 ft, Q = 3700 gpm, E = 72 %.

Given:

h_a = 52 ft   Q = 370 gpm E_p = 72%.

Solution:

Power added

$$P_a = Q \times \gamma \times h_a = \frac{370 \ gal}{min} \times \frac{8.34 \ lb}{gal} \times 52 \ ft \times \frac{hp.min}{550 \times 60 \ ft.lb} = 4.86 \ hp$$

Pump or shaft power

$$P_p = \frac{P_a}{E_p} = \frac{4.86 \ hp}{72\%} \times 100\% = 6.75 = \underline{6.8 \ hp}$$

EXAMPLE PROBLEM 29.19

A water treatment plant gets its water from the lake. The intake is below the water surface at an elevation of 230.4 m. The lake water is pumped to the plant influent at elevation 242.4 m. The total head losses are estimated to be 5.5 m when water is drawn at the rate of 340 L/s. Calculate pumping head, waterpower and the pump power, assuming the pump is 72% efficient.

Given:

Variable	Suction = 1	Discharge = 2
Pressure, p	$p_1 = 0$	$p_2 = 0$
Elevation, Z	$Z_1 = 230.4$ m	$Z_2 = 242.4$ m
Head loss, h_L	5.5 m	

Solution:

Head added

$$h_a = \frac{p_2}{\gamma} - \frac{p_1}{\gamma} + Z_2 - Z_1 + h_1$$

$$= 0 + (242.4 - 230.4)\ m + 5.5\ m = 17.5\ m = 18\ m$$

Power added

$$P_a = Q \times \gamma \times h_a = \frac{0.340\ m^3}{s} \times \frac{9.81\ kN}{m^3} \times 17.5\ m \times \frac{kW.s}{kN.m} = 58.3 = 58\ kW$$

Pump or shaft power

$$P_P = \frac{P_a}{E_P} = \frac{58.3\ kW}{72\%} \times 100\% = 80.9 = 81\,kW$$

EXAMPLE PROBLEM 29.20

In a water pumping system, pressure gauges attached to the suction side and discharge side read 2 in of mercury and 20 psi when the pump is discharged at the rate of 360 gpm. Determine

a) Head added and water horsepower
b) Pump or brake horsepower if 8.0 kW power is drawn and the electric motor is 80% efficient
c) Pump efficiency
d) Overall efficiency of the pumping unit

Given:

Variable	Suction = 1	Discharge = 2
Pressure/head	–2 in Hg	p_2 = 20 psi
Elevation	$(Z_2 - Z_1) = 0$	same
Pumping Rate	360 gpm = 6.0 gal/s	

Solution:

Head added

$$h_a = \frac{p_2}{\gamma} - \frac{p_1}{\gamma} + Z_2 - Z_1 + h_1 = \frac{p_2}{\gamma} - \frac{p_1}{\gamma}$$

$$= 20\ psi \times \frac{2.31\ ft}{psi} - -2\,in\ Hg \times \frac{13.6\ in\,water}{in\ Hg} \times \frac{ft}{12in} = 48.46 = 48\ ft$$

Power added or water horsepower

$$P_a = Q \times \gamma \times h_a = \frac{6.0\, gal}{s} \times \frac{8.341b}{gal} \times 48.46\, ft \times \frac{hp.s}{550\, ft.lb} = 4.409 = \underline{4.4\, hp}$$

Pump power or brake horsepower

$$P_P = P_I \times E_M = 8.0\ kW \times \frac{hp}{0.746\, kW} \times 0.80 = 8.579 = \underline{8.6\, hp}$$

Pump efficiency

$$E_p = \frac{P_a}{P_p} = \frac{4.409\, hp}{8.579\, hp} \times 100\% = 51.39 = \underline{51\%}$$

Overall or wire to water efficiency

$$E_O = \frac{P_a}{P_I} = \frac{4.409\ hp}{8.0\ kW} \times \frac{0.746\ kW}{hp} \times 100\% = 41.1 = \underline{41\%} \quad or$$

$$E_O = E_P \times E_M = 51.39\% \times 0.80 = 41.1 = \underline{41\%}$$

EXAMPLE PROBLEM 29.21

What discharge pressure is needed at the booster pumping station to pump water to a residential area with maximum height of 46 m? A minimum residual pressure of 220 kPa is needed for adequate supply of water. Assume the head losses in the watermain are 10 m.

Given:

Variable	Pump = 1	Faucet = 2
Pressure, kPa	?	$p_2 = 220\ kPa$
Elevation, m	$(Z_2 - Z_1) = 46$	
Head loss, m	10	

Solution:

Discharge pressure needed

$$p_1 = \gamma h_1 + \gamma(Z_2 - Z_1) + p_2 = \frac{9.81\ kPa}{m} \times 10\ m + \frac{9.81\ kPa}{m} \times 46\ m + 220\ kPa$$

$$= 769 = \underline{770\ kPa}$$

EXAMPLE PROBLEM 29.22

A pump is 72% efficient when pumping 25 L/s against a head of 35 m. What is the expected amperage drawn by the pump from 220-V supply? Assume the electric motor is 90% efficient.

Given:

$Q = 25$ L/s $h_a = 35$ m   emf = 220 V $E_p = 72\%$ $E_m = 90\%$ $I = ?$

Solution:

Power input

$$P_i = \frac{Q \times \gamma \times h_a}{E_p \times E_m} = \frac{0.025\,m^3}{s} \times \frac{9.81kN}{m^3} \times 35\,m \times \frac{1}{0.72 \times 0.9} \times \frac{kW.s}{kN.m}$$
$$= 13.2\,kW = 13\,kW$$

Current drawn

$$I = \frac{P_i}{emf} = \frac{13.2\,kW}{220\,V} \times \frac{1000\,W}{kW} \times \frac{V.A}{W} = 60.2 = \underline{60A}$$

29.5   FREE FLOW VELOCITY AND DISCHARGE Q

Under free flow conditions, the flow velocity through an opening will largely depend on the pressure head exerted. The actual flow velocity v would be less due to the fact that contraction and head loss across the opening.

Flowing hydrant discharge velocity/rate

$$\boxed{v = C\sqrt{2g\Delta h} = \quad Q = 0.785D^2 \times C\sqrt{2g \times \Delta h}}$$

29.6   HYDRANT TESTING

During hydrant testing, a reduction in pressure at the residual hydrant is observed by opening the hydrant in the vicinity of the hydrant being studied. This test allows us to estimate the flow capacity of a given hydrant during firefighting or other similar emergencies. Hydrant discharge from a flowing hydrant is found by observing the trajectory of the water jet or by reading the velocity pressure of the water jet, as explained earlier.

Hydrant flow rate

$$\boxed{Q = 0.785D^2\sqrt{2g \times \Delta h} = D^2\sqrt{\Delta p} \quad (SI)}$$

$p = kPa,\ D = m,\ Q = m^3/s$

Trajectory method

$$Q = \frac{\pi D^2 \times X}{4} \sqrt{\frac{g}{2y}}$$

X = *horizontal distance where the trajectory meets the horizontal surface*
Y = *vertical height of the jet*
D = *diameter of the hydrant nozzle*

Hydrant flow capacity at the residual pressure, which is usually 150 kPa, is calculated using the following formula.

Hydrant flow at the residual pressure

$$Q_R = Q_F \times \left(\frac{\Delta p_R}{\Delta p_F}\right)^{0.54} = Q_F \times \left(\frac{p_S - p_R}{p_S - p_F}\right)^{0.54}$$

Q_F = *Total flow from all the flowing hydrants during testing*
Δp_F = *drop in pressure at the residual hydrant* = $p_S - p_F$
Δp_R = *static pressure minus the residual pressure* = $p_S - p_R$
p_R = *Residual pressure* = *140–150 kPa*

EXAMPLE PROBLEM 29.23

During a fire flow test, the pitot gauge read a velocity pressure of 55 kPa. Applying the principle of hydraulics, estimate the discharge rate from a 60-mm hydrant nozzle. The coefficient of discharge can be assumed to be 85%.

Given:

D = 60 mm = 0.060 m   Q = ?   C = 0.85   Δp = 55 kPa

Solution:

Hydrant flow rate
$$Q = D^2 \sqrt{\Delta p} = (0.060)^2 \sqrt{55} = 2.66 \times 10^{-2} \, m^3/s = 26.6 = \underline{27 \, L/s}$$

EXAMPLE PROBLEM 29.24

Estimate the flow coming out of a hydrant connected to a 50-mm hose. The water jet coming out drops 90 cm before hitting the ground a distance of 1.6 m away from the hose. What is the rate of flow in L/s?

Given:

D = 50 mm = 0.050 m   X = 1.6 m   Y = 90 cm = 0.9 m   Q = ?

Solution:

Hydrant flow rate

$$Q = \frac{\pi D^2 X}{4} \sqrt{\frac{g}{2y}} = \frac{\pi}{4} \times (0.050m)^2 \times 1.6 \ m \times \sqrt{\frac{9.81 \ m}{s^2} \times \frac{1}{2 \times 0.9 \ m}}$$

$$= 7.33 \times 10^{-3} \ m^3/s = \underline{7.3 \ L/s}$$

EXAMPLE PROBLEM 29.25

During a hydrant flow test, the water jet coming out of a 2.5-in nozzle drops 3.0 ft before hitting the ground a distance of 4.5 ft away from the flowing hydrant. What is the rate of flow in gpm?

Given:

D = 2.5 in = 0.0208 ft   X = 4.5 ft   Y = 3.0 ft   Q = ?

Solution:

Hydrant flow rate

$$Q = \frac{\pi D^2 X}{4} \sqrt{\frac{g}{2y}} = \frac{\pi}{4} \times (0.208 \ ft)^2 \times 4.5 \ ft \times \sqrt{\frac{32.2 \ ft}{s^2} \times \frac{1}{2 \times 3.0 \ ft}}$$

$$= \frac{0.355 \ ft^3}{s} \times \frac{7.48 \ gal}{ft^3} \times \frac{60 \ s}{min} = 159 = \underline{160 \ gpm}$$

EXAMPLE PROBLEM 29.26

Applying the principle of flow through an orifice, estimate the leakage rate through a 1-mm-diameter crack in a pipe. The operating pressure is 400 kPa, and the coefficient of discharge can be assumed to be 60%.

Given:

D = 1 mm = 0.001 m   Q = ?   C = 0.60   Δp = 400 kPa

Solution:

Leakage rate

$$Q - Av = 0.785D^2 C \sqrt{2g\Delta h}$$

$$= 0.785(0.0010m)^2 \times 0.60 \times \sqrt{\frac{2 \times 9.81m}{s^2} \times \frac{400 \ kPa.m}{9.81kPa}}$$

$$= 1.3 \times 10^{-5} \ m^3/s = 0.013L/s = \underline{0.8 \ L/min}$$

EXAMPLE PROBLEM 29.27

In a fire flow test, four hydrants were opened to produce a total flow of 140 L/s. During the test, pressure at the residual hydrant dropped from 480 to 310 kPa. What hydrant flow can be expected at a residual pressure of 150 kPa?

Given:

$\Delta p_F = 480 - 310 = 170$ kPa $Q_F = 140$ L/s $\Delta p_R = 480 - 150 = 330$ kPa

Solution:

Hydrant flow rate at the residual pressure (max)

$$Q_R = Q_F \left(\frac{\Delta p_R}{\Delta p_F} \right)^{0.54} = 140\, L/s \times \left(\frac{330\, kPa}{170\, kPa} \right)^{0.54} = 200.3 = \underline{200\, L/s}$$

The hydrants are colour-coded based on the flow they can deliver at the residual pressure. Hydrants with maximum flows > 100 L/s are painted blue, 30–65 L/s range are orange, 65–100 L/s range are red and < 30 L/s are yellow.

EXAMPLE PROBLEM 29.28

In a fire flow test, four hydrants surrounding the test hydrant were opened to produce a total flow of 2500 gpm. During the test, pressure at the residual hydrant dropped from 62 psi to 42 psi. What hydrant flow can be expected at a residual pressure of 25 psi?

Given:

$\Delta p_F = 62 - 42 = 20$ psi $Q_F = 2500$ gpm $\Delta p_R = 62-25 = 37$ psi

Solution:

Hydrant flow rate at the residual pressure

$$Q_R = Q_F \left(\frac{\Delta p_R}{\Delta p_F} \right)^{0.54} = 2500\ gpm \times \left(\frac{37\ psi}{20\ psi} \right)^{0.54} = 3485 = \underline{3500\ gpm}$$

29.6.1 HYDRAULIC GRADIENT TEST

Hydraulic tests in water distribution system are carried out to determine the efficiency of water mains. With age, the pipe surface becomes rougher and results in more head losses, causing pressure loss for the same flow. If the water is scaling, it affects both the roughness and effective diameter of the water main. When significant head loss is experienced in a certain section of the main, it is recommended to carry out hydraulic gradient tests. Much like hydrant testing, during low demand, flow is caused by opening the flowing hydrant, and pressure gauge readings are taken in two hydrants on the same water main.

EXAMPLE PROBLEM 29.29

A hydraulic gradient test is performed in the field by opening the end hydrant. In the direction of flow, the pressure readings at hydrants 1 and 2, respectively, were observed to be 413 kPa and 395 kPa. From the map for the area, the elevations of hydrants 1 and 2 are 112.4 m and 111.8 m, respectively. Given that the two hydrants are connected by 610 m of 300-mm-diameter line, what is the hydraulic gradient for the test conditions?

Given:

Variable	Hydrant 1	Hydrant 2
Pressure, kPa	413	395
Elevation, m	112.4	111.8
Length, m	610	

Solution:

Head loss

$$h_1 = (Z_1 - Z_2) + \frac{(p_1 - p_2)}{\gamma}$$

$$= (112.4 - 111.8)\,m\,(413 - 395)\,kPa \times \frac{m}{9.81\,kPa} = 2.43 = \underline{2.4\,m}$$

Friction slope

$$S_f = \frac{h_1}{L} = \frac{2.43\,m}{600\,m} \times 100\% = 0.405 = \underline{0.41\%}$$

EXAMPLE PROBLEM 29.30

A hydraulic gradient field test was carried out. The pressure gauge readings at hydrants 1 and 2, respectively, were 65 psi kPa and 57 psi. Find friction hydraulic gradient. The two hydrants are 450 ft apart, and hydrant 2 is located 1.2 ft higher than hydrant 1.

Given:

Variable	Hydrant 1	Hydrant 2
Pressure, psi	65	57
Elevation, ft	0	1.2
Length, ft	610	

Solution:

Head loss

$$h_1 = (Z_1 - Z_2) + \frac{(p_1 - p_2)}{\gamma} = (65 - 57)\,psi \times \frac{2.31\,ft}{psi} + (-1.2)\,ft = 17.3\,ft = \underline{17\,ft}$$

Friction slope

$$S_f = \frac{h_1}{L} = \frac{17.3\,ft}{1200\,ft} \times 100\% = 1.44 = \underline{1.4\%}$$

PRACTICE PROBLEMS

PRACTICE PROBLEM 29.1

A 200-mm-diameter pipe is flowing at the rate of 45 L/s. The water pressure in the pipe at a point 15 m above the datum is read 350 kPa. Find the hydraulic head. (51 m)

PRACTICE PROBLEM 29.2

A 16-in-diameter water main carries a flow of 1500 gpm. The water pressure in the pipe at a point 15 ft above the datum is 55 psi. What is the total head? (142 ft)

PRACTICE PROBLEM 29.3

Determine the flow capacity of a 300-mm-diameter main for an allowable friction slope of 0.10%. Assume the Hazen–Williams coefficient of friction is 110. (31 L/s)

PRACTICE PROBLEM 29.4

Estimate the flow-carrying capacity of an 8-in water main for an allowable friction loss of 8 ft/mile length of pipe. Assume the Hazen–Williams roughness coefficient to be 120. (240 gpm)

PRACTICE PROBLEM 29.5

Applying the Darcy–Weisbach equation, find the flow capacity of a 300-mm-diameter main for an allowable friction slope of 1m/km. Assume friction factor f = 0.02. (39 L/s)

PRACTICE PROBLEM 29.6

Determine the flow-carrying capacity of an 8-in water main for an allowable friction loss of 0.15%. Assume the Darcy friction factor = 0.018. (300 gpm)

PRACTICE PROBLEM 29.7

A 16-in transmission pipe is quarter of a mile in length. Calculate the head loss when carrying a flow of 1200 gpm, assuming friction coefficient C =110. (11 ft)

PRACTICE PROBLEM 29.8

What friction slope can be expected in a 300-mm-diameter water main of C = 120 when carrying a flow of 50 L/s? (0.21%)

PRACTICE PROBLEM 29.9

After cleaning a pipe, it was intended to check the improvement in the coefficient by performing a pressure test. A pressure difference of 28 kPa was observed for a flow of 22 L/s. Compute the improved value of roughness coefficient C. (98)

PRACTICE PROBLEM 29.10

A field test was performed to check the performance of an old water main. A drop in pressure of 28 kPa is observed across 300 m length of 250-mm main line when a flow of 55 L/s is carried. What is the friction coefficient C? (94)

PRACTICE PROBLEM 29.11

The peak water demand is worked out to be 40 L/s. Select the size of the main (C =110) without exceeding the friction loss of 0.10%. (350 mm)

PRACTICE PROBLEM 29.12

A 12-in main carries a flow of 1200 gpm. At a junction, a reducer is placed to contract the diameter to 6 in. If the pressure before the junction is 35 psi, what will be the water pressure after the junction, assuming 2.5 ft head loss due to contraction? (33 psi)

PRACTICE PROBLEM 29.13

A 350-mm-diameter water main carries a flow of 130 L/s. At a junction, a reducer is placed to reduce the pipe diameter to 150 mm. If the pressure before the junction is 280 kPa, what will be the water pressure after the junction, assuming head loss due to contraction is equal to velocity head? (230 kPa)

PRACTICE PROBLEM 29.14

What pressure loss can be expected when a 12-in water main (f = 0.022) is carrying a flow of 1200 gpm? The length of the water main in question is 850 ft, and for losses due to fittings, assume an equivalent length of 50 ft. (1.5 psi)

PRACTICE PROBLEM 29.15

Determine the pressure loss in a 300-mm water main (f = 0.025) carrying a flow of 55 L/s. The length of the water main in question is 300 m, and for losses due to fittings, assume an equivalent length of 25 m. (8.2 kPa)

PRACTICE PROBLEM 29.16

A 400-mm-diameter water main is a steel pipe with a friction factor of 0.018. Calculate the pressure drop and friction slope between two hydrants 600 m apart when the pipe is carrying 210 L/s of water. Assume that hydrants 1 and 2 are at the same elevation. (38 kPa, 0.65%)

PRACTICE PROBLEM 29.17

As shown in Figure 29.2, water from the tower flows to the load centre during the peak hours. The pipeline from C to B is 900 m long and 50 mm in diameter, and C = 110. The elevations of junction C and load centre B, respectively, are 110.0 m and 105.0 m.

a) Calculate the hydraulic gradient in the water main section C to B. (2.0%)
b) Applying the Hazen–Williams equation, find how much demand is served by the tower. (96 L/s)

PRACTICE PROBLEM 29.18

Work out the pump power for the following pump data: H = 82 ft, Q = 500 gpm, E = 65%. (16 hp)

PRACTICE PROBLEM 29.19

In a pumping, a system suction gauge reads 46 kPa vacuum, and a discharge gauge located 1.0 m higher than the suction gauge reads 290 kPa. Find the head added by the pump to the water. (35 m)

PRACTICE PROBLEM 29.20

In a wastewater pumping system, pressure gauges attached to suction and discharge sides read 2.5 psi vacuum and 22 psi, respectively, when the pumping rate is 300 gpm. Determine

a) Head added and water horsepower. (57 ft, 4.3 hp)
b) Pump efficiency, if motor power is estimated to be 6.5 kW. (49%)

PRACTICE PROBLEM 29.21

A booster station is pumping water at a discharge pressure of 570 kPa. The pressure at a residence located at an elevation of 26 m above the pump is observed to be 190 kPa. Compute the head losses in the line. (13 m)

PRACTICE PROBLEM 29.22

During a pumping test, the power withdrawn by a pumping unit is observed to be 18.3 kW. The pumping rate gauge indicates a reading of 32 L/s, and the discharge pressure gauge reads 380 kPa. What is the overall efficiency of the pumping unit? (66%)

PRACTICE PROBLEM 29.23

A pitot gauge placed in front of a flowing hydrant jet reads 82 kPa of velocity pressure. Compute the discharge rate from a 60-mm nozzle. Assume a coefficient of 90%. (33 L/s)

PRACTICE PROBLEM 29.24

Water is flowing out of a 60-mm-diameter pipe. The pipe is 1.1 m above the point where the water jet hits the ground. The horizontal distance of the jet from the discharge point is 2.4 m. Estimate the flow of water. (14 L/s)

PRACTICE PROBLEM 29.25

Water is flowing out of a 2-in-diameter pipe. The pipe is 3.2 ft above the point where the water jet hits the ground. The horizontal distance of the jet from the discharge point is 4.8 ft. Estimate the flow of water. (130 gpm)

PRACTICE PROBLEM 29.26

Estimate the discharge rate of a 50-mm-diameter orifice flowing under a head of 5.5 m. Assume the coefficient of discharge is 0.60. (13 L/s)

PRACTICE PROBLEM 29.27

A fire flow test was conducted on a section of a water main in a residential area. Two hydrants, one on each side of the residual hydrant, were opened to produce a flow of 60 L/s. During the test, pressure at the residual hydrant dropped from 350 to 240 kPa. Compute the discharge at the residual pressure of 150 kPa. (83 L/s)

PRACTICE PROBLEM 29.28

A fire flow test was conducted on a section of the main in a residential area. Two hydrants, one on each side of the residual hydrant, were opened to produce a total flow of 980 gpm. During the test, pressure at the residual hydrant dropped from 75 psi to 52 psi. Compute the discharge at the residual pressure of 20 psi. (1600 gpm)

PRACTICE PROBLEM 29.29

A hydraulic gradient test is performed in the field. Two hydrants 750 m apart were chosen to observe the hydraulic head. For the maximum flow conditions, pressure reading at hydrants 1 and 2 were observed to be 477 kPa and 495 kPa. From the map for the area, the elevations of hydrants 1 and 2 are 112.4 m and 111.9 m, respectively. Find the hydraulic head at each of the hydrant and hence the head loss in the pipeline connecting the two hydrants. (0.18%)

PRACTICE PROBLEM 29.30

A hydraulic gradient test field test was carried out. The pressure gauge readings at hydrants 1 and 2, respectively, were 75 psi kPa and 62 psi. Determine the friction hydraulic gradient. The two hydrants are 2500 ft apart, and hydrant 2 is located 1.2 ft lower than hydrant 1. (1.30%)

30 Basic Wastewater Collection

Most of the concepts related to wastewater collection have been discussed in earlier chapters. It is strongly recommended that you review the material and specifically the units on chemical feeding and hydraulics. The key concepts related to wastewater collection are as follows.

30.1 FLOW VELOCITY

In the wastewater collection system, wastewater is collected by sanitary sewers and transported to the wastewater treatment plant. In the majority of cases, the sewer flows partially full; thus, the water area is only a fraction of the cross-sectional area of the pipe. In pressure mains, though, sewers pipes flow full, and the average flow velocity is given by the following equation:

Flow velocity

$$v = \frac{Q}{A}, \quad A = \frac{\pi}{4} \times D^2 = 0.785D^2$$

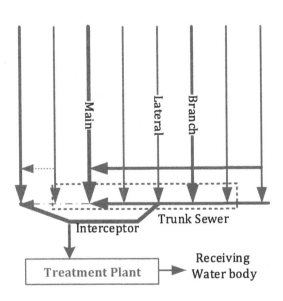

FIGURE 30.1 A typical wastewater collection system.

DOI: 10.1201/9781003468745-30

To work out the flow velocity, you would need to find the partial area. The relationship between full area and partial area is discussed in the next chapter.

30.2   DILUTION FORMULA

In chemical treatment of wastewater in sewers to neutralize odour and septicity, one of the most common concepts is dilution of chemical solutions and feeding rate of solutions. As has been talked about in many of the earlier units, it is reproduced here.

Dilution formula

$$\boxed{V_1 \times C_1 = V \times C_2}$$

The chemical feed rate can be found simply by multiplying the concentration of solution being fed and the volume of the wastewater being treated.

EXAMPLE PROBLEM 30.1

In a town, the daily average water supply is 12. 4 ML/d. If the wastewater production is assumed to be 70% of this, what is the daily average wastewater flow expected at the water pollution control plant?

Given:

Q = 12.4 ML/d   Q (wastewater) = 70%

Solution:

Daily average wastewater flow

$$Q = \frac{70\%}{100\%} \times 12.4 \ ML/d = 8.68 = \underline{8.7 \ ML/d}$$

EXAMPLE PROBLEM 30.2

At a given pumping station, the daily average pumping rate is 3000 gpm. What is the wastewater production in gal/c.d? Assume this station serves a community of 27,000 people.

Given:

Q = 200 L/s   Population = 27000   Q/Capita = ?

Solution:

Per capita flow

$$Q = \frac{3000 \ gal}{min} \times \frac{1440 \ min}{d} \times \frac{1}{27000 \ p} = \underline{160 \ gal/c \cdot d}$$

EXAMPLE PROBLEM 30.3

A sewer trench is 75 cm wide at the bottom and 4.0 m deep, and the side slope is 45°. How wide is the trench at the ground surface?

Given:

b_1 = 75 cm = 0.75 m   d = 4.0 m   Slope, S = tan 45° = 1

Solution:

Width of the trench
$$b_2 = b_1 + 2dS = 0.75\ m + 2 \times 4.0\ m \times 1 = 8.75 = \underline{8.8\ m}$$

EXAMPLE PROBLEM 30.4

On a city map, the distance between two manholes is measured to be 9.5 inches. If the map scale is 1:500, what is the linear distance between the manholes?

Given:

Map distance = 9.5 in   Ground distance = ?   Scale = 1:500

Solution:

Distance on the ground
$$= 9.5\ in \times \frac{500\ in\ ground}{1\ in\ map} \times \frac{ft}{12\ in} = 395.8 = \underline{400\ ft}$$

EXAMPLE PROBLEM 30.5

A 1.2 m × 2.4 m × 20 m trench is to be backfilled with sand. How many 9-m³-capacity truckloads are needed?

Given:

V = 1.2 m × 2.4 m × 20 m   Capacity = 9 m³/load

Solution:

Number of truckloads
$$\# = \frac{Volume\ of\ sand}{capacity/Load} = 1.2\ m \times 2.4\ m \times 20\ m \times \frac{load}{9\ m^3} = 6.4 = \underline{7\ loads}$$

Rounded up: a partial load has to be one more trip.

EXAMPLE PROBLEM 30.6

How many tons (2000 lb) of soil (SG = 2.4) are to be dug to excavate a 4.0 ft × 8.0 ft × 65 ft trench?

Given:

$$V = 4.0 \text{ ft} \times 8.0 \text{ ft} \times 65 \text{ ft} \qquad SG = 2.4 \qquad m = ?$$

Solution:

Mass density of soil

$$\rho = SG \times \rho_w = 2.4 \times \frac{62.4 \ lb}{ft^3} = 149.7 = 150 \ lb/ft^3$$

$$m = \rho \times V = \frac{149.7}{ft^3} \times 4.0 \ ft \times 8.0 \ ft \times 65 \ ft \times \frac{ton}{2000 \ lb} = 155.6 = \underline{160 \ tons}$$

EXAMPLE PROBLEM 30.7

If a sewer is laid on a 0.3% slope, estimate the drop in elevation over 1.2 km length of pipe.

Given:

$$L = 1.2 \text{ km} \qquad S = 0.30\% = 0.003 \text{ m/m} \qquad \Delta Z = ?$$

Solution:

Drop in elevation

$$\Delta Z = S \times L = \frac{0.30\%}{100\%} \times 1.2 \ km \times \frac{1000 \ m}{km} = 3.60 = \underline{3.6 \ m}$$

EXAMPLE PROBLEM 30.8

The upstream invert elevation for a 10-in sewer pipe is fixed at 330.58 ft. You are required to lay a 450-ft-long sewer at a slope of 0.40%. What must be the elevation of the downstream invert?

Given:

$$S = 0.40\% \qquad L = 450 \text{ ft} \qquad Z_1 = 330.58 \text{ ft} \qquad Z_2 = ?$$

Solution:

Downstream invert elevation

$$Z_2 = Z_1 - \Delta Z = 330.58 \ ft - \frac{0.40\%}{100\%} \times 450 \ ft = 328.780 = \underline{328.78 \ ft}$$

EXAMPLE PROBLEM 30.9

The interior of 90 m of 300-mm-diameter pipe is uniformly coated with 20 mm of grease. How many kL of wastewater can be held by this pipe when completely full?

Given

D = 300 mm   Effective D = 300 – 2 × 20 = 260 mm   L = 90 m   V = ?

Solution:

Effective capacity of sewer pipe

$$V = \frac{\pi}{4} \times (0.26\ m)^2 \times 90\ m = 4.77 = \underline{4.8\ m^3}$$

EXAMPLE PROBLEM 30.10

Estimate the full flow velocity in an 8-in-diameter sewer pipe carrying a flow of 350 gpm.

Given:

D = 8 in = 0.667 ft   Q = 350 gpm   v = ?

Solution:

Flow rate

$$Q = \frac{350\ gal}{min} \times \frac{min}{60\ s} \times \frac{ft^3}{7.48\ gal} = 0.779 = \underline{0.78\ ft^3/s}$$

Flow velocity

$$v = \frac{Q}{A} = \frac{0.779\ ft^3}{s} \times \frac{4}{\pi (0.667\ ft^2)} = 2.23 = \underline{2.2\ ft/s}$$

EXAMPLE PROBLEM 30.11

At a sewage pumping station, wastewater is pumped through a 200-mm-diameter force main at the rate of 30 L/s. What is the flow velocity in the force main?

Given:

Q = 30 L/s   D = 200 mm = 0.20 m   v = ?

Solution:

Flow velocity

$$v = \frac{Q}{A} = \frac{30\ L}{s} \times \frac{4}{\pi (0.20\ m)^2} \times \frac{m^3}{1000\ L} = 0.950 = \underline{0.95\ m/s}$$

EXAMPLE PROBLEM 30.12

How many minutes will it take for sewage to travel a distance of 850 m when flowing at an average speed of 0.75 m/s?

Given:

s = 850 m   v = 0.75 m/s   t = ?

Solution:

Time to travel

$$t = \frac{s}{v} = 850\ m \times \frac{s}{0.75\ m} \times \frac{min}{60\ s} = 18.8 = \underline{19\ min}$$

EXAMPLE PROBLEM 30.13

A flow of 750 gpm passes through a wet well measuring 50 ft long by 20 ft wide by 11 ft deep. What is the average theoretical detention time?

Given:

V = 50 ft × 20 ft × 11 ft   Q = 750 gpm   HDT = ?

Solution:

Hydraulic detention time

$$HDT = 50\ ft \times 20\ ft \times 11\ ft \times \frac{min}{750\ gal} \times \frac{7.48\ gal}{ft^3} = 109.7 = \underline{110\ min}$$

EXAMPLE PROBLEM 30.14

The wet well at a pump station measures 2.5 m × 2.1 m. With the inflow shut off, the water level with one pump running dropped by 0.50 m in 5 min. Find the pumping rate in L/s.

Given:

A = 2.5 m × 2.1 m   v = 0.50 m/5 min   Q = ?

Solution:

Flow rate

$$Q = Av = 2.5\ m \times 2.1\ m \times \frac{0.50\ m}{5.0\ min} \times \frac{min}{60\ s} \times \frac{1000\ L}{m^3} = 8.75 = \underline{8.8\ L/s}$$

EXAMPLE PROBLEM 30.15

Determine the pump capacity if an 8.0 ft × 6.0 ft wet well is lowered by 3.2 ft in 5.0 min. Assume no additional water flows into the wet well.

Given:

$A = 8.0 \text{ ft} \times 6.0 \text{ ft} = 48 \text{ ft}^2$ $d = 3.2 \text{ ft}$ $t = 5.0 \text{ min}$ $Q = ?$

Solution:

Flow rate

$$Q = Av = 48 \ ft^2 \times \frac{3.2 \ ft}{5.0 \ min} \times \frac{7.48 \ gal}{ft^3} = 229.7 = \underline{230 \ gpm}$$

EXAMPLE PROBLEM 30.16

A holding tank measuring 10 m × 3.0 m is buried in the ground. Find the uplifting pressure force at the bottom of the tank when the ground water is 1.8 m above the bottom of the tank.

Given:

$A = 10 \text{ m} \times 3.0 \text{ m} = 30 \text{ m}^2$ $h = 1.8 \text{ m}$ $F = ?$

Solution:

Force acting on the plug

$$F = p \times A = \gamma h \times A = 1.8 \ m \times \frac{9.81 \ kN}{m^3} \times 10 \ m \times 3.0 \ m = 529.7 = \underline{530 \ kN}$$

EXAMPLE PROBLEM 30.17

An 8-in-diameter inflated plug is holding water in a sewer pipe tested for leakage. Find the pressure force against the plug when the water depth in the upstream manhole is 5.0 ft above the centre of the plug.

Given:

$D = 8 \text{ in}$ $h = 5.0 \text{ ft}$ $F = ?$

Solution:

Uplifting force due to high water table

$$F = \gamma h A = \frac{62.4 \ lb}{ft^3} \times 5.0 \ ft \times \frac{\pi}{4} \times \left(\frac{8}{12} \ ft\right)^2 = 108.9 = \underline{110 \ lb}$$

EXAMPLE PROBLEM 30.18

A chemical cost is $55 per 100 kg. What is the chemical cost of treating a flow of 32 ML/d at a dosage of 12 mg/L?

Given:

$Q = 32 \text{ ML/d}$ $C = 12 \text{ mg/L (kg/ML)}$ Cost = $55/100 kg

Solution:

$$Cost = \frac{32\ ML}{d} \times \frac{12\ kg}{ML} \times \frac{\$55}{100\ kg} = 211.2 = \underline{\$210}$$

EXAMPLE PROBLEM 30.19

A chemical feeder is set to feed at the rate of 5.0 lb/h when treating a wastewater flow of 750 gpm. Find the chemical dosage applied in mg/L.

Given:

$M = 5.0$ lb/h $Q = 750$ gpm $C = ?$

Solution:

Chemical dosage

$$C = \frac{M}{Q} = \frac{5.0\ lb}{h} \times \frac{min}{750\ gal} \times \frac{h}{60\ min} \times \frac{gal}{8.34\ lb} \times \frac{1000000}{10^6} = 13.32 = \underline{13\ ppm}$$

EXAMPLE PROBLEM 30.20

You have 20 L of 5% liquid chlorine. What should be the diluted volume if you need a 2.0% solution?

Given:

$V_1 = 20$ L $C_1 = 5\%$ $V_2 = ?$ $C_2 = 2.0\%$

Solution:

$$V_2 = \frac{C_1}{C_2} \times V_1 = \frac{5.0\%}{2.0\%} \times 20\ L = 50.0 = \underline{50\ L}$$

EXAMPLE PROBLEM 30.21

Hypochlorite is supplied in a 5-gal container. The strength of the hypochlorite is 14%. For chemical feeding, you need to prepare a 2.0% solution. The feed tank has a capacity of 55 gal. If you pour one container of hypochlorite into the feed tank, how much water will you add to make the desired strength?

Given:

$V_1 = 5.0$ gal $C_1 = 14\%$ $V_2 = ?$ $C_2 = 2.0\%$

Solution:

Volume of final solution

$$V_2 = V_1 \times \frac{C_1}{C_2} = 5.0\ gal \times \frac{14\%}{2.0\%} = 35.0 = \underline{35\ gal}$$

Volume of water to be added

$$V_w = V_2 - V_1 = (35 - 5)\ gal = 30.0 = \underline{30\ gal}$$

PRACTICE PROBLEMS

PRACTICE PROBLEM 30.1

If the average hourly peak wastewater flow is assumed to be 25% of the daily average flow, estimate the peak flow. (6.3 m³/s)

PRACTICE PROBLEM 30.2

A city of 22,000 people contributes on average 2.1 MGD of wastewater. How much wastewater is produced per capita? (91 gal/c·d)

PRACTICE PROBLEM 30.3

Rework Example Problem 30.3, assuming a side slope of 0.75 horizontal to vertical. (6.8 m)

PRACTICE PROBLEM 30.4

For a ratio map scale of 1:100, how many ft of ground distance are represented by each in of distance on the map? (8.3 ft)

PRACTICE PROBLEM 30.5

Rework Example Problem 30.5, if the capacity of a dump truck is only 4.0 m³. (15 loads)

PRACTICE PROBLEM 30.6

A dump truck has a capacity of 11 yd³. How many tons of sand (SG = 2.3) are carried in each load? (21 tons)

PRACTICE PROBLEM 30.7

To provide a uniform slope of 0.35% to a 140-m-long, 200-mm-diameter sewer pipe, what must be the fall in elevation? (0.49 m)

PRACTICE PROBLEM 30.8

Find the slope of a 350-ft-long sewer pipe. The difference in invert elevations at the upstream and downstream manholes is observed to be 11 inches. (0.26%)

PRACTICE PROBLEM 30.9

What is the capacity of a new 90-m-long, 300-mm-diameter sewer pipe? (6.4 m³)

PRACTICE PROBLEM 30.10

Estimate the full flow velocity in a 12-in-diameter sewer pipe when carrying a wastewater flow of 550 gpm. (1.6 ft/s)

PRACTICE PROBLEM 30.11

If a 200-mm-diameter force main carries a metered flow rate of 1.6 ML/d, what is the average flow velocity in the sewer pipe? (0.59 m/s)

PRACTICE PROBLEM 30.12

Assuming the float travels 12% faster, how much time it will take to travel the same distance as in the previous example problem? (16 min 52 s)

PRACTICE PROBLEM 30.13

Repeat the previous problem assuming the wet well is a square section of each side of 6.5 m and the flow rate is 55 L/s. (44 min)

PRACTICE PROBLEM 30.14

A rectangular wet well is 10 ft × 5.0 ft. When no pumps were running, the level of the wet well was observed to be rising at the rate of 1.0 ft in 2 min and 30 s. What is the rate of wastewater flowing into the wet well? (150 gpm)

PRACTICE PROBLEM 30.15

If the lead pump is pumping at the rate of 150 gpm and the water in the 8.0 ft × 6.0 ft wet well is rising at a rate of 4 inches in 55 s, calculate the rate of inflow. (280 gpm)

PRACTICE PROBLEM 30.16

A 15-m-diameter storage tank is buried in the ground. What is the upward force at the bottom of the tank when the water table rises to a height of 2.1 m above the bottom? (3.6 MN)

PRACTICE PROBLEM 30.17

Repeat the previous problem, assuming the sewer pipe is of 300 mm diameter and the water depth is 1.5 m. (1.0 kN)

PRACTICE PROBLEM 30.18

Find the cost of treating a wastewater flow of 3.2 ML/d at a chorine dosage of 10 mg/L. Assume the cost of chlorine is $3.5/kg. ($112/d)

PRACTICE PROBLEM 30.19

Calculate the lb of chlorine required for dosing a flow of 0.50 MGD at the rate of 10 ppm. (42 lb)

PRACTICE PROBLEM 30.20

You have 25 L of 12.0% hypochlorite in a 200-L capacity solution tank. What is the strength of the solution when you add water to fill it up? (1.5%)

PRACTICE PROBLEM 30.21

Commercial bleach is supplied in a 5-gal container. The strength of hypochlorite is 12% available chlorine. For chemical feeding, you need to prepare a 1.5% solution. The feed tank has a capacity of 55 gal. If you pour one container of hypochlorite into the feed tank, how much water will you add to make the desired strength? (35 gal)

31 Advanced Wastewater Collection

For advanced wastewater collection systems, you need to apply the concepts of hydraulics. Most of these concepts were already discussed in the previous units. Some of the more advanced formulas and applications are discussed in the following sections.

31.1 CONTINUITY EQUATION

This concept of **continuity** or mass conservation is important during the filling and emptying of a given storage device. When water is being added or pumped out of a wet well or other structure, the rate of change in storage is given by the difference in inflow rate and outflow rate. Under a normal situation, inflow would refer to the rate of inflow in to the wet well, and the rate of outflow is the rate at which wastewater is pumped out (Figure 31.1).

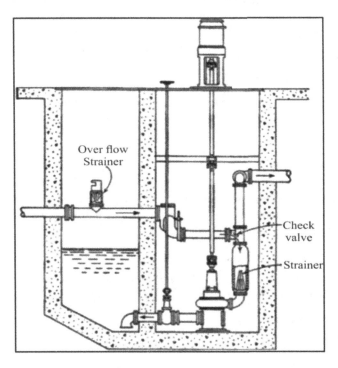

FIGURE 31.1 A drywell pumping system.

DOI: 10.1201/9781003468745-31

Pumping rate

$$Q_{out} = Q_{in} - Q_{sto}$$

When pumping rate exceeds inflow rate, the water level will drop to indicate that the water well is being emptied out.

31.2 PUMPING HEAD

In any pumping system, **pumping head** is the sum of the terms increase in pressure head from intake to discharge, vertical distance through which water is being lifted and head losses that occur in the piping system connecting the intake to the discharge point. In the case where you know the pressure at the suction and discharge side of the pump, the total dynamic head equals the pressure head added. The general equation for finding the total dynamic head is:

Total dynamic head (head added)

$$h_a = \frac{v_2^2}{2g} - \frac{v_1^2}{2g} + h_L + Z_2 - Z_1 + \frac{p_2}{\gamma} - \frac{p_1}{\gamma}$$

31.2.1 WATERPOWER

As the name suggests, waterpower is the actual power added or transferred to water by the pump. In other words, it represents the pump output. Knowing the pumping head, waterpower can be found.

Power added to wastewater

$$P_a = W \times h_a = Q \times \gamma \times h_a = Q \times p_a$$

P_a = pressure added, W = Weight flow rate, Q = pumping rate

31.2.2 PUMP EFFICIENCY

Pump power P_p is the power given to the pump by the prime mover. This is also called **brake power** or **shaft power**. Pump power is less than wire or input power by the amount indicated by the efficiency of the prime mover driving the pump. Knowing the power input and power output of the pump (waterpower), the efficiency of the pump can be worked out

Pump efficiency

$$E_p = \frac{Power\ added}{Shaft\ Power} = \frac{P_a}{P_m} \quad E_m = \frac{Pump\ Power}{Power\ Supplied} = \frac{P_p}{P_I}$$

$$E_o = \frac{Pump\ added}{Power\ Supplied} = \frac{P_a}{P_I}$$

31.2.3 PUMP POWER

Conversely, if the operating efficiency of the pump is read from the performance curves, the power required to operate the pump can be calculated.

31.2.4 POWER INPUT

Power input is the actual power required to pump water. This represents the wire power or power on which basis you are billed. If your supply system has a power factor of unity, this would represent the power input to the prime mover.

31.3 FLOW CAPACITY OF SEWER PIPES

Wastewater collection systems are primarily gravity flow systems. Under very few situations, wastewater is pumped under pressure, as in case of a **force main**. In gravity flow or open channel flow systems, water flow velocity is maximum when water depth is about 90% of the pipe diameter.

Manning's flow equation (SI)

$$v_F = \frac{0.4}{n} \times D^{2/3} \times \sqrt{S} \quad Q_F = \frac{0.312}{n} \times D^{8/3} \times \sqrt{S} \quad (SI)$$

Manning's flow equation (USC)

$$v_F = \frac{0.6}{n} \times D^{2/3} \times \sqrt{S} \quad Q_F = \frac{0.464}{n} \times D^{8/3} \times \sqrt{S} \quad (USC)$$

In this equation, the constant n represents the roughness coefficient of the sewer pipe, D is the diameter of the pipe in m, ft and S is slope of the pipe expressed as a decimal fraction. A typical value of n for sewer pipes is 0.013. If a value of n is not specified, use n = 0.013.

EXAMPLE PROBLEM 31.1

A dye is introduced at a stream manhole to estimate the flow velocity. The dye first appears after 3 minutes at a downstream manhole located at a distance of 390 ft. The dye disappears after 3 min and 40 seconds. Calculate the average flow velocity in m/s.

Given:

s = 390 ft t_1 = 3 min, 10 s t_2 = 3 min, 40 s   v = ?

Solution:

$$t_1 = 3.0 \ min \times \frac{60 \ s}{min} + 10 \ s = 190 \ s$$

$$t_2 = 3.0 \ min \times \frac{60 \ s}{min} + 40 \ s = 220 \ s$$

$$t_{avg} = \frac{1}{2}(190\ s + 220\ s) = 205\ s$$

Average flow velocity

$$\bar{v} = \frac{s}{t} = \frac{390\ ft}{205\ s} = 1.902 = \underline{1.9\ ft/s}$$

EXAMPLE PROBLEM 31.2

A float is dropped in a manhole to estimate the velocity of the wastewater. It takes 2 min and 35 s for the float to arrive at a manhole 440 ft downstream. Assume the average flow velocity is 85% of the float velocity.

Given:

s = 440 ft   C = 85% = 0.85   t = 2 min and 30 s   v = ?

Solution:

$$t = 2.0\ min \times \frac{60\ s}{min} + 35\ s = 155\ s$$

Float velocity

$$\bar{v}_F = \frac{s}{t} = \frac{440\ ft}{155\ s} = 2.83\ ft/s$$

Flow velocity

$$\bar{v} = C \times \bar{v}_F = 0.85 \times 2.838\ ft/s = 2.41 = \underline{2.4\ ft/s}$$

EXAMPLE PROBLEM 31.3

Estimate the full flow velocity in an 8-in-diameter sewer pipe carrying a flow of 350 gpm.

Given:

D = 8-in = 0.667 ft   Q = 350 gpm   v = ?

Solution:

Full flow velocity

$$v_F = \frac{Q}{A} = \frac{4Q}{\pi D^2} = \frac{4}{\pi} \times \frac{350\ gal}{min} \times \frac{1}{(0.667\ ft)^2} \times \frac{ft^3}{7.48\ gal} \times \frac{min}{60\ s}$$

$$= 2.23 = \underline{2.2\ ft/s}$$

EXAMPLE PROBLEM 31.4

Calculate the flow-carrying capacity of a 200-mm-diameter sewer pipe flowing half full when the average flow velocity is 0.70 m/s.

Given:

d/D = 0.5 (half full) $A/A_F = 0.5$ v = 0.70 m/s   D = 200 mm = 0.2 m

Solution:

Flow-carrying capacity (half full)

$$Q = v \times A = v \times \frac{\pi}{8} \times D^2 = \frac{0.7\, m}{s} \times \frac{\pi}{8} \times (0.2\, m)^2 \times \frac{1000\, L}{m^3}$$

$$= 11.0 = \underline{11\, L/s}$$

EXAMPLE PROBLEM 31.5

At a sewage pumping station, wastewater is pumped through a 6-in-diameter force main at the rate of 350 gpm. What is the flow velocity in the main?

Given:

Q = 350 gpm   D = 6-in = 0.50 ft   v = ?

Solution:

Full flow velocity

$$v_F = \frac{Q}{A} = \frac{4Q}{\pi D^2} = \frac{4}{\pi} \times \frac{450\, gal}{min} \times \frac{1}{(0.50\, ft)^2} \times \frac{ft^3}{7.48\, gal} \times \frac{min}{60\, s}$$

$$= 3.97 = \underline{4.0\, ft/s}$$

EXAMPLE PROBLEM 31.6

Wastewater is flowing into a wet well at the rate of 23 L/s. If the level of the 2.5 m × 2.0 m wet well is rising at the rate of 5.0 cm/min, determine the pump capacity.

Given:

A = 2.5 × 2.0 m = 5.0 m²   v = 5.0 cm/min Q_{in} = 23 L/s Q_{out} = ?

Solution:

Storage rate

$$Q_{stor} = v_{stor} \times A = \frac{5.0\, cm \times 5.0\, m^2}{min} \times \frac{m}{100\, cm} \times \frac{1000\, L}{m^3} \times \frac{min}{60\, s} = 4.167 = 4.2\, L/s$$

Outflow rate

$$Q_{out} = Q_{in} - Q_{sto} = 23 \ L/s - 4.167 \ L/s = 18.83 = \underline{19 \ L/s}$$

EXAMPLE PROBLEM 31.7

Water is flowing at the rate of 310 gpm into a wet well of an area of 55 ft². If the pumping rate is 290 gpm, what is the rise rate of wastewater in the well?

Given:

$Q_{in} = 310$ gpm $Q_{out} = 290$ gpm $A = 55$ ft²

Solution:

Storage rate

$$Q_{sto} = Q_{in} - Q_{out} = 310 - 290 = 20 \ gpm$$

Rise rate

$$v_{stor} = \frac{Q_{stor}}{A} = \frac{20 \ gal}{min} \times \frac{1}{55 \ ft^2} \times \frac{ft^3}{7.48 \ gal} \times \frac{12 \ in}{ft} = 0.583 = \underline{0.58 \ in/min}$$

EXAMPLE PROBLEM 31.8

The storage in a well between the high (pump starts) and low (pump stops) levels is 4.8 m³. If the pumping rate is 4.2 L/s and the inflow to the wet well is 2.3 L/s, what is the pumping operating time?

Given:

$V = 4.8$ m³ $Q_{in} = 2.3$ L/s $Q_{out} = 4.2$ L/s $t = ?$

Solution:

Storage rate

$$Q_{sto} = Q_{in} - Q_{out} = 2.3 \ L/s - 4.2 \ L/s = -1.9 \ L/s = Emptying$$

Time of pump operation

$$t = \frac{V}{Q} = \frac{4.8 \ m^3 . s}{1.9 \ L} \times \frac{1000 L}{m^3} \times \frac{min}{60 \ s} = 42.1 = \underline{42 \ min}$$

EXAMPLE PROBLEM 31.9

The suction and discharge pressure gauge readings of a given pump are 12 kPa vacuum and 45 kPa, respectively. The pressure gauges are placed at the same elevation. What is the dynamic head of the pump?

Given:

$Z_1 = Z_2$ $p_1 = -12$ kPa $p_2 = 45$ kPa $h_L = 0$

Pressure added, $p_a = p_2 - p_1$

Solution:

Head added by the pump

$$h_a = h_L + Z_2 - Z_1 + \frac{p_2}{\gamma} - \frac{p_1}{\gamma} = \left(\frac{p_2 - p_1}{\gamma}\right)$$

$$= (45 - -12)kPa \times \frac{m}{9.81\ kPa} = 5.81 = \underline{5.8\ m}$$

EXAMPLE PROBLEM 31.10

A pump has a capacity of 2500 gpm when pumping against a head of 25 ft (total head). Assuming the pump efficiency is 80%, what size (rated power) motor is required?

Given:

$Q = 2500$ gpm $h_a = 25$ ft $E_p = 80\%$ $P_p = ?$

Solution:

Power added by the pump

$$P_a = Q \times \gamma \times h_a = \frac{2500\ gal}{min} \times \frac{8.34\ lb}{gal} \times \frac{hp.s}{550\ lb.ft} \times \frac{min}{60\ s} \times 25\ ft = 15.79 = 16\ hp$$

Pump power

$$P_p = \frac{P_a}{E_p} = \frac{15.79}{80\%} \times 100\% = 19.74 = \underline{20\ hp}$$

EXAMPLE PROBLEM 31.11

A pumping unit draws 75 A at 220 V when pumping @ 35 L/s against a head of 31 m. What is the overall efficiency of the pumping unit?

Given:

$I = 75$ A $E = 220$ V $Q = 35$ L/s $h_a = 31$ m $E_o = ?$

Solution:

Power input

$$P_i = E \times I = 220\ V \times 75A \times \frac{W}{V.A} = 1.65 \times 10^4\ W = 16.5\ kW$$

Power added to water

$$P_a = Q \times \gamma \times h_a = \frac{0.035 \ m^3}{s} \times \frac{9.81 \ kN}{m^3} \times 31 \ m = 10.6 \ kW$$

Overall efficiency (wire to water)

$$E_o = \frac{P_a}{P_I} = \frac{10.6 \ kW}{16.5 \ kW} \times 100\% = 64.5 = \underline{65\%}$$

EXAMPLE PROBLEM 31.12

A pump motor draws 12 A of current when connected to a 220-V supply. What is the output of the motor, assuming the power factor is 80% and the efficiency of the motor is 90%?

Given:

I = 12 A   E = 220 V   PF = 80%   P = ?

Solution:

Power input to motor

$$P_i = E \times I = PF = 220 \ V \times 12A \times \frac{80\%}{100\%} \times \frac{kW}{KVA} = 2.11 \ kW$$

Power output of motor

$$P_o = P_i \times E_I = 2.11 \ kW \times \frac{90\%}{100\%} = 1.90 = \underline{1.9 \ kW}$$

EXAMPLE PROBLEM 31.13

An 18-in sub-main sewer is laid on a grade of 0.25%. Assuming a coefficient of pipe roughness of 0.013, determine the full flow velocity.

Given:

n = 0.013   D = 18-in = 1.5 ft v_F = ?   S = 0.25% = 0.0025

Solution:

Full flow velocity

$$v_F = \frac{0.6}{n} \times D^{2/3} \times \sqrt{S} = \frac{0.6}{0.013} \times (1.5)^{2/3} \times \sqrt{0.0025} = 3.02 = \underline{3.0 \ ft/s}$$

EXAMPLE PROBLEM 31.14

The slope of a 450-mm sub-main sewer is 0.25%. Assuming a coefficient of pipe roughness of 0.013, calculate the wastewater flow capacity when flowing full.

Given:

n = 0.013   D = 450 mm = 0.45 m $Q_F = ?$ S = 0.25% = 0.0025

Solution:

Full flow rate

$$Q_F = \frac{0.312}{n} \times D^{8/3} \times \sqrt{S} = \frac{0.312}{0.013} \times (0.45)^{8/3} \times \sqrt{0.0025}$$

$$= \frac{0.142 \ m^3}{s} \times \frac{1000 \ L}{m^3} = 142 = \underline{140 \ L/s}$$

31.3.1 PARTIAL FLOW CHARACTERISTICS

Based on full flow conditions, **partial flow** conditions (Figure 31.2) can be worked out. For a given ratio of one of the flow components, flow rate, flow velocity or flow depth, other flow component ratios are automatically fixed. Such relationships are given as shown in Table 31.1. For example, for a depth to diameter ratio of 0.40 or 40%, partial flow is 0.34 or 34% of full flow. At this depth of flow, flow velocity is 90% of the full flow velocity. It is important to note that flow velocity is at maximum when the depth of flow is 90% of the diameter of the pipe.

EXAMPLE PROBLEM 31.15

The slope of a 450-mm sub-main sewer is 0.25%. Assuming a coefficient of pipe roughness of 0.013, calculate the wastewater flow capacity when flowing 60% full.

Given:

n = 0.013   D = 450 mm = 0.45 m $Q/Q_F = 0.78$ S = 0.25% = 0.0025

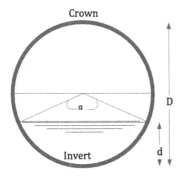

FIGURE 31.2 Partially filled sewer pipe.

TABLE 31.1

Partial Flow

d/D	rad	deg	P/P$_F$	P/P$_F$	A/A$_F$	v/v$_F$	Q/Q$_F$
0	0	0	0	0	0	0	0
0.10	1.287	73.7	0.205	0.254	0.052	0.401	0.021
0.15	1.591	91.1	0.253	0.372	0.094	0.517	0.049
0.20	1.855	106.3	0.005	0.482	0.142	0.615	0.088
0.25	2.094	120	0.006	0.587	0.196	0.701	0.137
0.30	2.319	132.8	0.006	0.684	0.252	0.776	0.196
0.35	2.532	145.1	0.007	0.774	0.312	0.843	0.263
0.40	2.739	156.9	0.008	0.857	0.374	0.902	0.337
0.45	2.941	168.5	0.008	0.932	0.436	0.954	0.417
0.50	3.142	180	0.009	1	0.5	1	0.5
0.55	3.342	191.5	0.009	1.06	0.564	1.039	0.586
0.60	3.544	203.1	0.01	1.111	0.626	1.072	0.672
0.65	3.751	214.9	0.01	1.153	0.688	1.099	0.756
0.70	3.965	227.2	0.011	1.185	0.748	1.12	0.837
0.75	4.189	240	0.012	1.207	0.804	1.133	0.912
0.80	4.429	253.7	0.012	1.217	0.858	1.14	0.977
0.85	4.692	268.9	0.013	1.213	0.906	1.137	1.03
0.90	4.996	286.3	0.014	1.192	0.948	1.124	1.066
0.95	5.381	308.3	0.015	1.146	0.981	1.095	1.075
1	6.283	360	0.017	1	1	1	1

Solution:

Full flow rate

$$Q_F = \frac{0.312}{n} \times D^{8/3} \times \sqrt{S} = \frac{0.312}{0.013} \times (0.45)^{8/3} = \sqrt{0.0025}$$

$$= \frac{0.142 \ m^3}{s} \times \frac{1000 \ L}{m^3} = 142 = \underline{140 \ L/s}$$

Partly full flow rate

$$Q = 0.78Q_F = 0.78 \times 142 \ L/s = 110.7 = \underline{110 \ L/s}$$

EXAMPLE PROBLEM 31.16

The slope of a 400-mm sub-main sewer is 0.30%. Assuming a coefficient of pipe roughness of 0.013, calculate the flow velocity during minimum flow period, which is 20% full (v/v$_F$ = 0.62).

Given:

n = 0.013   D = 400 mm = 0.40 m   v = ?   v/v$_F$ = 0.62   S = 0.30% = 0.003

Solution:

Full flow velocity

$$v_F = \frac{0.4}{n} \times D^{2/3} \times \sqrt{S} = \frac{0.4}{0.013} \times (0.40)^{2/3} \times \sqrt{0.003} = 0.914 = 0.91 \; m/s$$

Partly full flow velocity

$$v = 0.62 \times 0.9149 \; m/s = 0.567 = \underline{0.57 \; m/s}$$

EXAMPLE PROBLEM 31.17

Calculate the minimum slope on which a rectangular channel with 1.2 m of width should be laid to maintain a velocity of 1.0 m/s while flowing at a depth of 0.90 m. The channel is made of formed, unfinished concrete.

Given:

b = 1.2 m   d = 0.90 m   v = 1.0 m/s   n = 0.014   S = ?

Solution:

Hydraulic radius

$$R_h = \frac{A}{P_w} = \frac{1.2 \; m \times 0.90 \; m}{(1.2 \; m + 2 \times 0.90 \; m)} = 0.36 \; m$$

Slope of the channel

$$S = \frac{n^2 v^2}{R_h^{4/3}} = 0.014^2 \times \frac{1.0^2}{0.36^{4/3}} = 7.65 \times 10^{-4} = \underline{0.077\%}$$

EXAMPLE PROBLEM 31.18

A 12-in-diameter sanitary sewer with n of 0.013 is laid on a grade of 0.10%. Determine the full flow capacity of the sewer pipe.

Given:

n = 0.013   D = 400 mm   S = 0.1% = 0.001

Solution:

Full flow capacity

$$Q_F = \frac{0.464}{n} \times D^{8/3} \times \sqrt{S} = \frac{0.464}{0.013} \times 1.0^{8/3} \times \sqrt{0.001} = 1.128 \; ft^3/s$$

$$= \frac{1.128 \; ft^3}{s} \times \frac{7.48 \; gal}{ft^3} \times \frac{60 \; s}{min} = 506 = \underline{510 \; gpm}$$

EXAMPLE PROBLEM 31.19

A 300-mm-diameter sanitary sewer with n of 0.012 is required to carry a flow of 22 L/s when flowing full. What is the minimum grade required?

Given:

$n = 0.012$ $D = 300$ mm $= 0.30$ m $S = ?$ $Q_F = 22$ L/s $= 0.022$ m³/s

Solution:

Slope of the sewer pipe

$$S = \left(\frac{Qn}{0.312D^{8/3}}\right)^2 = \left(\frac{0.022 \times 0.012}{0.312 \times (0.3)^{8/3}}\right)^2 = 4.40 \times 10^{-4} = \underline{0.044\%}$$

EXAMPLE PROBLEM 31.20

What diameter sewer pipe with n of 0.012 is required to carry 400 gpm flowing full? The available grade is 0.060%.

Given:

$n = 0.012$ $D = ?$ $S = 0.06\% = 0.006$ $Q_F = 450$ gpm

Solution:

Full flow capacity

$$Q_F = \frac{400 \ gal}{min} \times \frac{ft^3}{7.48 \ gal} \times \frac{min}{60 \ s} = 0.891 \ ft^3/s$$

Diameter of sewer pipe

$$D = \left(\frac{Qn}{0.464\sqrt{S}}\right)^{3/8} = \left(\frac{0.891 \times 0.013}{0.464\sqrt{0.0006}}\right)^{3/8} \times \frac{12 \ in}{ft} = \underline{18 \ in}$$

PRACTICE PROBLEMS

PRACTICE PROBLEM 31.1

The average time taken by a dye to travel a distance of 520 ft between two manholes is 4 min and 15 seconds. What is the average sewage flow velocity? (2.0 ft/s)

PRACTICE PROBLEM 31.2

It took 2 min and 20 s for a float to travel a distance of 125 m between two manholes. Estimate the velocity of the wastewater in the sewer line assuming flow velocity to be 88% of the float velocity. (0.79 m/s)

PRACTICE PROBLEM 31.3

Estimate the full flow velocity in a 12-in sewer pipe when carrying a flow of 600 gpm. (1.7 ft/s)

PRACTICE PROBLEM 31.4

Estimate the flow in the 300-mm server pipe flowing at a depth 2/3rd of the diameter at the rate of 0.75 m/s. (38 L/s)

PRACTICE PROBLEM 31.5

If a 200-mm-diameter force main carries a metered flow rate of 1.6 ML/d, what is the average flow velocity in the sewer pipe? (0.59 m/s)

PRACTICE PROBLEM 31.6

Water in wet well measuring 2.5 m × 2.0 m is rising at the rate of 10 cm per 40 s when the well is pumped at the rate of 10 L/s. What is the rate of inflow? (23 L/s)

PRACTICE PROBLEM 31.7

When a wet well is pumped at the rate of 1.7 ML/d, the water level drops at the rate of 4.2 L/s. What is the water inflow into the well? (15 L/s)

PRACTICE PROBLEM 31.8

Rework the previous problem assuming the inflow rate is 1.8 L/s. (33 min)

PRACTICE PROBLEM 31.9

Rework the previous problem assuming there is positive pressure of 12 kPa at the suction side. (3.4 m)

PRACTICE PROBLEM 31.10

Work out the pump power for the following data: H = 50 ft, Q = 450 gpm, E = 73%. (7.8 hp)

PRACTICE PROBLEM 31.11

A pump is 72% efficient when pumping 25 L/s against a head of 35 m. What is the expected amperage drawn by the pump from 220-V supply? Assume the motor is 90%. (60 A)

PRACTICE PROBLEM 31.12

For the data of the previous example problem, the power factor is improved to 88%. What will be the motor output? (2.4 kW)

PRACTICE PROBLEM 31.13

A 24-in sewer main is laid on a grade of 0.20% slope. Find the full flow velocity. (3.3 ft/s)

PRACTICE PROBLEM 31.14

A 300-mm sub-main sewer is laid on a grade of 0.15%. Assuming a coefficient of pipe roughness of 0.013, work out the full flow capacity of the sewer pipe. (37 L/s)

PRACTICE PROBLEM 31.15

Repeat Example Problem 31.15 assuming the pipe is flowing 75% full. (130 L/s)

PRACTICE PROBLEM 31.16

The slope of a 300-mm sub-main sewer is 0.250%. Assuming a coefficient of pipe roughness of 0.013, calculate the flow velocity during minimum flow period, which is 25% full. (0.48 m/s)

PRACTICE PROBLEM 31.17

Calculate the minimum slope on which a rectangular channel with 1.5 m of width should be laid to maintain a velocity of 1.0 m/s while flowing at a depth of 1.0 m. Assume n = 0.015. (0.07%)

PRACTICE PROBLEM 31.18

An 18-in-diameter sanitary sewer with n of 0.013 is laid on a grade of 0.050%. Compute the full flow capacity of the sewer pipe. (1100 gpm)

PRACTICE PROBLEM 31.19

A 375-mm-diameter sanitary sewer with n of 0.013 is required to carry a flow of 28 L/s when flowing full. What is the minimum grade required? (0.025%)

PRACTICE PROBLEM 31.20

What minimum diameter sewer pipe with n of 0.013 is required to carry 450 gpm when flowing full? The available grade is 0.10%. (12 in)

32 Preliminary Treatment

As the name suggests, preliminary treatment is responsible for the removal of material from wastewater that might otherwise impair or harm head works or the operation of downstream processes, including primary and secondary treatment. This material usually consists of the removal of wood, rags, plastic and grit. Some preliminary treatment methods and equipment include pumping, screening, shredding and grit removal. In some cases, prechlorination and aeration are used as part of pretreatment. Though flow measurement is not a treatment process, it is very common to have a flow measuring device at the head works of a plant.

32.1 WET WELL

In general, flow in sanitary sewers is due to gravity. However, in low-lying areas, it may be required to lift the flow to a higher sewer in which wastewater may again flow by gravity. Wastewater collected in sanitary sewers flows into large wet wells from where it is pumped directly into the plant for gravity flow through the plant. The capacity of a wet well refers to its volume. For a rectangular-shaped wet well, the pumping rate Q can be found by observing the drop rate due to pumping with the flow into the wet well stopped. This is illustrated in Figure 32.1.

Pumping rate

$$Q = A_s \times \frac{\Delta H}{\Delta t}$$

$A_s = L \times W = $ *Surface Area of the wet well* $\Delta H = $ *Drop of water level in time interval* Δt

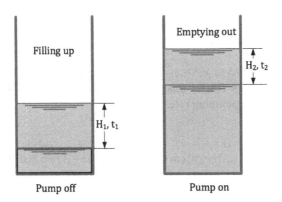

FIGURE 32.1 Wet well pumping rate.

DOI: 10.1201/9781003468745-32

This provides an easy way of conducting a pump test. This concept can also be used to estimate the influent rate by noting the rise rate before the lead pump comes on. In fact, by observing the rise rate before pumping starts and drop rate when the wet well is being emptied out by pumping, the pumping rate, and hence the flow into the plant from a given pumping station, can be calculated. This information is useful in calibrating or comparing the accuracy of flow measuring equipment and devices.

Influent rate

$$Q = A_s \times \left(\frac{H_1}{t_1} + \frac{H_2}{t_2} \right)$$

H_1, H_2 = rise and drop of water level in the wet well
t_1, t_2 = time taken for water level to rise and drop respectively.

EXAMPLE PROBLEM 32.1

A wet well is 12 ft × 10 ft. The difference between the minimum water level (pump shuts off) and the water level when the lead pump comes on is 2.0 ft. How much water flows into the well between two consecutive pumping cycles? If the time interval between two consecutive pumping cycles is 3 min and 35 s in the morning hours, calculate the instantaneous inflow rate.

Given:

$A_s = 12\,\text{ft} \times 10\,\text{ft}$ $\Delta H = 2.0\,\text{ft}$ $\Delta t = 3\,\text{min and } 35\,\text{s} = 115\,\text{s}$ $Q = ?$

Solution:

Instantaneous flow rate

$$Q = \frac{\Delta V}{\Delta t} = \frac{A_s \times d}{\Delta t} = \frac{12\,ft \times 10\,ft \times 2.0\,ft}{115\,s} \times \frac{17.48\,gal}{m^3} \times \frac{60\,s}{min} = 936 = \underline{940\ gpm}$$

EXAMPLE PROBLEM 32.2

A wet well measures 3.5 m × 3.0 m. When the pump is on, the water level drops by 50 cm in 2 min and 30 s. During this period, inflow to the wet well is estimated to be 12 L/s. Determine the pumping rate.

Given:

$A_s = 3.5\,\text{m} \times 3.0\,\text{m}$ $\Delta H = 50\,\text{cm} = 0.50\,\text{m}$ $\Delta t = 2\,\text{min and } 30\,\text{s} = 150\,\text{s}$
$Q_i = 12\,\text{L/s}$ $Q = ?$

Solution:

Pumping rate

$$Q = \frac{\Delta V}{\Delta t} + Q_i = \frac{A_s \times d}{\Delta t} = \frac{3.5\ m \times 3.0 \times 0.50\ m}{150\ s} \times \frac{1000\ L}{m^3} + 12\ L/s = 47.0 = \underline{47\ L/s}$$

32.2 SCREENS

Screening is the first and foremost step in the preliminary treatment. The volume of screenings such as rags, cans and cardboard ranges from 10 L–100 L/ML of wastewater flow. This ratio is primarily affected by the characteristics of wastewater but may remain stable for a plant.

Screens need to be disposed of separately. Thus, it is important to keep track of the volume of screenings removed from the wastewater flow. Based on the rate of removal, the time required to fill a disposal pit of known capacity can be calculated.

$$\boxed{Fill\ Time = \frac{Volume\ of\ the\ pit}{Rate\ of\ filling}}$$

EXAMPLE PROBLEM 32.3

During the month of April, an average of 36.6 gallons of screenings were removed each day. If the screening pit has a capacity of 25 ft³, after how many days should the pit be emptied out?

Given:

$$V = 25\ ft^3 \qquad Q = 36.6\ gal/d \qquad t = ?$$

Solution:

Time to empty

$$t = \frac{V}{Q} = \frac{25\ ft^3.d}{36.6\ gal} \times \frac{7.48\ gal}{ft^3} = 5.11 = \underline{5\ d}$$

EXAMPLE PROBLEM 32.4

In a wastewater plant, on average, 120 L screenings are removed each day. If it is desired to empty the pit on a weekly basis, what should be the minimum capacity of the screening pit?

Given:

$$V = ? \quad Q = 120\ L/d \quad t = 7\ d$$

Solution:

Capacity of the screening pit

$$V = Q \times t = \frac{320\ L}{d} \times 7\ d \times \frac{m^3}{1000\ L} = 2.24 = \underline{2.2\ m^3}$$

EXAMPLE PROBLEM 32.5

You are asked to size a new screening pit that can hold screenings for as long as 60 days. If the maximum screening removal rate experienced at the plant is 66 gal/day in the worst case scenario, determine the required capacity of the pit.

Given:

$$V = ? \qquad Q = 66\ gal/d \qquad t = 60\ d$$

Solution:

Required capacity of the pit

$$V = Q \times t = \frac{66\ gal}{d} \times 60\ d \times \frac{ft^3}{7.48\ gal} = 529.4 = \underline{530\ ft^3}$$

EXAMPLE PROBLEM 32.6

The screenings at a given plant are removed at an average rate of 28 L/ML of wastewater flow. How long will a disposal pit of 10 m³ capacity last for this plant with a daily flow of 12 ML/d?

Given:

$$Q = 12\ ML/d \qquad Screening = 28\ L/ML \qquad V = 10\ m^3$$

Solution:

Screenings removal rate

$$Q = \frac{28L}{ML} \times \frac{12\ ML}{d} = 336\ L/d$$

Time to last

$$t = \frac{V}{Q} = \frac{10\ m^3 . d}{336\ L} \times \frac{1000\ L}{m^3} = 29.7 = \underline{30\ d}$$

32.3   GRIT REMOVAL

Grit consists of relatively heavy (SG = 2.6) solids, primary inorganic in nature. If not removed in preliminary treatment, they may damage mechanical equipment and unnecessarily occupy useful space in aeration tanks and sludge digesters. The removal of grit is practised by slowing down the flow velocity to about 30 cm/s (1 ft/s). An acceptable range is 20–40 cm/s (0.6–1.2 ft/s). Too-high velocities can cause washout of grit, and too-slow flow velocities will result in settling of organics with the grit.

Grit removal is done in grit chambers, which may be grit channels, grit tanks or grit clarifiers. Preaeration of wastewater prior to settling helps to freshen the water, scrub out entrained gases and improve settleability due to decreased specific gravity. In the case of grit channels, normal operation consists of measuring velocities at high, average and low flows. Velocity in the grit channel can be determined by the use of the continuity equation, or it can be estimated by observing the distance travelled by floats in a given amount of time.

Horizontal flow velocity

$$v_H = \frac{Q}{A_X} = \frac{Q}{b \times d}$$

32.3.1 HYDRAULIC DETENTION TIME

Average flow velocity is typically 80% to 90% of the float velocity. The time interval between entry and exit is the hydraulic detention time. Knowing the flow-through velocity v_H, detention time can be determined for a given length of the channel.

Detention time

$$t_d = \frac{L}{v_H} = \frac{V}{Q}$$

32.3.2 SETTLING VELOCITY

Settling rate will depend on the size and specific gravity of the grit particle. The settling velocity of grit (0.2 mm) is about 2.3 cm/s. This means that if the flow depth is 0.5 m, a grit particle of average size 0.2 mm introduced at the surface will settle in time:

$$t_s = \frac{d}{v_s} = \frac{0.5 \, m.s}{2.3 \, cm} \times \frac{100 \, cm}{m} = 22 \, s$$

Hydraulic detention time (HDT) should be greater than or equal to settling time. As shown earlier, the flow velocity for a given length of channel controls HDT. Velocity in grit channels may be controlled by varying the number of channels under operation. Flow velocity in grit channels is usually regulated by providing a proportional weir at the end of the channel or changing the shape of grit channel. Figure 30.2 shows the relationship between settling and horizontal flow velocity.

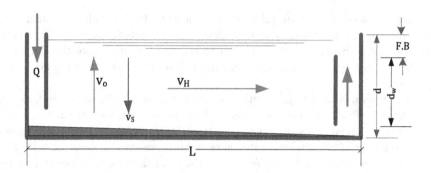

FIGURE 32.2   Flow velocity and settling velocity.

32.3.3   FLOAT VELOCITY

A simple method for estimating the velocity is to place a float like a ping pong ball or a stick in the channel and measuring the time it takes to travel a given distance.

Float velocity

$$v_F = \frac{distance}{time} = \frac{s}{t}$$

32.4   GRIT QUANTITIES

For separate sanitary sewer systems, the amount of grit may range from 10–30 L/ML of wastewater processed. During wet weather, it will be much higher. Grit is normally disposed of by burial. Over a given period, the average quantity of grit removed at a given facility can be calculated. Based on this information, you can estimate the required disposal capacity.

EXAMPLE PROBLEM 32.7

In a grit channel, surface velocity is observed by a float. It takes 21 seconds to travel 5 m length of the channel. Calculate the average flow velocity and hydraulic detention time if the total length of the channel is 18 m.

Given:

$$L = 18\,m \quad v_F = 5\,m/21\,s \quad HDT = ?$$

Solution:

Horizontal flow velocity

$$v_H = 0.85 v_F = 0.85 \times \frac{5.0\,m}{21\,s} = 0.202 = 0.2\,m/s$$

Detention time

$$t_d = \frac{L}{v_H} = \frac{18 \ m.s}{0.202 \ m} \times \frac{min}{60 \ s} = 1.50 = \underline{1.5 \ min}$$

EXAMPLE PROBLEM 32.8

A grit channel is 3.5 ft wide. The flow depth in the channel is observed to be 2.0 ft at a time when the flow gauge reads 3.4 MGD. Estimate the flow velocity.

Given:

Q = 3.4 MG/d   W = 3.5 ft   d = 2.0 ft

Solution:

Horizontal flow velocity

$$v_H = \frac{Q}{A_X} = \frac{Q}{W \times d} = \frac{3.5 \times 10^6 \ gal}{d} \frac{ft^3}{7.48 \ gal} \times \frac{1}{3.5 \ ft \times 2.0 \ ft}$$

$$= \frac{66844 \ ft}{d} \times \frac{d}{24 \ h} \times \frac{h}{3600 \ s} = 0.773 = \underline{0.77 \ ft/s}$$

EXAMPLE PROBLEM 32.9

A wastewater plant has an average flow of 8.0 ML/d. An average of 110 L of grit is removed each day. a) How many litres of grit are removed per ML of wastewater flow? b) What disposal capacity must be available for a period of 90 d?

Given:

Q = 8.0 ML/d   Grit production = 110 L/d

Solution:

Grit production rate

$$Grit = \frac{110 \ L}{d} \times \frac{d}{80 \ ML} = 14 \ L/ML$$

Disposal capacity required

$$V = \frac{110 \ L}{d} \times \frac{m^3}{1000 \ L} \times 90 \ d = 9.9 = \underline{10 \ m^3}$$

EXAMPLE PROBLEM 32.10

A new screening pit is being designed to hold enough screenings for exactly 60 days. If daily average removal rate of grit is 77 gal/day in the worst-case scenario, determine the desired capacity of the pit. If the pit depth is 6.0 ft, what size should each side of the square-shaped pit be?

Given:

t = 60 d   Grit production = 77 gal/d

Solution:

Capacity of the pit

$$V = \frac{77\, gal}{d} \times 60\, d \frac{ft^3}{7.48\, gal} = 617.6 = \underline{620\, ft^3}$$

Side of the square pit

$$s = \sqrt{\frac{V}{d}} = \sqrt{\frac{617.6\, ft^3}{6.0\, ft}} = 10.1 = \underline{10\, ft}$$

32.5   FLOW MEASUREMENT

Although flow measuring devices have no direct role in the treatment of wastewater, they are an integral part of preliminary treatment. By knowing the quantity of flow entering the plant, adjustments are made to the pumping rate, chlorination rate and other processes within the plant. The self-cleaning characteristics of the Parshall flume make it the preferred choice for flow measuring in wastewater treatment plants. Another measuring device used in open channels is a weir. The main disadvantage of the weir is the tendency of solids to settle upstream of the weir. Too much accumulation may affect the flow accuracy of the flow reading.

The necessity of accurate flow measurement cannot be overemphasized. In many situations, it may be necessary to check the accuracy of flow measurement from time to time. This can be done by comparing the flow reading with the flow rate measured by other methods. It may include volumetric methods, velocity methods and wet well pumping rates. For a given installation of weir or a flume, the flow rate is calculated by observing the depth of flow, which is measured by a secondary device installed at a specified point upstream of the contraction. Any error in the measurement of depth will be magnified many times in the estimation of flow rate. For example, if the depth measurements are read higher, the flow rate in the case of triangular weirs will be significantly overestimated.

Weir formula

$$\boxed{Q = K \times H^n}$$

The exponent n is 2.5 in the case of triangular weirs. If the head over a triangular weir increases by 20%, it indicates a 58% increase in flow rate.

$$\frac{Q_2}{Q_1} = \left(\frac{H_2}{H_1}\right)^{2.5} = 1.2^{2.5} = 1.58 = 58\%\uparrow$$

It is the operator's job to make sure that depth sensing devices are properly functioning and calibrated.

EXAMPLE PROBLEM 32.11

If a 11-m-diameter clarifier drops 48 cm in exactly 6 hours, what is the pumping rate out of the tank in L/s?

Given:

$D = 11$ m $\qquad t = 6$ h $\qquad d = 48$ cm $= 0.48$ m $\qquad Q = ?$

Solution:

Pumping rate

$$Q = \frac{V}{t} = \frac{\pi D^2}{4} \times \frac{d}{t} = \frac{\pi (11\,m)^2}{4} \times \frac{0.48\,m}{6.0\,h} \times \frac{h}{3600\,s} \times \frac{1000\,L}{m^3} = 2.11 = \underline{2.1\,L/s}$$

32.6 SAMPLING

There are three basic aspects of sampling:

1. Sample type (grab versus composite)
2. Flow measurement (flow volume versus instantaneous)
3. Sample technique (manual versus automatic)

However, regardless of the kind of sample and technique used to collect it, the sample should be as representative as possible.

32.6.1 GRAB SAMPLE

A grab sample is a discrete individual sample taken within a short period of time. Analysis of the grab sample will characterize the quality of water at a given time. Composite samples reflect *average concentrations*, whereas a grab or discrete sample reflects conditions at the time the sample was collected.

32.6.2 COMPOSITE SAMPLE

A composite sample is a mixed or combined sample that is formed by combining a series of individual and discrete samples of specific volumes at specified intervals.

Composite samples can be developed based on time or flow. There are four types of composite samples.

CONSTANT TIME—CONSTANT VOLUME

Discrete samples of constant volume are collected at a fixed time interval. This type of composite will fail to provide a good average if the quality varies highly from hour to hour. This is also called a **simple composite** sample.

CONSTANT TIME—VOLUME PROPORTIONAL TO FLOW

Samples are taken at equal increments of time and are composited proportionally to *volume of flow* since the last sample was taken.

CONSTANT TIME—VOLUME PROPORTIONAL TO FLOW RATE

Samples are taken at equal increments of time and are composited proportionally to flow rate at the time each sample was taken.

CONSTANT VOLUME—TIME PROPORTIONAL TO FLOW

Samples of equal volume are taken at equal increments of flow volume and composited. This method is quite suited to using automatic samplers paced with flow. The frequency of sampling will change proportionally to the flow.

32.6.3 SAMPLE TECHNIQUE

Sample technique refers to the method by which a grab or composite sample is actually collected—manually or by using an automatic sampler. To make a flow proportion sample, hourly grabs are generally taken manually or using an automatic sampling system. Portions (aliquots) are then measured out in direct proportion to

---

TABLE 32.1
Flow-Proportioned Sample

Volume Proportional to	Equation
Flow rate	$A_i = V_c \times Q_i / \Sigma Q_i$
Flow volume	$A_i = V_c \times \Delta Q / \Sigma(\Delta Q)$

i = *number ID of the discrete sample*
V_c = *volume of the composite sample*
Q_i = *ith hourly flow rate or totalized flow volume*
$\Delta Q = Q_i - Q_{i-1}$
Σ = *sum of all the values*
Δ = *difference between two values*
A_i = *portion (aliquot) of the grab sample to be taken*

---

the flow rate at a given hour or flow volume since the last hour. To calculate the aliquot volumes to be taken from different hourly grabs to make up a composite sample of a desired volume, equations shown in Table 32.1 are used.

EXAMPLE PROBLEM 32.12

Hourly grabs were taken of the wastewater entering a treatment plant. The recorded flow rates are shown as follows. Tabulate the volume of aliquots to be used from the hourly grabs to produce a 5.0-L flow-proportioned composite sample.

Solution:

Time, t (h)	Sample, i #	Flow, Q_i (ML/d)	Aliquot, A_i (mL)
6:00	1	0.8	85
7:00	2	1.7	180
8:00	3	2.8	300
9:00	4	4.0	430
10:00	5	5.1	540
11:00	6	5.9	630
12:00	7	6.3	670
1:00	8	5.0	530
2:00	9	4.0	430
3:00	10	4.0	430
4:00	11	3.8	410
5.00	12	3.5	370
		$\Sigma = 46.9$	5005

Volume of aliquot (fifth discrete sample)

$$A_s = V_c \times \frac{Q_5}{\Sigma Q_i} = 5000 \ mL \times \frac{5.1}{46.9} = 543 = \underline{540 \ mL}$$

32.6.4 MATHEMATICAL COMPOSITING

In some situations, it may be desirable to study the variation of a given parameter over a certain period. To do this, each discrete (grab) sample is analysed separately. This will require more laboratory work but will result in more information. This information is useful in studying the hourly variation. The analytical results for all the discrete samples are then mathematically composited to yield weighted average value, as follows:

Composite C

$$\overline{C} = \frac{\Sigma(Q_i \times C_i)}{\Sigma Q_i}$$

Subscript i represents flow and concentration for the ith sample.

32.6.5 COMPOSITE CONCENTRATION

The previous formula can also be used to calculate the composite concentration of a given parameter when different flow streams are contributing to flow (Figure 32.4). This is especially true when industrial discharge makes up a significant portion of municipal wastewater.

Some industries, particularly the food processing industries, may have a greater impact on BOD loading in municipal wastewater. It is important to study the impact before a given industrial discharger is allowed to discharge waste into sanitary sewer systems.

EXAMPLE PROBLEM 32.13

At the head end of a treatment plant, grab samples were analysed separately. The analytical results and the flow rate data are shown in Table 32.2. Calculate the flow-weighted average of suspended solids concentration for the day.

Solution:

Sample calculations (fifth sample)

$$= \frac{60\ L}{s} \times \frac{120\ mg}{L} \times \frac{g}{1000\ mg} = 7.2\ g/s$$

Average daily low

$$\bar{Q} = \frac{\Sigma(Q_i)}{n} = \frac{880\ L/s}{12} = 73.3 = 73\ L/s$$

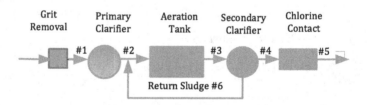

FIGURE 32.3 Sampling locations in a wastewater treatment plant.

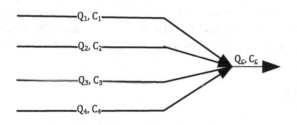

FIGURE 32.4 Composite concentration.

TABLE 32.2
Data Worksheet (Example Problem 32.13)

Time, t	i, #	Flow, Q_i, L/s	SS_i, mg/L	$Q_i \times SS_i$, g/s
0	1	50	55	2.75
2	2	40	45	1.8
4	3	30	45	1.35
6	4	40	80	3.2
8	5	60	120	7.2
10	6	100	150	15
12	7	120	220	26.4
14	8	110	250	27.5
16	9	100	180	18
18	10	90	160	14.4
20	11	80	180	14.4
22	12	60	220	13.2
24		50	55	2.75
	$\Sigma =$	880	1705	145.2

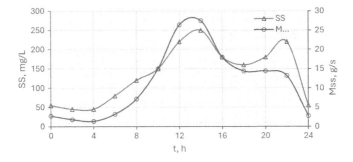

FIGURE 32.5 Hourly plot of SS concentration.

Average SS for the day

$$\overline{SS} = \frac{\Sigma(SS_i)}{n} = \frac{1705 \, mg/L}{12} = 142.0 = 140 \; mg/L$$

Weighted average SS concentration

$$\overline{C} = \frac{\Sigma(Q_i \times SS_i)}{\Sigma Q_i} = \frac{145.2 \, g}{s} \times \frac{s}{880 \, L} \times \frac{1000 \, mg}{g} = 165 = \underline{170 \; mg/L}$$

As seen in Figure 32.5, SS concentration and mass rate of SS follow similar trends; however, mass BOD has two distinct peaks. The weighted average of SS is 170 mg/L compared to a simple average of 140 mg/L.

EXAMPLE PROBLEM 32.14

At the head end of a wastewater pollution control plant, discrete samples of raw wastewater were collected over a period of an 8-h shift. Samples were analysed separately for determination of BOD. The analytical results and the flow rate data are tabulated in Table 32.3. Calculate the average flow, average BOD, and weighted average of BOD for the morning shift period.

Solution:

Calculations are made using MS Excel and shown in Table 32.3. In the last column of Table 32.3, the mass flow rate of BOD is shown. Accumulated values for the 8-h period are shown in the last row.

Average flow for the morning shift

$$\bar{Q} = \frac{\Sigma(Q_i)}{n} = \frac{10820 \; gpm}{8} = 1352 = 1400 \; gpm$$

Average BOD for the morning shift

$$\overline{BOD} = \frac{\Sigma(BOD_i)}{n} = \frac{1150 \; ppm}{8} = 143.7 = 140 \; ppm$$

Weighted average BOD concentration

$$\overline{BOD} = \frac{\Sigma(Q_i \times BOD_i)}{\Sigma Q_i} = \frac{1603300 \; ppm}{10820} = 148.1 = \underline{150 \; ppm}$$

The weighted average of the BOD of 150 ppm is greater than the simple average of 140 ppm.

TABLE 32.3
Data Worksheet (Example Problem 32.14)

Time, t	i, #	Flow Q_i, gpm	BOD_i, ppm	$Q_i \times BOD_i$
8:00	1	750	110	82500
9:00	2	900	120	108000
10:00	3	1120	140	156800
11:00	4	1500	170	255000
12:00	5	1600	200	320000
13:00	6	1800	150	270000
14:00	7	1650	140	231000
15:00	8	1500	120	180000
	Σ=	10820	Σ=	1603300

EXAMPLE PROBLEM 32.15

You are going to program an automatic sampler to collect a 2-gal composite sample over an 8-h period. If the discrete sample is collected every hour, what volume of aliquot (mL) should you program the sampler for?

Given:

$V = 2.0 \text{ gal} \qquad A_i =?$

Solution:

$$A_i = \frac{V}{\#} = 2.0 \text{ } gal \times \frac{h}{8h \times aliquot} = \frac{0.25 \text{ } gal}{aliquot} \times \frac{3.78 \text{ } L}{gal} \times \frac{1000 \text{ } mL}{L} = 945 \text{ } mL$$

PRACTICE PROBLEMS

PRACTICE PROBLEM 32.1

The water level in a given wet well is sufficiently low to allow shutting off all pumps. Before the pumps are restarted, the rise rate is observed to be 1.5 ft in 166 s. Estimate the inflow rate. The surface of the wet well is a square of 12 ft each side. (580 gpm).

PRACTICE PROBLEM 32.2

A wet well measures 3.0 m × 3.0 m. When the pump is on, the water level drops by 50 cm in 1 min and 50 s. During this period, inflow to the wet well is estimated to be 15 L/s. Determine the pumping rate. (56 L/s)

PRACTICE PROBLEM 32.3

During the month of April, an average of 36.6 gallons of screenings were removed each day. Find the total volume of screenings produced in the month of April. (150 ft³)

PRACTICE PROBLEM 32.4

In a sewage treatment plant, on average, 240 L of screenings are removed each day. If it is desired to empty the pit on a bi-weekly basis, what should be the minimum capacity of the screening pit? (3.4 m³)

PRACTICE PROBLEM 32.5

A new screening pit is being designed to replace the old one at a wastewater treatment plant. It is desired to size the new pit with a holding capacity of 30 d. If the screenings normally removed by the plant are 75 gal/day in the worst-case scenario, determine the size in ft³ the pit should be. (300 ft³)

PRACTICE PROBLEM 32.6

The average rate of screenings produced at a given facility is 190 L/d. The effective capacity (excluding soil cover) of the disposal pit is 38 m³. How many days will it take for the pit to be full? (20 d)

PRACTICE PROBLEM 32.7

Estimate the velocity of a grit channel if a stick travels 10 m in 30 s. What should be the length of the channel to provide a detention time of 1 min? (0.28 m/s, 17 m)

PRACTICE PROBLEM 32.8

What will be the flow velocity in a 3.0-ft-wide channel during the peak flow of 3.7 MGD when the depth of flow is 1.8 ft? (1.1 ft/s)

PRACTICE PROBLEM 32.9

A grit channel removed 800 L of grit during a period when the total flow is recorded to be 40,000 m³. Express the quantity of grit in L/1000 m³ of flow. Considering this average rate of grit generation, how long will a disposal pit of 50 m³ capacity last if the average daily flow is 12,000 m³/d? (20 L/1000 m³, 210 d)

PRACTICE PROBLEM 32.10

Repeat Example Problem 32.10, assuming the pit is of rectangular shape with length twice the width and the pit is emptied out every 8 weeks. (14 ft × 7 ft)

PRACTICE PROBLEM 32.11

It is intended to make a 2.5-L flow proportioned composite sample during the 8-h period of peak loading. Calculate the portions to be used from the noon hourly grab sample based on the given flow rate data. (350 mL)

Time	9:00	10:00	11:00	noon	1:.00	2:00	3:00	4:00	
Flow	90	104	113	116	112	106	100	95	$\Sigma = 836$

PRACTICE PROBLEM 32.12

How long will it take for a pump to fill an empty 3.0-m-radius tank to the 2.5 m level if the pumping rate is 5.0 L/s? (3.9 h)

PRACTICE PROBLEM 32.13

For the data of Example Problem 32.11, find the SS loading in kg for the 2-h period from 10 AM to 12 noon. (300 kg)

PRACTICE PROBLEM 32.14

Discrete samples collected every couple of hours were analysed for BOD, as shown. Calculate the flow-weighted BOD concentration for that day. (150 mg/L)

#	1	2	3	4	5	6	7	8	9	10	11	12
(mg/L)	67	60	56	90	120	180	190	190	180	150	140	100
(gpm)	5.0	4.0	3.0	4.0	6.0	10	12	11	10	9.0	8.0	6.0

PRACTICE PROBLEM 32.15

You are going to program an automatic sampler to collect a 5-L composite sample over a 24-h period. If a discrete sample is collected every hour, what volume of aliquot should you enter into the sampler? (210 mL)

33 Clarification

Sedimentation is the removal of settleable solids, as shown in Figure 33.1. This is a physical process which allows the suspended solids to settle under the force of gravity. This operation in wastewater treatment is more commonly referred to as **clarification** and is used as part of both primary and secondary treatment. The purpose of primary treatment is to remove the raw solids (primary or raw sludge), and the role of secondary treatment is to separate biological solids (activated sludge) and produce a clear effluent.

33.1 HYDRAULIC LOADING

The removal of solids by the clarifier depends on the solid content, nature of solids and volume of the wastewater being treated (hydraulic loading). The commonly used parameters to describe the hydraulic loading include hydraulic detention time, HDT, surface overflow rate, v_0 and weir loading WL. All solids with settling velocity greater than the overflow rate will settle out.

Hydraulic detention time

$$t_d = \frac{V}{Q} = \frac{d}{v_o} \mid (1-3h)$$

Overflow rate or surface loading

$$v_o = \frac{Q}{A_s} \quad (20-40 \ m^3/m^2.d) \ (500-1000 \ gal/ft^2.d)$$

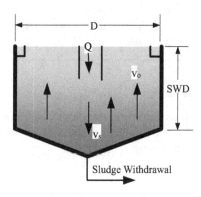

FIGURE 33.1 Hydraulic loading on a clarifier.

 DOI: 10.1201/9781003468745-33

The weir loading rate is used to determine the velocity of wastewater over the weir. The velocity informs the operator about the efficiency of the sedimentation process. At constant wastewater flow, the shorter the length of the weir, the faster the water velocity out of the basin. Conversely, the longer the weir length, the slower the velocity out of the basin. Excessive weir loading indicates high overflow velocities near the weir. As a result of a high weir overflow rate, light solids are pulled over the weirs and into the effluent troughs.

Weir loading

$$WL = \frac{Q}{L_W}(125 - 250\ m^3/m.d),\ (10000 - 20000\ gal/ft.d)$$

EXAMPLE PROBLEM 33.1

A circular clarifier has a diameter of 98 ft. If the entire circumference acts as a weir and the flow is 2.9 MGD, what is the weir overflow rate in gpd/ft?

Given:

D = 98 ft $\qquad$ Q = 2.9 MG/d $\qquad$ WL =?

Solution:

Length of the weir

$$L_w = \pi D = \pi \times 98\ ft = 307.8\ ft$$

Weir loading rate

$$WL = \frac{Q}{L_w} = \frac{2.9 \times 10^6\ gal}{d} \times \frac{1}{307.8\ ft} = 9419 = \underline{9400\ gal/ft.d}$$

EXAMPLE PROBLEM 33.2

A circular clarifier has a diameter of 55 ft. If the influent flow rate is 1.5 MG/d, compute the surface overflow rate in gal/ft².d and ft/h.

Given:

D = 55 ft $\qquad$ Q =1.5 MG/d $\qquad$ v_0=?

Solution:

Surface area

$$A = \frac{\pi D^2}{4} = \frac{\pi}{4} \times (55\ ft)^2 = 2375.84 = 2400\ ft^2$$

Overflow rate

$$v_o = \frac{Q}{A_s} = \frac{1.5 \, MG}{d} \times \frac{1}{2375.8 \, ft^2} = 631.3 = \underline{630 \, gal/ft^2.d}$$

$$= \frac{631.3 \, gal}{ft^2.d} \times \frac{ft^3}{7.48 \, gal} \times \frac{d}{24 \, h} = 3.51 = \underline{3.5 \, ft/h}$$

EXAMPLE PROBLEM 33.3

A clarifier capacity is designed to provide a surface overflow rate of 24 m³/m².d. If the side water depth is 2.5 m, what is the detention time in hours?

Given:

$$v_0 = 24 \, m^3/m^2.d \quad d = 2.5 \, m \quad HDT = ?$$

Solution:

Overflow rate is essentially the upward flow velocity with which water will rise in the clarifier before it overflows the weir. Detention time is the time water will take to rise and exit.

Hydraulic detention time

$$t_d = \frac{d}{v_o} = 2.5 \, m \times \frac{d}{24 \, m} \times \frac{24 \, h}{d} = 2.50 = \underline{2.5 \, h}$$

EXAMPLE PROBLEM 33.4

A circular clarifier receives a flow of 10 ML/d. A primary clarifier is 20 m in diameter, and the side water depth is 3.5 m. Calculate the hydraulic loading in terms of hydraulic detention time and weir loading.

Given:

$$Q = 13 \, ML/d \quad d = 3.5 \, m \quad D = 20 \, m \quad L_w = \pi \times 20 \, m = 62.8$$

Solution:

Detention time

$$t_d = \frac{V}{Q} = \frac{\pi}{4} \times (20 \, m)^2 = 3.5 \, m \times \frac{d}{13 \, ML} \times \frac{ML}{1000 \, m^3} \times \frac{24 \, h}{h} = 2.63 = \underline{2.6 \, h}$$

Weir loading

$$WL = \frac{Q}{L_w} = \frac{Q}{\pi D} = \frac{10 \, ML}{d} \times \frac{1000 \, m^3}{ML} \times \frac{1}{\pi \times 40 \, m} = 159.2 = \underline{160 \, m^3/m.d}$$

EXAMPLE PROBLEM 33.5

A rectangular clarifier 25 m long and 10 m wide receives a flow of 8.0 ML/d. Determine the overflow rate and horizontal flow velocity when depth of flow is 3.0 m.

Given:

$L = 25$ m $W = 10$ m $d = 3.0$ m $Q = 8.0$ ML/d

Solution:

Overflow rate

$$v_o = \frac{Q}{A_s} = \frac{A}{L \times W} = \frac{8000\,m^3}{d} \times \frac{1}{25\,m \times 10\,m} = 32.0 = \underline{32\ m^3/m.d}$$

Horizontal flow velocity

$$v_H = \frac{Q}{A_X} = \frac{A}{d \times W} = \frac{8000\,m^3}{d} \times \frac{1}{3.0\,m \times 10\,m} = 266.6 = \underline{270\,m/d}$$

EXAMPLE PROBLEM 33.6

In a wastewater treatment plant, it is desired that overflow rate not exceed 24 m³/m².d, and weir loading is not to exceed 180 m³/m·d. If the daily flow to the plant is 10 ML/d, what is the minimum diameter of the clarifier that would satisfy both conditions?

Given:

$Q = 10$ ML/d $D = ?$ WL = 180 m³/m.d $v_o = 24$ m³/m².d

Solution:

Based on overflow rate

$$D_{min} = \sqrt{\frac{4}{\pi} \times \frac{Q}{v_o}} = \sqrt{\frac{4}{\pi} \times \frac{10\,ML}{d} \times \frac{d}{24\,m} \times \frac{1000\,m^3}{ML}} = 23.03 = \underline{23\,m}$$

Based on weir loading

$$D_{min} = \frac{Q}{\pi \times WL} = \frac{1}{\pi} \times \frac{10\,ML}{d} \times \frac{m.d}{180\,m^3} \times \frac{1000\,m^3}{ML} = 17.68 = \underline{18\,m}$$

Hence, the minimum diameter is largest of the two values, that is, 23 m.

33.2 REMOVAL OF BOD AND SS

33.2.1 PRIMARY CLARIFICATION

The main purpose of unit operations is to remove contaminants. The percent removal of a given contaminant in terms of influent and effluent concentrations is given by the following equation:

Percent removal efficiency

$$PR = \frac{C_r}{C_i} = \frac{(C_i - C_e)}{C_i} \times 100\%$$

EXAMPLE PROBLEM 33.7

In an activated sludge plant, the final effluent BOD is 10 mg/L when the raw wastewater BOD is 190 mg/L. The activated sludge operation is carried out to remove 90% of BOD. What is the BOD removal by the primary treatment?

Given

$$PR_{II} = 90\% \qquad BOD_{RAW} = 190 \text{ mg/L} \qquad BOD_{FE} = 10 \text{ mg/L}$$

Solution:

Primary effluent BOD

$$BOD_{PE} = \frac{BOD_{FE}}{(1 - PR_{II})} = \frac{10 \text{ mg/L}}{(1 - 0.90)} = 100.0 = 100 \text{ mg/L}$$

Primary removal of BOD

$$PR_I = \frac{BOD_{RAW} - BOD_{PE}}{BOD_{RAW}} = \frac{(190 - 100) \text{ mg/L}}{190 \text{ mg/L}} = 0.473 = \underline{47\%}$$

33.3 VOLUME OF SLUDGE

The quantity of solids removed ends up in sludge. Based on the principle of mass conservation, the mass of dry solids in the sludge will equal the dry solids removed from the wastewater. The volume of the sludge produced depends on the solids removed SS_r and the concentration of solids SS_{sl} in the wet sludge, as seen in Figure 33.2.

Mass of solids in the sludge

$$M_{ss(r)} = Q \times SS_r = Q \times SS_i \times PR/100$$

33.3.1 PRIMARY OR RAW SLUDGE

If the concentration of solids in the sludge SS_{sl} is known, the daily production of sludge volume can be calculated. Concentration of SS in the sludge is generally expressed as percentage of mass. Multiplying by the density of the sludge, it can be converted to the mass of dry solids per unit volume of the sludge.

Volume of primary sludge

$$\boxed{Q_{Sl} = \frac{M_{ss(r)}}{SS_{sl}} = \frac{Q \times SS_r}{SS_{sl}} = \frac{Q \times SS_i \times PR/100}{SS_{sl}}}$$

33.3.2 VOLUME OF RAW SLUDGE

Another way of estimating the volume of raw sludge is by running the Imhoff cone test on the influent and effluent streams to the primary clarifier. The Imhoff cone test is discussed in Section 38.1 of the unit on laboratory tests. Based on the Imhoff cone test, the volume of settled solids in a given sample is read as mL/L of the sample. The difference in volume readings in influent and effluent indicates the volume of sludge, as shown.

Volume of primary sludge

$$\boxed{Q_{Sl} = Q \times (V_i - V_e)}$$

where V is the Imhoff cone reading, and Q is the daily wastewater flow rate.

EXAMPLE PROBLEM 33.8

Estimate the sludge pumped to a digester from a clarifier that treats a flow of 2.4 MGD. The influent Imhoff cone test indicates 12 mL/L and the effluent 0.5 mL/L.

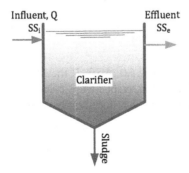

FIGURE 33.2 Solids removal and sludge production.

Given:

Q = 2.4 MGD V_i = 12 mL/L V_e = 0.50 mL/L

Solution:

Volume of sludge removed

$$V_r = V_i - V_e = 12 - 0.5 = 11.5 \ mL/L$$

Sludge pumping rate

$$Q_{Sl} = Q \times V_r = \frac{2.4MG}{d} \times \frac{11.5\,mL}{L} \times \frac{L}{1000\,mL} \times \frac{10^6\,gal}{MG} = 27600 = \underline{28000\,gal/d}$$

EXAMPLE PROBLEM 33.9

The flow to a primary clarifier is 800 m³/d. If the solid concentrations in the influent and effluent are 190 mg/L and 90 mg/L, respectively, how many kg of solids are removed by the primary treatment?

Given:

Q = 6800 m³/d SS_i = 190 mg/L SS_e = 90 mg/L

Solution:

SS removed by primary clarification

$$SS_r = SS_i - SS_e = 190 - 90 = 100.0 = 100 \ mg/L$$

Mass of SS removed by primary clarification

$$M_{ss(r)} = Q \times SS_r = \frac{6800\,m^3}{d} \times \frac{100\,g}{m^3} \times \frac{kg}{1000\,g} = \underline{680\,kg/d}$$

EXAMPLE PROBLEM 33.10

The flow to a primary clarifier is 5.0 MGD. If the SS concentrations in the influent and effluent are 190 and 110 ppm, respectively, calculate the removal efficiency and estimate the volume of raw sludge produced daily, assuming SS concentration in the sludge to be 4.0%.

Given:

Q = 5.0 MG/d SS_i (influent) = 190 mg/L SS_e = 110 mg/L   SS (sludge) = 4.0%

Solution:

SS removed by primary clarification

$$PR = \frac{(SS_i - SS_e)}{SS_i} \times 100\% = \frac{(190 - 110)}{110} \times 100\% = 72.7 = \underline{73\%}$$

Volume of primary or raw sludge

$$Q_{sl} = \frac{Q \times SS_r}{SS_{sl}} = \frac{5.0 \times 10^6 \, gal}{d} \times \frac{80}{10^6} \times \frac{100\%}{4.0\%} = 40\,000 \, gal/d$$

EXAMPLE PROBLEM 33.11

The SS content of the primary influent is 360 mg/L, and removal in the primary is 40%. How many kg of solids are removed as primary sludge if the average daily flow is 16800 m³/d? Sludge is pumped every 2 hours at the rate of 200 L/min. How long should the sludge pump be run each operation? Assume the solids concentration in the sludge to be 5.0%.

Given:

$Q = 16800 \, m^3/d$ $SS_i = 360$ mg/L $PR_I = 40\%$ $SS_{SL} = 5.0\% = 50 \, kg/m^3$
Sludge pumping cycle is every 2 hours @ 200 L/min.

Solution:

Mass of SS removed by primary clarification

$$M_{SS(r)} = Q \times SS_r = \frac{16800 \, m^3}{d} \times \frac{360 \, g}{m^3} \times \frac{kg}{1000 \, g} \times \frac{40\%}{100\%} = 2419 = \underline{2400 \, kg/d}$$

Volume of primary or raw sludge

$$V_{sl} = \frac{M_{SS}}{SS_{sl}} = \frac{2419 \, kg}{d} \times \frac{d}{24 \, h} \times \frac{m^3}{50 \, kg} = 2.015 \, m^3/h$$

Time period to operate sludge pump every hour

$$t = \frac{V}{Q} = \frac{2 \times 2.015 \, m^3}{2 \, h} \times \frac{1000 \, L}{m^3} \times \frac{min}{200 \, L} = 20.0 = \underline{20 \, min/2h}$$

EXAMPLE PROBLEM 33.12

In a wastewater treatment plant, the average daily flow to each primary unit is 1.3 MGD containing 250 mg/L of solids. Assuming a solids removal of 45% by settling, work out the volume of primary sludge produced, assuming the consistency of raw sludge is 3.5%.

Given:

$Q = 1.3$ MG/d   PR = 45%   SS = 250 mg/L $SS_{sl} = 3.5\%$

Solution:

Mass of SS removed by primary clarification

$$M_{SS} = Q \times SS_i \times PR = \frac{1.3\,MG}{d} \times \frac{250\,mg}{L} \times \frac{8.34\,lb/MG}{mg/L} \times \frac{45\%}{100\%} = 1219\,lb/d$$

Volume of primary or raw sludge produced

$$Q_{Sl} = \frac{M_{SS}}{SS_{sl}} = \frac{1219.7\,lb}{d} \times \frac{100\%}{3.5\%} \times \frac{gal}{8.34\,lb} = 4178 = \underline{4200\ gal/d}$$

EXAMPLE PROBLEM 33.13

Calculate the operating rate in minutes per hour for a primary sludge pump if the influent flow to the clarifier is 7.2 ML/d and SS in the primary influent and effluent, respectively, are 260 mg/L and 100 mg/L. The solid content of raw sludge is 4%, and the sludge pumping rate is 150 L/min.

Given:

$Q = 7.2$ ML/d $SS_i = 260$ mg/L , $SS_e = 100$ mg/L   SS= 4% = 40 kg/m³ $t = ?$

Solution:

Mass of SS removed by primary clarification

$$M_{SS(r)} = Q \times SS_r = \frac{7.2\,ML}{d} \times \frac{(260-100)\,kg}{ML} = 1152\ kg/d$$

Volume of primary sludge produced

$$V_{Sl} = \frac{M_{SS}}{SS_{sl}} = \frac{1152\,kg}{d} \times \frac{d}{24\,h} \times \frac{m^3}{40\,kg} = 1.2\ m^3/h$$

Time period to operate sludge pump every hour

$$t = \frac{V}{Q} = \frac{1.2\,m^3}{h} \times \frac{1000\,L}{m^3} \times \frac{min}{150\,L} = \underline{8.0\ min/h}$$

33.4   SOLIDS LOADING RATE (SLR)

The **solids loading rate** parameter is used to determine solids loading on activated sludge secondary clarifiers (Figure 33.3). It is defined as the mass flow rate of activated sludge solids contained in the mixed liquor per unit surface area of the clarifier. A typical solids loading rate is around 100 kg/m².d.

Solids rate loading

$$SLR = \frac{M_{SS}}{A_S} = \frac{(Q+Q_{RS}) \times MLSS}{A_S}$$

In activated sludge processes, the wastewater stream entering the secondary clarifier is large, as it also includes the return sludge flow. As seen in Figure 33.3, the suspended solids concentration in this flow stream is the same as of mixed liquor (MLSS) as this flow exits from the aeration tank.

EXAMPLE PROBLEM 33.14

A secondary clarifier is 25 m in diameter. The return sludge flow is 50% of the average flow of 10 ML/d. If the MLSS concentration is 2800 g/m³, calculate the solids loading rate on the clarifier.

Given:

$Q = 10$ ML/d   MLSS $= 2800$ g/m³ $R = 50\%$ $D = 25$ m

Solution:

Solids loading rate

$$SLR = \frac{(Q+Q_{RS}) \times MLSS}{A_S} = \frac{1.5 \times 10\, ML}{d} \times \frac{2800\, kg}{ML} \times \frac{4}{\pi(25\,m)^2} \times \frac{1000\, g}{kg}$$

$$= 85.6 = \underline{86\ kg/m^2.d}$$

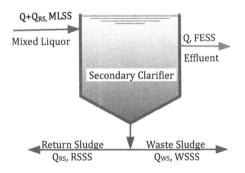

FIGURE 33.3   Solids loading rate.

EXAMPLE PROBLEM 33.15

A secondary clarifier handles a flow of 1.5 MGD with SS content (MLSS) of 3400 mg/L. The clarifier is 8.5 ft deep and 48 ft in diameter. In total flow of 1.5 MGD, one-third, is the return sludge flow. Find the detention time and solids loading rate.

Given:

$Q + Q_{RS} = 1.5$ MG/d $\qquad Q = 1.0$ MGD $\qquad$ MSSS $= 3400$ mg/L

$D = 48$ ft $\qquad\qquad\qquad$ SWD $= 8.5$ ft

Solution:

Detention time

$$t_d = \frac{V}{Q} = (48\,ft)^2 \times 8.5\,ft \times \frac{d}{1 \times 10^6\,gal} \times \frac{7.48\,gal}{ft^3} \times \frac{24\,h}{d} = 2.76 = \underline{2.8\,h}$$

Solids loading rate

$$SLR = \frac{(Q + Q_{RS}) \times MLSS}{A_s} = \frac{1.5\,MG}{d} \times \frac{3400\,mg}{L} \times \frac{8.34\,lb}{MG} \times \frac{4}{\pi(48\,ft)^2}$$

$$= 23.5 = \underline{24\ lb/ft^2.d}$$

EXAMPLE PROBLEM 33.16

A secondary clarifier handles a flow of 8.5 ML/d. To prevent solids leaving with the effluent, weir loading is kept low by providing an inboard channel. The clarifier is 3.5 m deep, and the inboard channel is set at radial distance of 7.5 m. Find the weir loading rate.

Given:

$Q = 8.5$ ML/d $\qquad$ $r_w = 7.5$ m $\qquad$ $D_w = 15$ m $\qquad$ WL =?

Solution:

In the case of the inboard channel, effluent enters from both sides such that the length of the weir is twice the circumference.

Weir loading rate

$$WL = \frac{Q}{L_W} = \frac{8.5\,ML}{d} \times \frac{1}{2\pi \times 15\,m} \times \frac{1000\,m^3}{ML} = 90.01 = \underline{90\,m^3/m.d}$$

33.5 SECONDARY CLARIFIER SLUDGE PRODUCTION

In secondary treatment, BOD in colloidal and dissolved form is converted to new biological growth, which is separated in the secondary clarifier. The amount of sludge wasted from the secondary clarifier is keyed in proportion to the new biological growth rate. The solids produced during secondary treatment depend upon many factors, including the organic matter (BOD) removed by the system and growth rate of the biomass. For a known BOD load, the biogrowth will depend upon the type of process or, more precisely, the F/M ratio. For most treatment plants, every kg of BOD broken down by the microorganism produces between 0.3 to 0.7 kg of new biological growth; these are solids that have to be removed (wasted) from the secondary treatment system. In the following example, calculations of the sludge waste rate are shown.

EXAMPLE PROBLEM 33.17

Determine the waste activated sludge (WAS) pumping rate in gpm if 3200 lb/day are to be wasted and the concentration of solids in waste sludge (WSSS) is 0.45%.

Given:

$WSSS = 0.45\%$ $M_{WS} = 3200$ lb/d $Q_{WS} = ?$

Solution:

Volume of secondary sludge to be wasted

$$Q_{ws} = \frac{M_{WS}}{WSSS} = \frac{3200\,lb}{d} \times \frac{gal}{8.34\,lb} \times \frac{100\%}{0.45\%} \times \frac{d}{1440\,min} = 59.2 = \underline{59\ gpm}$$

33.5.1 BOD Removed by Secondary Treatment

Based on removal efficiencies, the mass of BOD removed in the secondary treatment process can be worked out as follows:

Mass of BOD removed in secondary treatment

$$\boxed{M_{BOD(r)} = Q \times BOD_r = Q \times BOD_{raw} \times (1 - PR_I / 100) \times PR_{II} / 100}$$

PR_{II} = *percent BOD removal efficiency of the secondary treatment*
BOD_r = *BOD removed in the secondary treatment*
PR_I = *percent BOD removal by the primary treatment*

33.5.2 Mass of Solids Removed

Use the conversion factor (Y) to determine kg of solids (biomass). Y represents the fraction of BOD converted to new growth.

Mass of SS in the secondary sludge

$$M_{SS(sl)} = M_{BOD(r)} \times Y = Y \times Q \times BOD_{raw} \times (1 - PR_I/100) \times PR_{II}/100$$

EXAMPLE PROBLEM 33.18

4.5 ML of plant influent contains 200 mg/L of BOD. The average BOD removal efficiencies of primary and secondary treatment are 35% and 85% respectively. If the new growth rate is 0.35 kg of SS/kg of BOD removed, how many kg of dry sludge solids are produced each day?

Given:

$Q = 4.5$ ML/d $BOD_{raw} = 200$ mg/L $PR_I = 35\%$ $PR_{II} = 85\%$ $Y = 0.50$

Solution:

Mass of solids in the sludge

$$M_{SS(sl)} = Y \times Q \times BOD_{raw} \times (1 - PR_I/100)PR_{II}/100$$

$$= 0.35 \times \frac{4.5\,ML}{d} \times \frac{200\,mg}{ML} \times (1 - 0.35) \times 0.85 = 174.0 = \underline{170\,kg/d}$$

EXAMPLE PROBLEM 33.19

6.5 ML of plant influent contains 250 mg/L of BOD. The average BOD removal efficiencies of primary and secondary treatment are 30% and 80% respectively. If the new growth rate $Y = 0.40$ kg of SS/kg of BOD removed, how many kg of dry sludge solids are produced each day by the secondary treatment? Also calculate the volume of new sludge based on an SS content of 1.0%.

Given:

$Q = 6.5$ ML/d $BOD_{raw} = 250$ mg/L $PR_I = 30\%$ $PR_{II} = 80\%$
$Y = 0.40$ $SS_{SL} = 1\% = 10$ kg/m^3

Solution:

Mass of BOD removed by the secondary process

$$M_{SS(sl)} = Y \times Q \times BOD_{raw} \times (1 - PR_I/100)PR_{II}/100$$

$$= 0.40 \times \frac{6.5\,ML}{d} \times \frac{250\,mg}{ML} \times (1 - 0.30) \times 0.80 = 364 = \underline{360\,kg/d}$$

Volume of secondary sludge

$$V_{Sl} = \frac{M_{SS}}{SS_{sl}} = \frac{364\,kg}{d} \times \frac{m^3}{10\,kg} = 36.4 = \underline{36\,m^3/d}$$

EXAMPLE PROBLEM 33.20

The BOD content of the plant influent of 2.5 MGD is 220 mg/L. The average BOD removal efficiencies of primary and secondary treatment are 35% and 85% respectively. If the new growth rate Y = 0.35 lb of SS/lb of BOD removed, how many lb of dry sludge solids are produced each day by the secondary treatment? Also estimate the volume of new activated sludge produced, assuming solids concentration is 0.85%.

Given:

$Q = 2.5$ MG/d $BOD_{raw} = 220$ mg/L $PR_I = 35\%$ $PR_{II} = 85\%$

$Y = 0.35$ $SS_{SL} = 0.85\%$

Solution:

Mass of BOD removed by the secondary

$$M_{SS(sl)} = Y \times Q \times BOD_{raw} \times (1 - PR_I / 100) PR_{II} / 100$$

$$= 0.35 \times \frac{2.5\,MG}{d} \times \frac{220\,mg}{L} \times \frac{8.34\,lb/MG}{mg/L}(1 - 0.35) \times 0.85$$

$$= 887.0 = \underline{890\,lb/d}$$

Volume of secondary sludge

$$V_{Sl} = \frac{M_{SS}}{SS_{sl}} = \frac{887.0\,lb}{d} \times \frac{100}{0.85} \times \frac{gal}{8.34\,lb} = 12512 = \underline{13\,000\,gal/d}$$

PRACTICE PROBLEMS

PRACTICE PROBLEM 33.1

Wastewater flow during peak hours is 2.4 MGD. If the entire circumference of a 75-ft-diameter clarifier acts as a weir, what is the maximum weir overflow rate? (8900 gpd/ft)

PRACTICE PROBLEM 33.2

A wastewater treatment plant has daily average flow of 5.0 MGD and consists of two circular primary clarifiers, each with a radius of 40 ft. What is the surface overflow rate? (500 gal/ft²·d)

PRACTICE PROBLEM 33.3

A 30-m circular clarifier has a capacity of 2.5 ML. If the daily average flow to the clarifier is 20 ML/d, calculate the detention time and weir loading. (3 h, 210 m³/m·d)

PRACTICE PROBLEM 33.4

A secondary clarifier is 3.0 m deep on the side. It is large enough to allow a minimum of 2.0 h detention time during periods of peak flow. Estimate the overflow rate during the peak flow periods. (36 m³/m²·d)

PRACTICE PROBLEM 33.5

For the data of Example Problem 33.5, calculate detention time using: a) depth and surface overflow rate and b) volume and flow rate. (2.3 h)

PRACTICE PROBLEM 33.6

Rework Example Problem 33.6 for a daily wastewater flow of 15 ML/d. (28 m)

PRACTICE PROBLEM 33.7

In a trickling filter plant, it is desired to have a final effluent BOD not exceeding 20 mg/L when the raw wastewater BOD is 190 mg/L. On average, a primary removal of 40% is achieved at this plant. What should be the minimum removal by the trickling filter operation? (82%)

PRACTICE PROBLEM 33.8

A primary effluent flow is 15.6 ML/d containing 170 mg/L of BOD. If the secondary effluent BOD is 20 mg/L, how many kg of BOD are removed in the secondary treatment? (2340 kg/d)

PRACTICE PROBLEM 33.9

The SS removal efficiency of a primary clarifier is as high as 50%. The average daily flow to the plant is 2.5 MGD, with an SS concentration of 220 g/m³. Estimate the volume of raw sludge based on SS in the sludge to be 3.0%. (9200 gal/d)

PRACTICE PROBLEM 33.10

Calculate kg of dry solids removed by primary clarification in a primary treatment plant with an average daily flow of 7.5 ML/d containing 260 mg/L of suspended solids. Assume the removal efficiency of the primary clarifier is 60%. Also, calculate the volume of sludge (containing 4% dry solids) pumped in each shift of 8 hours. (1170 kg/d, 9.8 m³/shift)

PRACTICE PROBLEM 33.11

A primary sludge pump is operated every 2 hours. The pumping rate is 1.5 L/s. On average, a primary clarifier removes solids at the rate of 120 g/m³ from wastewater entering the clarifier. What will be the time of operation of the sludge pump in min/cycle if the daily average flow is 5.5 ML/d and the solid content of the primary sludge is 4.5%? (14 min/cycle)

PRACTICE PROBLEM 33.12

In a wastewater treatment plant, the daily flow is 4.5 MGD. This plant consists of three primary clarifiers. The suspended solids concentration in the raw wastewater entering the plant is 230 mg/L, and 50% is removed by primary clarification. Estimate the total production of raw sludge, assuming the raw sludge contains 2.5% of dry solids. (21,000 gal/d)

PRACTICE PROBLEM 33.13

Rework Example Problem 33.13 assuming the primary removal of suspended solids is 150 mg/L and the sludge consistency is 3.2%.

PRACTICE PROBLEM 33.14

Solve Example Problem 33.14 using a recirculation rate of 100%. (110 kg/m²·d)

PRACTICE PROBLEM 33.15

A secondary clarifier provided with an inboard channel for weir overflow treats a flow of 15 ML/d. The clarifier is 3.5 m deep, and the inboard channel is set at a radial distance of 10. m. Find the weir loading rate. (120 m³/m·d)

PRACTICE PROBLEM 33.16

A secondary clarifier handles a flow of 2.0 MGD with SS content (MLSS) of 2100 mg/L. The clarifier is 9 ft deep and 50 ft in diameter. In total flow, one half is the return flow. Find the detention time and solids loading rate. (3.2 h, 18 lb/ft²·d)

PRACTICE PROBLEM 33.17

In the operation of an activated sludge plant, 3650 lb of solids are wasted daily. Determine the waste activated sludge (WAS) pumping rate in gpm if the concentration of solids in waste sludge (WSSS) is 0.38%. (80 gpm)

PRACTICE PROBLEM 33.18

Solve Example Problem 33.15 assuming $Y = 0.30$ and the BOD removal efficiency of the secondary system is 90%. (160 kg/d)

PRACTICE PROBLEM 33.19

Solve Example Problem 33.16 assuming $Y = 0.30$, BOD removal efficiency = 90% and SS in the secondary sludge = 1.5%. (20 m³/d)

PRACTICE PROBLEM 33.20

Solve Example Problem 33.15 assuming $Y = 0.30$, BOD removal efficiency of the secondary treatment is 80% and secondary sludge contains 1.0% of dry solids (8600 gal/d)

34 Activated Sludge Process

34.1 INTRODUCTION

In Figure 34.1, a schematic of an activated sludge plant is shown. An activated sludge process is the most common type of secondary treatment used for treatment of wastewater. If properly operated, activated sludge should be able to produce plant effluent with a BOD concentration of 15 mg/L or less. Depending on the size of the plant and BOD loading, there are many variations of activated sludge, including conventional, tapered aeration, step aeration, contact stabilization, extended aeration and high rate.

As shown in Figure 34.1, The three main components of an activated sludge process are the aeration tank, secondary clarifier and return pump.

The **aeration tank** is the bioreactor. The contents of the tank are known as mixed liquor since they contain activated sludge and primary effluent containing unoxidized BOD. The activated sludge in the tank liquid biochemically oxidizes the organics not removed in the primary treatment. As a result of this, new biological growth takes place, and part of the BOD is converted to carbon dioxide and water.

Secondary clarification allows settling of solids in the mixed liquor and produces clear effluent. Its main purpose is to separate the activated sludge, which settles at the bottom, and supernatant is the secondary effluent.

A **return sludge pump** returns the activated sludge to the aeration tank to maintain a proper level of mixed liquor solids (microorganisms) in the aeration tank.

A **waste sludge pump**'s function is to pump some of the activated sludge to the sludge processing unit. The quantity of waste sludge depends on the new growth and level of solids maintained in the mixed liquor. As more growth takes place, there is a net increase in MLSS concentration in the aeration tank. Thus, at frequent intervals, some of the activated sludge is wasted to maintain a set level of MLSS concentration.

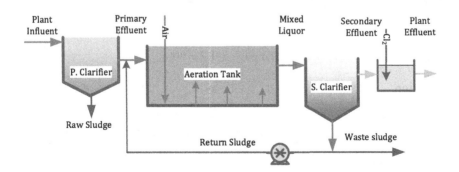

FIGURE 34.1 Schematic of an activated sludge plant.

DOI: 10.1201/9781003468745-34

443

34.2   HYDRAULIC LOADING

Hydraulic loading on the aeration tank is expressed in terms of **aeration period**, which is essentially the hydraulic detention time in the aeration tank. Return flow is not considered in the calculations. Aeration period is commonly expressed in hours.

Aeration period

$$AP = \frac{Volume\,under\,aeration}{Daily\,average\,flow\,rate} = \frac{V_A}{Q}$$

34.3   BOD LOADING

The mass of BOD going into the aeration tank is calculated by knowing the BOD concentration in the secondary influent. If the activated sludge process is proceeded by primary treatment with a percent removal of PR_1:

Mass rate of BOD

$$M_{BOD} = Q \times \frac{BOD_{Raw}\left(100 - PR_1\right)}{100}$$

BOD loading rate (BODLR) is expressed as kg of BOD per unit aeration volume per day. BODLR is expressed as $g/m^3 \cdot d$ and can range from as low as 200 for extended aeration to as high as 1500 for high-rate systems. (12–90 lb/1000 ft².d)

BOD loading rate

$$BODLR = \frac{M_{BOD}}{VA} = \frac{Q \times BOD_{PE}}{V_A} = \frac{BOD_{PE}}{AP}$$

EXAMPLE PROBLEM 34.1

An activated sludge plant has a daily flow of 4.0 MGD, and raw wastewater has a BOD content of 220 mg/L. Assuming 30% removal of BOD by the primary treatment, find the lb of BOD entering the aeration tanks.

Given:

$Q = 4.0\,MG/d$, $BOD_{raw} = 220$ mg/L $PR = 30\%$

Solution:

BOD loading rate

$$M_{BOD} = Q \times BOD_{raw}\left(1 - \frac{PR}{100\%}\right) = \frac{4.0\,MG}{d} \times \frac{220\,mg}{L} \times \left(1 - \frac{30\%}{100\%}\right) \times \frac{8.34\,lb/MG}{mg/L}$$
$$= 5137 = \underline{5100\,lb/d}$$

EXAMPLE PROBLEM 34.2

In a conventional activated sludge plant, aeration capacity is 5000 m³. Primary effluent BOD content is 150 g/m³. Daily average flow is 20 ML/d. What is the BOD loading rate?

Given:

$Q = 20$ ML/d $= 20\,000$ m³/d   BOD $= 150$ g/m³ $V_A = 5000$ m³

Solution:

BOD loading rate

$$BODLR = \frac{M_{BOD}}{V} = \frac{20000\,m^3}{d} \times \frac{150\,g}{m^3} \times \frac{1}{5000\,m^3} = 600.0 = \underline{600\,g/m^3.d}$$

EXAMPLE PROBLEM 34.3

A rectangular aeration basin is 25 m × 10 m × 5.0 m. Calculate the aeration period for a daily flow of 3.0 ML/d.

Given:

Capacity $= 25$ m × 10 m × 5.0 m $Q = 3.0$ ML/d   AP $= ?$

Solution:

Volume under aeration

$$V_A = 25\,m \times 10\,m \times 5\,m = 750.0 = 750\,m^3$$

Aeration period

$$AP = \frac{V}{Q} = 750\,m^3 \times \frac{d}{3.0\,ML} \times \frac{24\,h}{d} \times \frac{ML}{1000\,m^3} = \underline{6.0\,h}$$

EXAMPLE PROBLEM 34.4

It has been decided to provide three aeration tanks, each 3 m deep, to treat a flow of 10 ML/d. If the length to width ratio is kept 3 to 1, what should be the length of each tank to provide an aeration period of 8 hours?

Given:

d $= 3$ m   # $= 3$ $Q = 10$ ML/d   AP $= 8$ h   L/W $= 3$

Solution:

Capacity of each aeration tank

$$V_A = AP \times Q = \frac{8.0\,h}{3\,tanks} \times \frac{10\,ML}{d} \times \frac{d}{24\,h} \times \frac{1000\,m^3}{ML} = \underline{1110\,m^3/tank}$$

Length of the aeration tank

$$L = \sqrt{\frac{3V}{d}} = \sqrt{\frac{3 \times 1110\,m^3}{3.0\,m}} = 33.3 = \underline{33\,m}$$

If the length is 33 m, then width would be one-third, that is, 11 m.

EXAMPLE PROBLEM 34.5

The BOD concentration of the raw wastewater is 250 g/m³, and the BOD removal efficiency of the primary treatment is 40%. If the daily plant flow is 6.6 ML/d and the total aeration volume is 2.2 ML, calculate the BOD loading rate on the aeration in g/m³. d.

Given:

$BOD_{raw} = 250$ g/m³ $PR_I = 40\% = 0.40$ $Q = 6.6$ ML/d $V_A = 2.2$ ML

Solution:

BOD of primary effluent

$$BOD_{PE} = \frac{250g}{m^3} \times \left(1 - \frac{40\%}{100\%}\right) = 150.0 = 150\,g/m^3$$

BOD loading rate

$$BODLR = \frac{M_{BOD}}{V} = \frac{150\,kg}{ML} \times \frac{6.6\,ML}{d} \times \frac{1}{2200\,m^3} = \underline{450\,g/m^3.d}$$

Aeration period

$$AP = \frac{V}{Q} = 2.2\,ML \times \frac{d}{6.6\,ML} \times \frac{24h}{d} = 8.00 = \underline{8.0h}$$

34.4   FOOD TO MICROORGANISMS RATIO (F/M)

The food to microorganisms ratio (**F/M ratio**) is one of the important process control parameters. For a particular system, the F/M ratio depends mainly on the type of activated sludge process and the characteristics of the wastewater entering the system. As the name indicates, the F/M ratio determines the balance between food entering the aeration tank (M_{BOD}) and the microorganisms as measured by the mass of solids in the mixed liquor (m_{MLSS}).

Food to microorganism ratio

$$\frac{F}{M} = \frac{M_{BOD}}{m_{MLSS}} = \frac{Q \times BOD_{PE}}{MLSS \times V_A} = \frac{BOD_{PE}}{MLSS} \times \frac{1}{AP}$$

34.4.1 COMMENTS

- COD test being matter of hours as compared to BOD test which takes five days, COD value is sometimes used as a measure of organic load. This is only valid if the BOD/COD ratio is constant for a given wastewater. The BOD/COD ratio is typically 0.6 for municipal wastewater.
- It is sometimes preferred to use a volatile fraction of mixed liquor solids (MLVSS) to represent microorganisms. The MLVSS concentration can be calculated by knowing MLSS and percent volatile solids content.
- Because the F/M ratio is mass rate divided by mass, its units are that of the inverse of time, usually per day. The F/M ratio of the conventional activated sludge process is typically in the range of 0.2–0.5/d.
- When operating an activated sludge process, it must be maintained within the recommended F/M range by controlling the solids in the aeration basin. Failure to do so may result in poor settleability.
- The F/M ratio should preferably be calculated based on moving average.
- Changes should not be made abruptly. At one time, do not adjust a given parameter more than 10–15%.
- A high F/M ratio is indicative of young sludge, whereas a low F/M ratio produced is the mark of well-oxidized sludge.
- The F/M ratio directly affects the settleability of sludge. A low F/M ratio produces quick-settling sludge, whereas a high F/M ratio causes sludge bulking.

EXAMPLE PROBLEM 34.6

A conventional activated sludge treatment plant receives a daily flow of 20 000 m³/d. Given that the BOD of the primary effluent is 170 mg/L, calculate the F/M ratio for an aeration volume of 3500 m³ and MLSS of 2000 g/m³.

Given:

$$Q = 20\ 000\ m^3/d \quad BOD = 170\ g/m^3 \quad MLSS = 2000\ g/m^3 \quad F/M = ?$$

Solution:

Food to microorganism ratio

$$\frac{F}{M} = \frac{Q \times BOD_{PE}}{MLSS \times V_A} = \frac{170\ g/m^3}{2000\ g/m^3} \times \frac{20000\ m^3}{d} \times \frac{1}{3500\ m^3} = 0.485 = \underline{0.49/d}$$

EXAMPLE PROBLEM 34.7

An activated sludge aeration tank receives a primary effluent flow of 5.5 MGD with a BOD of 140 mg/L. The mixed liquor solids concentration is 1900 mg/L, and the aeration tank capacity is 0.65 MG. Calculate the F/M ratio.

Given:

$Q = 5.5 \text{ MG/d}$ $V_A = 0.65 \text{ MG}$ $MLSS = 1900 \text{ g/m}^3$ $BOD = 140 \text{ g/m}^3$

Solution:

Food to microorganism ratio

$$\frac{F}{M} = \frac{Q \times BOD_{PE}}{MLSS \times V_A} = \frac{140 \, g/m^3}{1900 \, g/m^3} \times \frac{5.5 \, MG}{d} \times \frac{1}{0.65 \, MG} = 0.623 = \underline{0.62 \, / \, d}$$

EXAMPLE PROBLEM 34.8

The flow to a 3200-m³ high-rate aeration system is 3.0 ML/d. The BOD concentration of the primary effluent is 350 g/m³. What should be the concentration of MLSS to achieve an F/M ratio of 1.0/d?

Given:

$Q = 3.0 \text{ ML/d}$ $V_A = 320 \text{ m}^3 = 0.32 \text{ ML}$ $BOD = 350 \text{ mg/L}$ $F/M = 1.0/d$

Solution:

Mixed liquor solids concentration

$$MLSS = \frac{Q \times BOD_{PE}}{(F/M) \times V_A} = \frac{3.0 \, ML}{d} \times \frac{1}{0.32 \, ML} \times \frac{350 \, mg}{L} \times \frac{d}{1.0} = 3281 = \underline{3300 \, mg/L}$$

EXAMPLE PROBLEM 34.9

An extended aeration plant has an influent BOD of 200 g/m³ and an aeration period of 36 hours. What size tank would be required to serve a trailer park with 15 hook-ups, averaging 2.0 people per trailer? Also find the operating F/M ratio if the system is to be operating by maintaining MLSS of 1500 mg/L. Due to less convenience available at the trailers, assume a per capita wastewater production of 200 L/person.d.

Given:

$BOD = 200 \text{ mg/L}$ $AP = 36 \text{ h}$ $Pop = 15 \times 2 = 30 \text{ p}$ $Q = 200 \text{ L/p·d}$

Solution:

Daily flow rate

$$Q = \frac{200 \, L}{p.d} \times \frac{2 \, p}{trailer} \times 15 \, trailers \times \frac{m^3}{1000 \, L} = 6.00 = 6.0 \, m^3/d$$

Volume under aeration

$$V_A = AP \times Q = 36\,h \times \frac{6.0\,m^3}{d} \times \frac{d}{24\,h} = 9.00 = \underline{9.0\,m^3}$$

Food to microorganism ratio

$$\frac{F}{M} = \frac{1}{AP} \times \frac{BOD_{PE}}{MLSS} = \frac{1}{36\,h} \times \frac{200\,g\,/\,m^3}{1500\,g\,/\,m^3} \times \frac{24\,h}{d} = 0.0888 = \underline{0.089\,/\,d}$$

34.5   SOLIDS RETENTION TIME

Solids retention time (SRT) represents the average length of time for which bio-solids are held in the system before being discharged as part of secondary effluent or waste-activated sludge flow stream. For this reason, this parameter is also called **mean cell residence time** (MCRT). Much like hydraulic detention time is volume per unit volume flow rate, SRT is solids in the system divided by the rate at which solids exit the system, as shown in Figure 34.2.

SRT is much larger than HDT because the microorganisms (activated sludge) stay longer in the system than the flowing wastewater. Whereas water has only one pass, activated sludge has numerous passes before being discharged out. SRT is expressed in days and can range from 5–15 days for the majority of activated sludge treatment systems.

For efficient BOD removal, conventional and contact stabilization plants should be operated in an SRT range of 2–5 days, with the upper limit used for winter operation. If nitrification is required, then processing SRT over 5 days must be maintained. Like the F/M ratio, SRT can be adjusted by varying the quantity waste sludge. Since a biological system can be easily upset, any changes in the waste sludge pumping rate should not be greater than 10–15% of the previous day rate. For the same reason, pumping more frequently or continuously

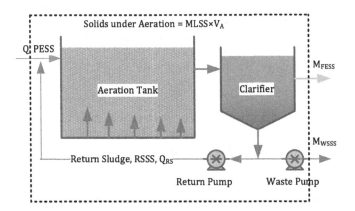

FIGURE 34.2   Solids retention time.

is preferred to wasting in one or two batches. To make SRT calculations, you should know:

1. Total solids in the system, that is, bio-solids in the aeration tank and secondary clarifier.
2. Rate at which solids are leaving the system, which includes daily discharge of solids in the final effluent and waste-activated sludge.

34.5.1   SRT FORMULA

SRT can be determined by knowing the mass of bio-solids in the system and the rate at which these solids exit the system. Mostly solids are wasted as part of the recirculated or returned sludge. In that case, the solids concentration in the waste sludge is the same as that of return sludge.

Solids retention time

$$SRT = \frac{Solids\,in\,the\,system}{Solids\,exiting\,rate} = \frac{(V_A + V_C) \times MLSS}{WSSS \times Q_{WS} + FESS \times Q_{FE}}$$

V_C = *volume of the clarifier*
V_A = *volume of the aeration tank*
$MLSS$ = *mixed liquor solids concentration*
$WSSS$ = *SS in the waste activated sludge*
$FESS$ = *solids concentration in the final effluent*
Q = *daily wastewater flow rate*
Q_{WS} = *waste-activated sludge pumping rate*

34.5.2   GENERAL FORM

The previous expression is the most general formula for calculating SRT. The terms in the numerator and the denominator, respectively, give the mass of solids in the system and the mass rate at which solids exit the system. Dividing mass (kg) by mass rate (kg/d) yields the unit of time (d). The general form of the SRT formula is rarely used. Usually, the terms V_C, FESS or both are relatively small in magnitude and are ignored.

34.5.3   MODIFIED SRT EQUATIONS

The general form of the equation is not warranted in most cases. The terms like volume of mixed liquor in the final clarifier and concentrations of solids in the final effluent are relatively small and are ignored.

34.5.3.1   Long Form

Usually, the solids in the final clarifier are not considered. This is true when the volume of the clarifier is relatively small.

Solids retention time

$$SRT = \frac{V_A \times MLSS}{WSSS \times Q_{WS} + FESS \times Q_{FE}}$$

34.5.3.2 Short Form

SRT calculations become easy if the solids leaving as part of the final effluent are ignored. This will be more valid if the FESS concentration is relatively low. Ignoring FESS:

Solids retention time

$$SRT = \frac{V_A \times MLSS}{WSSS \times Q_{WS}}$$

34.5.4 POINTS TO KEEP IN MIND

- When the denominator consists of two terms, evaluate each term first before adding them together. The final units of both the terms should be the same, for example, lb/d or kg/d.
- When Q and V are expressed in ML, write all the concentrations expressed in mg/L as kg/ML.
- Similarly, when volume and flow units are in m³, write concentrations in g/m³. This will allow the direct cancellation of volume and concentration units, leaving you with units of time.
- When concentration of solids in the return sludge (RSSS) is specified, replace WSSS with RSSS.

EXAMPLE PROBLEM 34.10

Determine the solids retention time in days considering the solids in the secondary clarifier.

Given:

$V_A = 6.0$ ML   MLSS = 2500 mg/L $V_C = 0.42$ ML   WSSS = 8100 mg/L
$Q = 12$ ML/d   FESS = 14 mg/L $Q_{WS} = 0.24$ ML/d

Solution:

Solids retention time

$$SRT = \frac{(V_A + V_C) \times MLSS}{WSSS \times Q_{WS} + FESS \times Q_{FE}}$$

$$= \frac{(6.0 + 0.42) ML \times 2500 \ kg}{ML} \times \frac{d}{(8100 \times 0.24 + 14 \times 12) kg}$$

$$= 1.207 = \underline{1.2 \, d}$$

EXAMPLE PROBLEM 34.11

An aerator is 23 ft deep with a holding capacity of 1.3 MG. The MLSS concentration is maintained at 2500 mg/L, and the return activated sludge solids concentration (RSSS) is 5000 mg/L. If 120,000 gal of the activated sludge is wasted daily, find the solids retention time.

Given:

$V_A = 1.3$ MG   MLSS $= 2500$ g/m^3 WSSS $=$ RSSS $= 5000$ g/m^3
$Q_{WS} = 120,000$ gal/d

Solution:

Solids retention time

$$SRT = \frac{MLSS}{WSSS} \times \frac{V_A}{Q_{WS}} = \frac{1.3 \times 10^6 \, gal}{120000 \, gal/d} \times \frac{2500 \, mg/L}{5000 \, mg/L} = 5.41 = \underline{5.4 \, d}$$

EXAMPLE PROBLEM 34.12

The aeration tank is 7.0 m deep with a surface area of 650 m^2. The MLSS concentration is 2500 mg/L, and the solids concentration in the return activated sludge is 6000 mg/L. If 500 m^3 of activated sludge is wasted per day and the final effluent has a SS concentration of 10 mg/L at 7000 m^3/d, calculate SRT in days.

Given:

$d = 7.0$ m $A_s = 650$ m^2 MLSS $= 2500$ g/m^3 WSSS $= 6000$ g/m^3
$Q = 7000$ m^3/d $Q_{WS} = 500$ m^3/d   FESS $= 10$ g/m^3

Solution:

Solids retention time

$$SRT = \frac{V_A \times MLSS}{WSSS \times Q_{WS} + FESS \times Q_{FE}}$$

$$= \frac{650 \, m^2 \times 7.0 \, m \times 2500 \, g}{m^3} \times \frac{d}{(6000 \times 500 + 10 \times 7000) \, g}$$

$$= 3.705 = \underline{3.7 d}$$

34.6   ACTIVATED SLUDGE PROCESS CONTROL

The quality of the effluent of an activated sludge system will depend on the characteristics and quality of wastewater as well as how the actual process is controlled. Two important operational parameters are:

1. Return sludge (Q_{RS}, RSSS)
2. Waste sludge (Q_{WS}, WSSS)

The important thing to remember is that each of these factors is related to the other, and the impact on all process variables must be considered before changing one variable. For conventional activated sludge process, the return rate is generally about 20%–50% of the incoming wastewater flow. Based on the settling characteristics of the mixed liquor solids, you may need to change the return rate of the activated sludge. Return rate is generally expressed as a percentage of influent flow.

34.6.1 RETURN SLUDGE RATE

The return of activated sludge from the secondary clarifier to the aeration tank makes it possible for the microorganisms to be in the system longer than the flowing wastewater.

Based on the **settleability test**, the return rate of the activated sludge can be estimated. This assumes that whatever sludge is settling down needs to be returned to the aeration tank and that the settleability test completely represents the secondary clarifier. Usually, actual return rates are higher than hypothetical rates. Assuming the settleability test accurately simulates the settling of mixed liquor solids in the clarifier, the return rate of activated sludge should be in proportion to the volume of settled sludge, as shown in Figure 34.3.

Hypothetical recirculation

$$R_{hyp} = \frac{Q_{RS}}{Q} = \frac{V_{sl}}{1000 - V_{sl}}$$

V_{Sl} = *volume occupied by the sludge in mL*
R_{HYP} = *hypothetical recirculation ratio*

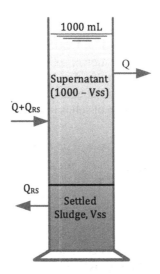

FIGURE 34.3 Settleability test and recirculation ratio.

Caution: Return rate based on a settleability test may not be very accurate, as the graduated cylinder in some types of activated sludge processes may not truly represent the final clarifier.

EXAMPLE PROBLEM 34.13

The settleability test indicates a settling volume of 180 mL/L. Based on this, what recirculation ratio would you recommend?

Given:

$V_{SL} = 180$ mL/L $R = ?$

Solution:

Hypothetical recirculation ratio

$$R_{hyp} = \frac{Q_{RS}}{Q} = \frac{V_{sl}}{1000 - V_{sl}} = \frac{180}{1000 - 180} = 0.219 = \underline{0.22 = 22\%}$$

34.6.2 SLUDGE VOLUME INDEX

Sludge volume index (SVI) is a better way of defining the settleability of activated sludge. It is defined as the volume of sludge containing 1 gram of dry solids. The volume of sludge as measured in the test contains solids as indicated by the MLSS concentration. Dividing V_{SL} from the settleability test by the MLSS and converting the units to mL/g yields the value of the SVI.

Sludge volume index

$$\boxed{SVI = \frac{V_{sl}}{MLSS} \, expressed \, as \, g/mL}$$

SVI basically is the volume occupied by 1 gram of dry solids in the sludge. In that sense, SVI is inverse of the concentration of solids in the return sludge. For the same reason, some people prefer to use **sludge density index** (SDI), which is the reciprocal of SVI. When the return ratio is set equal to a hypothetical value, SDI should represent the expected concentration of solids in the return sludge.

Sludge density index

$$\boxed{SDI = \frac{1}{SVI} \quad or \quad SDI\% = \left(\frac{100}{SVI}\right)}$$

$$\boxed{R = R_{hyp}, \, RSSS = SDI \quad R < R_{hyp}, \, RSSS < SDI}$$

When R is not equal to R_{HYP}, RSSS concentration is based on the actual return ratio. When the actual rate of return is significantly higher than the hypothetical rate, the return sludge will be relatively thin.

EXAMPLE PROBLEM 34.14

The volume of settled solids in a 30-min settling test was read to be 300 mL/L. A solids test on the same sample of mixed liquor yielded a concentration of total solids of 3000 mg/L. Calculate SVI, SDI and minimum return rate Q_{RS}.

Given:

V_{ssl} = 300 mL/L   MLSS = 3000 mg/L = 3.0 g/L

Solution:

Sludge volume index

$$SVI = \frac{V_{SI}}{MLSS} = \frac{300\ mL}{L} \times \frac{L}{3000\ mg} \times \frac{1000\ mg}{g} = \underline{100\ mL/g}$$

Sludge density index

$$SDI = \frac{1}{SVI} = \frac{g}{100\ mL} \times \frac{mL}{1.0\ g} \times 100\% = 1.00 = \underline{1.0\%}$$

Hypothetical return ratio

$$R_{hyp} = \frac{V_{sl}}{1000 - V_{sl}} = \frac{300\ ml}{L} \times \frac{L}{(1000 - 300)\ mL} \times 100\% = 42.8 = \underline{43\%}$$

34.6.3 SOLIDS MASS BALANCE

Based on the mass balance around the final clarifier, the concentration of RSSS may be estimated by knowing the MLSS concentration and return ratio (Figure 34.4).

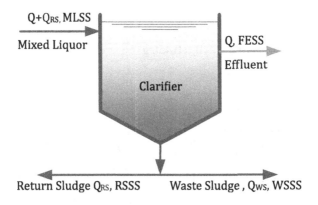

FIGURE 34.4 Solids mass balance.

Mass balance equation

$$(Q+Q_{RS})MLSS = (Q_{RS}+Q_{WS})RSSS+(Q-Q_{WS})FESS$$

Under the steady state conditions, the wastage rate is such that the solids discharged balance the new solids produced in the aeration basin; thus, we need not consider Q_{WS}. Dropping the terms Q_{WS} and FESS (negligible solids leave the system in the clarifier effluent flow):

> *The real value of SVI is to indicate the stability of the activated sludge process.*

Mass balance yields

$$\frac{R}{1+R} = \frac{MLSS}{RSSS} \quad or \quad R = \frac{Q_{RS}}{Q} = \frac{MLSS}{RSSS-MLSS}$$

The above expression shows that, return rate can be set independently from the settleability test. However, there is an inherent assumption that all the solids in the mixed liquor are separated by settling at the bottom of the clarifier. For a given concentration of MLSS, the recirculation ratio will be high for a low concentration of solids in the return sludge and vice versa. In other words, high recirculation or return rates will result in the relatively diluted return activated sludge.

Note: This equation will not hold true during big surges, as steady state conditions will be disturbed. During peak flow conditions due to high influx of mixed liquor into the secondary clarifier, solids loading is increased suddenly.

EXAMPLE PROBLEM 34.15

A solids test is performed on a 50-mL sample of mixed liquor using a dish weighing 0.300 g. After drying, the sample weighs 0.450 g. If the return rate is maintained at 40%, calculate the expected solids concentration in the return sludge.

Given:

V = 50 mL   A = 0.450 g   B = 0.450 g

Solution:

Mixed liquor solids concentration

$$MLSS = \frac{m_{ss}}{V} = \frac{(0.450-0.300)g}{50.0\,mL} \times \frac{1000\,L}{m^3} \times \frac{1000\,mL}{L} = 3000\,g/m^3$$

Return sludge solids concentration

$$RSSS = MLSS = \frac{(1+R)}{R} \times \frac{3000\,mg}{L} \times \frac{(1+0.40)}{0.40} = 11000\,mg/L = \underline{1.1\%}$$

EXAMPLE PROBLEM 34.16

An SVI test performed on a sample of mixed liquor from a step-aeration activated sludge treatment system yields a sludge volume of 220 mL/L after 30 min of settling. A suspended solids test run on the same sample indicates an MLSS concentration of 2400 mg/L. What is the SVI? Determine the recirculation ratio based on the settling test and the predicted concentration of RSSS.

Given:

V_{SL} = 220 mL/L MLSS = 2400 mg/L

Solution:

Sludge volume index

$$SVI = \frac{V_{sl}}{MLSS} = \frac{220 \ mL}{L} \times \frac{L}{2400 \ mg} \times \frac{1000 \ mg}{g} = 92.0 = \underline{92 \ mL/g}$$

Hypothetical return ratio

$$R_{hyp} = \frac{V_{sl}}{1000 - V_{sl}} = \frac{220 \ ml}{L} \times \frac{L}{(1000 - 220) \ mL} = 0.28 = \underline{28\%}$$

Return sludge solids concentration

$$RSSS = MLSS \left(\frac{1+R}{R} \right) = \frac{2400 \ mg}{L} \times \frac{(1+0.28)}{0.28}$$
$$= 10971 = 11000 \ mg/L = 11 \ g/L = \underline{1.1\%}$$

EXAMPLE PROBLEM 34.17

Given the following, calculate the return sludge pumping rate in m³/h in a conventional activated sludge treatment system.

Given:

MLSS = 2000 mg/L   RSSS = 7900 mg/L   Q = 2.5 ML/d

Solution:

Return ratio

$$R = \frac{MLSS}{RSSS - MLSS} = \frac{2000}{(7900 - 2000)} = 0.34 = 34\%$$

Rate of return sludge

$$Q_R = 2.5 \ ML/d \times 0.34 = 0.85 = \underline{0.85 \ ML/d}$$

34.6.4 Sludge Blanket

One way of controlling the return sludge rate is by observing the sludge blanket in the final clarifier. Another method is based on the settleability test. Settleability is defined as the percentage of volume occupied by the sludge after 30 minutes of settling. A sample of mixed liquor for suspended solids testing is withdrawn from the aeration basin effluent.

34.6.5 Sludge Wasting Rate

In order to maintain balance between mixed liquor as indicators of microorganisms, it will be necessary to waste some activated sludge. The wasting can be done on a continuous basis or in batches, which is usually done during low flow periods. The volume of sludge to be wasted can be estimated based on SRT control, F/M control and MLSS control.

34.6.6 SRT Control

The activated sludge process is controlled by calculating the amount of activated sludge that should be wasted for a target value of SRT. Setting the parameter SRT also fixes the type of microorganisms predominant in the system and the extent to which nitrification may occur in the activated sludge process. The amount of sludge to be wasted can be determined if the suspended solids in the mixed liquor (MLSS) and return sludge (RSSS) are measured. If the return ratio is known, the RSSS can be estimated using the mass balance equation, as discussed in the previous section. For example, if the return sludge rate is 25% of the wastewater flow rate, return sludge will be five times thicker than mixed liquor.

Concentration factor

$$\frac{RSSS}{MLSS} = \frac{(1+R)}{R} = \frac{(1+0.25)}{0.25} = 5$$

That is to say, when the return rate is one-fourth of the influent flow, the clarifier can be expected to concentrate the mixed liquor by a factor of five. Under these operating conditions, the return sludge solids concentration is five times thicker than that of mixed liquor. An increase in the return rate will cause the return sludge to be thinner.

34.6.7 Expressions for Calculating Sludge Wastage Rate

When employing SRT control, you can figure out the amount of sludge that needs to be wasted to achieve a set or target value of SRT. The general form of the SRT equation is:

Solids retention time

$$SRT = \frac{(V_A + V_C) \times MLSS}{WSSS \times Q_{WS} + FESS \times Q_{FE}}$$

Using this equation, the sludge wastage rate can be expressed in terms of MLSS and RSSS. Based on variations of the SRT formula, the sludge wastage rate can be found using the following expressions.

34.6.7.1   General Form

Sludge wasting rate

$$Q_{WS} = \frac{(V_A + V_C)}{SRT} \times \frac{MLSS}{RSSS} - Q \times \frac{FESS}{RSSS}$$

34.6.7.2   Commonly Used Form

Since the volume of the final clarifier is small compared to the volume of the aeration tank, the previous equation is modified by dropping V_C:

Sludge wasting rate

$$Q_{WS} = \frac{V_A}{SRT} \times \frac{MLSS}{RSSS} - Q \times \frac{FESS}{RSSS}$$

34.6.7.3   Short Form

Since final effluent solids are negligibly small compared to MLSS, so the term FESS can be dropped. When FESS is negligible, it becomes:

Sludge wasting rate

$$Q_{WS} = \frac{V_A}{SRT} \times \frac{MLSS}{RSSS}$$

EXAMPLE PROBLEM 34.18

For the following operating data, calculate the amount of sludge to be wasted in m^3/d to attain an SRT of 8 d.

Given:

MLSS = 2500 g/m³   RSSS = 1.1% = 11 g/L   Q = 30 ML/d
FESS = 15 g/m³ V_A = 4.5 ML

Solution:

Rate of sludge wastage

$$Q_{WS} = \frac{V_A}{SRT} \times \frac{MLSS}{RSSS} - Q \times \frac{FESS}{RSSS} = \frac{4.5\,ML}{8.0\,d} \times \frac{2500}{11000} - \frac{30\,ML}{d} \times \frac{15}{11000}$$

$$= \frac{0.0867\,ML}{d} \times \frac{1000\,m^3}{ML} = 86.7 = \underline{87\,m^3/d}$$

EXAMPLE PROBLEM 34.19

In a conventional activated sludge plant, MLSS is determined to be 3200 ppm, and the activated sludge is returned at the same rate as the daily wastewater flow rate of 7.5 MGD. Calculate the rate at which activated sludge should be wasted to attain a SRT of 10 d, given that the aeration volume is 2.0 MG.

Given:

$$MLSS = 3200 \, ppm \qquad V_A = 2.0 \, MG \qquad SRT = 10 \, d \qquad R = 1$$

Solution:

Rate of sludge wastage

$$Q_{WS} = \frac{V_A}{SRT} \times \frac{MLSS}{RSSS} = \frac{V_A}{SRT} \times \frac{R}{(1+R)} = \frac{2.0 \times 10^6 \, gal}{10 \, d} \times \frac{1}{2} = \underline{100\,000 \, gal/d}$$

> *Using the short form would yield a conservative value of the sludge wasting rate.*

34.6.8 F/M CONTROL

F/M control is used to ensure that the activated sludge process is loaded at the rate microorganisms in the mixed liquor are able to consume. Too much or too little may upset the process. In a given plant, the loadings vary widely on a day-to-day basis. For this reason, the wastage rate should be worked out on 7-day moving averages of Q, MLSS, BOD and RSSS.

BOD test results for a given sample are available after 5 days. An estimate of the BOD concentration can be made by running a COD test or total carbon test, which are a lot quicker. For most municipal wastewater, BOD is about 60% of COD concentration and 250% of total organic carbon (TOC). To calculate the amount of activated sludge to be wasted based on F/M control, the mass of the desired MLSS or MLVSS, as the case may be, is calculated. Compared to the current value of MLSS, the amounts of excess solids in the aeration tank are calculated.

Excess of solids

$$\boxed{M_{ss(extra)} = V_A \times \Delta MLSS = V_A (MLSS_{act} - MLSS_{des})}$$

Additional volume of waste sludge

$$\boxed{Q_{Sl(extra)} = \frac{M_{ss(extra)}}{WSSS} = \frac{V_A \times \Delta MLSS}{WSSS}}$$

Daily waste sludge rate

$$\boxed{Q_{WS(tot)} = Q_{WS(current)} + Q_{WS(additional)}}$$

To prevent any shock to the system, the additional volume of sludge should be wasted over a period of 5–10 days. The new wastage rate should be within 10% of the current rate.

EXAMPLE PROBLEM 34.20

An activated sludge plant's optimum F/M ratio is 0.65/d based on COD and volatile solids in the mixed liquor. Given the following, work out the MLSS that should be maintained in the aeration tank.

$Q = 4.0\ \text{MGD}$ $V_A = 1.25\ \text{MG}$ $\text{MLSS} = ?$ $VF = 70\%$ $COD_{PE} = 300\ \text{mg/L}$

Solution:

Mixed liquor volatile solids concentration

$$MLVSS = \frac{COD_{PE}}{(F/M)} \times \frac{Q}{V_A} \times \frac{300\ mg}{L} \times \frac{d}{0.65} \times \frac{4.0\ MG/d}{1.25\ MG} = 1476 = \underline{1480\ mg/L}$$

Mixed liquor solids concentration

$$MLSS = \frac{MLVSS}{VF} = \frac{1476.9\ mg/L}{0.70} = 2109 = \underline{2100\ mg/L}$$

EXAMPLE PROBLEM 34.21

Given the following, calculate the additional volume of waste sludge to maintain a F/M ratio of 0.30/d.

Given:

$Q = 2.5\ \text{ML/d}$ $V_A = 0.45\ \text{ML}$ $\text{MLSS} = 2200\ \text{g/m}^3$ $\text{WSSS} = \text{RSSS} = 1.0\%$
$BOD_{PE} = 100\ \text{g/m}^3$ $F/M = 0.30/\text{d}$

Solution:

Mixed liquor solids concentration

$$MLSS = \frac{BOD_{PE}}{(F/M)} \times \frac{Q}{V_A} \times \frac{100\ mg}{L} \times \frac{d}{0.30} \times \frac{2.5\ ML/d}{0.45\ ML} = 1852 = \underline{1850\ mg/L}$$

Additional sludge wastage

$$V_{add} = \frac{\Delta MLSS}{WSSS} \times V_A = \frac{2200 - 1850}{10000} \times 4.5\ ML \times \frac{1000\ kL}{ML} = 156.6 = \underline{160\ kL}$$

34.6.9 MLSS Control

MLSS control is very simple and straightforward. For the same mass of solids in the aeration tank, the F/M ratio will vary directly with the organic loading to the activated sludge process. Thus, this control may not be very effective where the loadings show a wide range of fluctuations. For many of the plants where BOD loadings are relatively constant on a day-to-day basis, MLSS concentration can be effectively used as a control parameter for the activated sludge process.

Additional mixed liquor solids to be wasted

$$\Delta m_{MLSS} = V_A \Delta MLSS = V_A \left(MLSS_{current} - MLSS_{desired} \right)$$

Further, given the concentration of WSSS/RSSS, the additional volume of waste sludge can be calculated

Additional sludge to be wasted

$$V_{WS} = \frac{\Delta m_{MLSS}}{WSSS} = \frac{V_A \Delta MLSS}{WSSS}$$

Based on this, the change required in the waste-activated sludge pumping rate can be calculated. Remember, sudden changes can easily upset the activated sludge process, so the new pumping rate should be changed in small steps of not more than 10–15% at a given time.

EXAMPLE PROBLEM 34.22

The desired F/M ratio of a particular activated sludge plant is 0.40/d. If the 13 ML/d of wastewater has a BOD of 160 mg/L after settling, how many kg of MLSS should be maintained in the aeration tank?

Given:

F/M = 0.4/d   Q = 30 ML/d   BOD = 160 mg/L   WSSS = 10 kg/m³

Solution:

Mass of mixed liquor solids

$$m_{MLSS} = \frac{Q \times BOD_{PE}}{(F/M)} = \frac{13\,ML}{d} \times \frac{160\,kg}{ML} \times \frac{d}{0.40} = 5200.0 = \underline{5200\,kg}$$

Comparing this with the actual mass of MLSS in the aeration tank, the additional amount of sludge to waste can be calculated. For example, if it is desired to maintain

5200 kg of MLSS as compared to the current 6000 kg, this will mean 800 kg need to be wasted in addition to existing sludge wastage.

Volume of waste sludge

$$V_{WS} = \frac{\Delta m_{MLSS}}{WSSS} = 800\,kg \times \frac{m^3}{10.\,kg} = 80.0 = \underline{80\ m^3}$$

EXAMPLE PROBLEM 34.23

The desired F/M ratio of an activated sludge plant is 0.25/d. If the daily average wastewater flow is 4.5 MGD and raw wastewater BOD is 220 mg/L, find the concentration of MLSS that should be maintained. Assume 35% of BOD is removed by primary treatment.

a) How many lb of MLSS should be maintained in the aeration tank?
b) If the aeration volume is 1.5 MG, find the desired MLSS concentration.

Given:

F/M = 0.25/d   Q = 4.5 MGD BOD_{raw} = 220 mg/L

Solution:

a) Mass of mixed liquor solids in the aeration tank

$$m_{MLSS} = \frac{Q \times BOD_{PE}}{(F/M)} = \frac{4.5\ MG}{d} \times \frac{220(1-0.35)\,mg}{L} \times \frac{8.34\,lb/MG}{mg/L} \times \frac{d}{0.25}$$

$$= 21467 = \underline{21000\ lb}$$

b) Concentration of mixed liquor solids

$$MLSS = \frac{m_{MLSS}}{V_A} = \frac{21467\,lb}{1.5\ MG} \times \frac{mg/L}{8.34\ lb/MG} = 1716 = \underline{1700\ mg/L}$$

EXAMPLE PROBLEM 34.24

Given return sludge and mixed liquor suspended solids concentrations of 9800 and 4800 g/m³, respectively, and aeration capacity of 1250 m³, calculate the volume of sludge to be wasted to reduce the MLSS to 3700 g/m³. Also determine the pumping rate in L/s.

a) If wasted continuously over a 24-h period
b) If wasted in a batch over a 4-h period

Given:

Current MLSS = 4800   WSSS = 9800   Desired MLSS = 3700 V_A = 1250 m³

Solution:

Change in mixed liquor solids

$$\Delta m_{MLSS} = V_A \Delta MLSS = 1250\,m^3 \times (4800 - 3700)\,kg\,/\,m^3 = 1375.0 = 1380\,kg$$

Additional volume of waste sludge

$$V_{WS} = \frac{\Delta m_{MLSS}}{WSSS} = 1375\,kg \times \frac{m^3}{9800\,g} \times \frac{1000\,g}{kg} = 140.3 = 140\,m^3$$

Additional wastage rate over 24-h period

$$Q_{WS+} = \frac{140.3\,m^3}{24\,h} \times \frac{h}{3600\,s} \times \frac{1000\,L}{m^3} = 1.624 = \underline{1.6\,L/s}$$

Over 4-h period

$$Q_{WS+} = \frac{140.3\,m^3}{4.0\,h} \times \frac{h}{3600\,s} \times \frac{1000\,L}{m^3} = 9.74 = \underline{9.7\,L/s}$$

It should be preferred to waste over a longer time period to avoid shock to the system. If wasted over 2 days, the wastage rate must be increased by 0.8 L/s.

EXAMPLE PROBLEM 34.25

An extended aeration plant has an influent BOD of 180 g/m³. It is desired to operate the plant at an F/M ratio of 0.10/d, maintaining an MLVSS of 1500 mg/L. What size aeration tank would be required, assuming 80% of the mixed liquor solids are volatile? This plant is serving a small community with daily wastewater production of 60 m³/d. What should be the diameter of the aeration tank when operated to maintain a mixed liquor depth of 3.5 m?

Given:

BOD = 180 mg/L   Q = 60 m³/d   VF = 80%   MLVSS = 1500 g/m³

Solution:

Mixed liquor solids concentration

$$MLSS = \frac{MLSS}{VF} = \frac{1500\,mg/L}{0.80} = 1875\,g/m^3$$

Volume under aeration

$$V_A = \frac{Q \times BOD}{F/M \times MLSS} = \frac{60 \ m^3}{d} \times \frac{d}{0.1} \times \frac{180 \ g/m^3}{1875 \ g/m^3} = 57.6 = \underline{60 \ m^3}$$

Diameter of the tank

$$D = \sqrt{\frac{4}{\pi} \times \frac{V}{d}} = \sqrt{\frac{4}{\pi} \times \frac{57.6 \ m^3}{3.5 \ m}} = 4.57 = 5.0 \ m$$

PRACTICE PROBLEMS

PRACTICE PROBLEM 34.1

An activated sludge plant receives a daily flow of 4.0 MGD, and raw wastewater has BOD content of 220 mg/L. Assuming 35% removal of BOD by the primary treatment, find the lb of BOD entering the aeration tanks. (4800 lb/d)

PRACTICE PROBLEM 34.2

In a conventional activated sludge plant, the aeration capacity is 4.8 ML. The BOD of the primary effluent is 170 g/m³. For a daily average flow of 15 ML/d, calculate the BOD loading rate. (530 g/m²·d)

PRACTICE PROBLEM 34.3

A water pollution control plant's daily average flow is 4.5 MGD. The total aeration capacity is 1.2 MG and is operated by maintaining mixed liquor suspended solids (MLSS) of 2000 mg/L. Determine the aeration period. (6.4 h)

PRACTICE PROBLEM 34.4

Repeat Example Problem 34.4 assuming the length is twice the width. (27 m)

PRACTICE PROBLEM 34.5

The flow to a 2.0-ML aeration unit is 72 L/s. If the BOD concentration of the primary effluent is 130 g/m³, what is the BOD loading rate on the aeration unit? (400 g/m³·d)

PRACTICE PROBLEM 34.6

A plant has a daily flow of 4500 m³ and an average primary effluent BOD of 150 mg/L, MLVSS of 1500 mg/L and aeration capacity 900 m³. Calculate the F/M ratio. (0.5/d)

PRACTICE PROBLEM 34.7

An aeration tank receives a primary effluent flow of 5.0 MGD with a BOD concentration of 150 g/m³. The mixed liquor suspended solids concentration is 2500 mg/L, and the aeration volume is 1.3 MG. Find the F/M ratio. (0.33/d)

PRACTICE PROBLEM 34.8

An extended aeration plant has an influent BOD of 200 g/m³. At what level should MLSS concentration be maintained to achieve an F/M ratio of 0.10/d? The aeration period for the existing flow is 36 hours. (1300 mg/L)

PRACTICE PROBLEM 34.9

Due to a 50% increase in hydraulic loading, the aeration period is reduced to 24 h, and BOD of the influent is found to be 190 g/m³. What should the MLSS concentration be to achieve an F/M ratio of 0.1/d? (1900 g/m³)

PRACTICE PROBLEM 34.10

For the data of Ex. Prob. 34.10, calculate SRT, neglecting a) V_C and b) FESS and V_C. (7.1d, 7.7d)

PRACTICE PROBLEM 34.11

Repeat assuming sludge wastage rate of 0.1 MGD. (6.5 d)

PRACTICE PROBLEM 34.12

To increase the SRT, wastage of sludge is reduced from 500 to 400 m³ per day. What is the new SRT? (4.6 d)

PRACTICE PROBLEM 34.13

The volume occupied by sludge after 30 min of settling is observed to be 26%. Calculate the return sludge rate as a percentage of the wastewater flow. (35%)

PRACTICE PROBLEM 34.14

Rework the Example Problem 34.14 for V_{ss} of 25%. (83 mL/g, 1.2%, 33%)

PRACTICE PROBLEM 34.15

For the data of Ex. Prob. 34.15, work out the expected concentration of SS if sludge return rate is increased to 50%. (0.9%)

PRACTICE PROBLEM 34.16

In the previous example, if the actual recirculation ratio is kept at 50% instead of 28%, calculate the concentration of solids in the return sludge and SDI. (0.72%, 1.1%)

PRACTICE PROBLEM 34.17

It is planned to operate an activated sludge treatment system with solids concentrations of 1600 mg/L in the mixed liquor and 4800 mg/L in the return sludge. What should the recirculation ratio be? (0.5 or 50%)

PRACTICE PROBLEM 34.18

Repeat Example Problem 34.18 for MLSS = 2000 g/m³ and RSSS = 0.8%. (85 m³/d)

PRACTICE PROBLEM 34.19

In a conventional activated sludge plant, MLSS is determined to be 2200 ppm, and the aeration volume is 0.75 MG. The activated sludge is returned at the rate of 50% of the daily wastewater flow rate of 3.0 MGD. Calculate the rate at which activated sludge should be wasted to attain a SRT of 7.0 d. (36000 gal/d)

PRACTICE PROBLEM 34.20

An activated sludge plant operates well at an F/M ratio of 0.50/d based on COD and suspended solids in the mixed liquor. Given the following, work out the

MLSS that should be maintained in the aeration tank. Average daily flow = 3.0 MGD, aeration capacity = 1.0 MG and COD of the primary effluent = 250 mg/L. (1500 mg/L)

PRACTICE PROBLEM 34.21

Rework the EX. Prob. 34.21 if it is desired to lower MLSS to 2030 g/m^3. Calculate the continuous rate of additional waste sludge in L/min. (53 L/min)

PRACTICE PROBLEM 34.22

The desired F/M ratio of a particular activated sludge plant is 0.35/d. The daily average wastewater flow is 10 ML/d, with BOD content of 220 mg/L. Assuming 30% BOD removal by primary clarification, find the mass of MLSS to be kept in the aeration tank. (4400 kg)

PRACTICE PROBLEM 34.23

It is desired to operate an activated sludge plant maintaining a F/M ratio of 0.22/d for an daily average wastewater flow of 3.0 MGD with BOD content of 200 mg/L. This plant has four aeration tanks, each measuring 140 ft × 25 ft, with a liquor depth of 10 ft. Assuming primary BOD removal of 30%, find the concentration of MLSS that should be maintained. (1800 mg/L)

PRACTICE PROBLEM 34.24

For an activated sludge plant, on a given day, MLSS and RSSS, respectively, are 4500 g/m^3 and 1.2%. The total aeration capacity of the plant is 1300 m^3, and the desired MLSS is 4000 mg/L Determine the waste sludge pumping rate to lower MLL to desired value of 4000 mg/L.

a) If wasted continuously over a 24-h period. (3.8 L/min)
b) If wasted in a batch over a 4-h period. (23 L/min)

PRACTICE PROBLEM 34.25

A contact stabilization plant has an influent BOD of 160 g/m^3. It is desired to operate the plant at an F/M ratio of 0.15/d maintaining MLSS of 1800 mg/L. What capacity aeration tank would be required? This prefabricated plant is expected to treat a daily wastewater flow of 85 kL/d. (50 m^3)

35 Trickling Filter

Trickling filters and **biological rotating contactors** (RBCs) are biological systems in which biological growths are attached to the surfaces of supporting media. In trickling filters, the media can be crushed rock or synthetic. The term filter is a misnomer, as the purpose of the media is not to strain the water. Although the physical structures differ, the biological process is essentially the same in all attached culture systems.

35.1 HYDRAULIC LOADING

Much like clarifying, hydraulic loading is the volume of wastewater applied to the filter surface, including both untreated wastewater and recirculation flows. This surface loading rate is commonly expressed in $m^3/m^2 \cdot d$.

Hydraulic loading rate

$$HLR = \frac{(Q + Q_R)}{A_s}$$

The important thing to remember is that recirculated flows must be included. Typical hydraulic loadings for high-rate trickling filters may vary from 10 to 90 $m^3/m^2 \cdot d$. Too low hydraulic loading can cause plugging of the filter, thus resulting in septic conditions. Recirculation may be necessary. The **recirculation ratio** is the ratio of recirculated flows to the wastewater entering the plant.

Recirculation ratio

$$R = \frac{Q_{RS}}{Q} = \frac{(Q_{DR} + Q_{IR})}{Q}$$

As shown in Figure 35.1, recirculation may be **direct** or **indirect**. By applying direct recirculation, greater organic loading can be achieved. In addition, BOD removal efficiency is enhanced by passing wastewater through a filter more than once. Indirect recirculation is the return of underflow from the final clarifiers to the

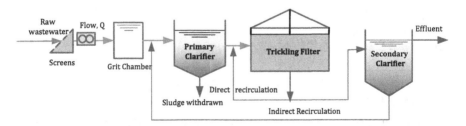

FIGURE 35.1 A trickling filter plant.

DOI: 10.1201/9781003468745-35

wet well for removal of solids and to prevent stalling of the distributor arm during low wastewater flow periods. Indirect recirculation ratios are generally limited to 0.5. Direct circulation has the advantage of influencing neither the primary nor the secondary clarifier. The recirculation ratio may be as high as 5.0.

EXAMPLE PROBLEM 35.1

A treatment plant receives a flow of 3.5 MGD. If the trickling filter effluent is recirculated at the rate of 3.7 MGD, what is the recirculation ratio?

Given:

$Q = 3.5$ MGD $Q_R = 3.7$ MGD $R = ?$

Solution:

Return ratio

$$R = \frac{Q_R}{Q} = \frac{3.7}{3.5} = 1.05 = \underline{1.1}$$

EXAMPLE PROBLEM 35.2

A high-rate trickling filter receives a primary effluent flow of 6.5 ML/d. What is the hydraulic loading rate if the filter is 30 m in diameter?

Given:

$Q = 5.5$ ML/d $D = 30$ m   HLR $= ?$

Solution:

Hydraulic loading rate

$$HLR = \frac{Q}{A_S} = \frac{4Q}{\pi D^2} = \frac{5.5\,ML}{d} \times \frac{1000\,m^3}{ML} \times \frac{4}{\pi (30\,m)^2} = 7.78 = \underline{7.8\,m^3/m^2.d}$$

35.2   ORGANIC LOADING

The BOD load on a trickling filter is calculated using the BOD in the primary effluent applied to the filter, without regard to any BOD in the recirculated flows.

BOD loading rate

$$\boxed{BODLR = \frac{M_{BOD}}{V_F} = \frac{Q \times BOD_{PE}}{V_F} \quad Sub\ F = Filter}$$

EXAMPLE PROBLEM 35.3

Raw municipal wastewater entering a trickling filter plant contains 220 mg/L of BOD. If the average primary removal is 35%, what is the minimum secondary removal required to produce plant effluent not exceeding 30 mg/L?

Given:

$SS_i = 220$ mg/L $PR_I = 35\%$ $PR_{II} = ?$ $BOD_e = 30$ mg/L

Solution:

Primary effluent BOD

$$BOD_e = BOD_i \left(1 - \frac{PR}{100}\right) = \frac{220 \; mg}{L} \times \left(1 - \frac{35\%}{100\%}\right) = 143.0 = 140 \; mg/L$$

Secondary removal

$$PR = \frac{\left(BOD_i - BOD_e\right)}{BOD_i} \times 100\% = \frac{(143 - 30) \, mg/L}{143 \, mg/L} \times 100\% = 0.790 = \underline{79\%}$$

EXAMPLE PROBLEM 35.4

A trickling filter 70 ft in diameter with 6.5 ft of media depth receives a flow of 1.2 MGD. If the BOD content of the primary effluent is 150 mg/L, what is the organic loading on the filter?

Given:

$D = 70$ ft $d = 6.5$ ft $Q = 1.2$ MGD $BOD = 150$ mg/L

Solution:

Volume of filter media

$$V_F = A_S \times d = \frac{\pi D^2}{4} \times d = \frac{\pi (70 \, ft)^2}{4} \times 6.5 \, ft = 25014 \, ft^3$$

Organic loading rate

$$BODL = \frac{Q \times BOD_{PE}}{V_F} = \frac{1.2 \, MG}{d} \times \frac{150 \, mg}{L} \times \frac{1}{25014 \, ft^3} \times \frac{8.34 \, lb/MG}{mg/L}$$

$$= 0.060 \, lb/ft^3 . d = \underline{60 \, lb/1000 \, ft^3 . d}$$

EXAMPLE PROBLEM 35.5

A standard rate trickling filter has a diameter of 26 m and an average media depth of 2.1 m. The daily wastewater flow is 5200 m³/d with an average BOD of 180 mg/L. During periods of low influent flow, 2.5 ML/d of underflow from the final clarifier is returned to the wet well. Calculate hydraulic loading, return ratio and BOD loading, assuming 35% BOD removal by the primary treatment.

Given:

$D = 26$ m $d = 2.1$ m $Q_R = 2.5$ ML/d $= 2500$ m³/d $Q = 5200$ m³/d

Solution:

Hydraulic loading rate

$$HLR = \frac{(Q+Q_R)}{A_s} = \frac{(2500+5200)\,m^3}{d} \times \frac{4}{\pi(26\,m)^2} = 14.5 = \underline{15\,m^3/m^2.d}$$

Return ratio

$$R = \frac{Q_R}{Q} = \frac{2500}{5200} = 0.481 = \underline{0.48}$$

Primary effluent BOD

$$BOD_{PE} = BOD_{Raw}(1-PR) = \frac{180\,g}{m^3}\left(1-\frac{35\%}{100\%}\right)$$
$$= 117 = \underline{120\,g/m^3}$$

Organic loading rate

$$BODL = \frac{Q \times BOD_{PE}}{V_F} = \frac{5200\,m^3}{d} \times \frac{117\,g}{m^3}\frac{4}{\pi(26\,m)^2 \times 2.1\,m}$$

$$= 546 = \underline{550\,g/m^3.d}$$

EXAMPLE PROBLEM 35.6

A standard rate trickling filter is 75 ft in diameter, with an average media depth of 7.0 ft. The daily wastewater flow is 1.3 MGD, with an average BOD of 190 mg/L. During periods of low influent flow, 0.65 MGD of underflow from the final clarifier is returned to the wet well. Calculate hydraulic loading, return ratio and BOD loading, assuming 30% BOD removal by the primary treatment.

Given:

D = 75 ft   d = 7.0 ft Q_R = 0.65 MGD   Q = 1.3 MGD   BOD = 190 mg/L

Solution:

Hydraulic loading rate

$$HLR = \frac{(Q+Q_R)}{A_S} = \frac{(1.3+0.65)\times10^6\,gal}{d} \times \frac{4}{\pi(75\,ft)^2} = 441 = \underline{440\,gal/ft^2.d}$$

$$R = \frac{Q_R}{Q} = \frac{0.65}{1.3} = \underline{0.50}$$

Primary effluent BOD

$$BOD_{PE} = BOD_{Raw}\left(1 - PR/100\right) = \frac{190\ mg}{L}\left(1 - \frac{30\%}{100\%}\right) = 133 = 130\ mg/L$$

BOD loading rate

$$BODL = \frac{Q \times BOD_{PE}}{V_F} = \frac{1.3\ MG}{d} \times \frac{133\ mg}{L} \times \frac{8.34\ lb/MG}{mg/L} \times \frac{4}{\pi\left(75\ ft\right)^2 \times 7.0\ ft}$$
$$= 0.0466\ lb/ft^3.d = \underline{47\ lb/1000\ ft^3.d}$$

35.3 ESTIMATING BOD REMOVAL

BOD removal efficiency of biological filtration primarily depends on such factors as depth of bed, kind of media, temperature, recirculation and organic loading. Empirical equations have been developed to predict the BOD removal efficiency based on organic loading and recirculation ratios. One of the most popular formulations evolved from filter plants at military installations in the US. The following equation is applicable to single-stage stone media filters, followed by a final clarifier and treating settled domestic wastewater with a temperature of 20°C.

BOD percent removal (SI)

$$PR_{20} = \frac{100\%}{\left(1 + 0.014\sqrt{BODLR/F}\right)} \qquad F = \frac{\left(1 + R\right)}{\left(1 + 0.1R\right)^2}$$

In this formula, BODLR is expressed in g/m³·d, and F is dimensionless. If the loading is expressed in kg/m³·d, the value of the constant is 0.44. When BODLR is expressed in lb/ft³·d, the value of the constant is 1.77.

BOD percent removal (USC)

$$PR_{20} = \frac{100\%}{\left(1 + 1.77\sqrt{BODLR/F}\right)} \qquad F = \frac{\left(1 + R\right)}{\left(1 + 0.1R\right)^2}$$

As discussed before, BOD removal is strongly dependent on wastewater temperature. Filters in cold climates operate at lower efficiencies for the same loadings. Removal efficiency must be adjusted for the operating temperature. The following empirical formula can be used to find BOD removal at temperatures other than 20°C.

Percent removal at temperature T

$$PR_T = PR_{20}(1.035)^{T-20}$$

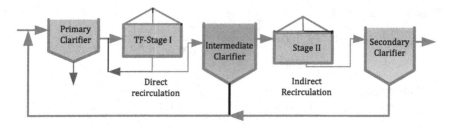

FIGURE 35.2   Staging in tricking filter operation.

Based on the actual operating date, BOD removal efficiency can be calculated by knowing the BOD concentration of the influent and effluent from a process or treatment plant, as the case may be. When a high-quality effluent is required, two tricking filters are connected in series, as seen in Figure 35.2.

EXAMPLE PROBLEM 35.7

In a trickling filter plant, the operating temperature is 15°C. What is the reduction in BOD percent removal?

Given:

$T = 15°\text{C}$ $PR = ?$

Solution:

Reduction in BOD removal

$$\frac{PR_{15}}{PR_{20}} = (1.035)^{T-20} = (1.035)^{15-20} = (1.035)^{-5} = 0.84$$

$$PR_{15} = 0.84PR_{20} \quad \text{or} \quad 16\% \, drop$$

EXAMPLE PROBLEM 35.8

A treatment plant receives a flow of 3.5 MGD. If the trickling filter effluent is recirculated at the rate of 3.7 MGD, what is the value of factor F in the BOD removal formula?

Given:

$Q = 3.5 \, \text{MGD}$ $Q_R = 3.7 \, \text{MGD}$ $F = ?$

Solution:

Return ratio

$$R = \frac{Q_R}{Q} = \frac{3.7}{3.5} = 1.05 = 1.1$$

Recirculation factor F

$$F = \frac{(1+R)}{(1+0.1R)^2} = \frac{1+1.05}{(1+0.105)^2} = 1.68 = \underline{1.7}$$

EXAMPLE PROBLEM 35.9

A trickling filter plant has a filter tank with a diameter of 21 m and media depth of 2.0 m. The wastewater influent is 3000 m³/d with a BOD of 200 mg/L. The plant is operated with indirect recirculation during low flow equal to 1500 m³/d and constant direct recirculation at 3300 m³/d. Calculate BOD removal efficiency of the plant at a wastewater temperature of 16°C. Assume 35% of BOD is removed by the primary treatment.

Given:

D = 21 m   d = 2.0 m   E = 35%   T = 16°C   Q = 3000 m³/d
BOD_{raw} = 200 g/m³ Q_{DR} = 1500 m³/d Q_{IR} = 3300 m³/d

Solution:

Primary effluent BOD

$$BOD_{PE} = BOD_{Raw}\left(1 - \frac{PR}{100}\right) = \frac{200\,g}{m^3}\left(1 - \frac{35\%}{100\%}\right) = 130.0 = 130\,g/m^3$$

BOD loading rate

$$BODLR = \frac{Q \times BOD_{PE}}{V_F} = \frac{3000\,m^3}{d} \times \frac{117\,g}{m^3}\frac{4}{\pi(21m)^2 \times 2.0\,m} = 557 = 560\,g/m^3.d$$

Return ratio

$$R = \frac{Q_R}{Q} = \frac{(1500+3300)}{3000} = 1.60 = 1.6$$

Factor F

$$F = \frac{(1+R)}{(1+0.1R)^2} = \frac{1+1.6}{(1+0.16)^2} = 2.35 = 2.4$$

BOD removal efficiency

$$E_{20} = \frac{100\%}{\left(1+0.014\sqrt{BODLR/F}\right)} = \frac{100\%}{\left(1+0.014\sqrt{560/2.35}\right)} = 82.2 = \underline{82\%}$$

$$E_{16} = E_{20}(1.35)^{T-20} = 82\%(1.035)^{(16-20)} = 72.1 = \underline{72\%}$$

EXAMPLE PROBLEM 35.10

A trickling filter is sized so that BOD loading is 500 g/m³·d and the wastewater is recirculated to provide a recirculation ratio of 1.0. Calculate the maximum allowable concentration of BOD in the primary effluent (filter influent) to produce an effluent BOD of 30 mg/L.

Given:

BODL = 500 g/m³·d   R = 1.0 BOD_e = 30 mg/L BOD_i = ?

Solution:

Factor F

$$F = \frac{(1+R)}{(1+0.1R)^2} = \frac{1+1.0}{(1+0.10)^2} = 1.650 = 1.65$$

BOD removal efficiency

$$PR_{20} = \frac{100\%}{\left(1+0.014\sqrt{\dfrac{BODL}{F}}\right)} = \frac{100\%}{\left(1+0.014\sqrt{\dfrac{500}{1.65}}\right)} = 80.4 = \underline{80\%}$$

Primary effluent BOD

$$BOD_i = \frac{BOD_e}{(1-PR/100)} = \frac{30\ mg/L}{(1-0.80)} = 150.0 = \underline{150\ mg/L}$$

EXAMPLE PROBLEM 35.11

A trickling filter plant has a filter tank with a diameter of 70 ft and media depth of 6.5 ft. The wastewater influent is 0.80 MGD with a BOD of 200 mg/L. The plant is operated with indirect recirculation during low flow equal to 0.40 MGD and constant direct recirculation at 1.0 MGD m³/d. Calculate BOD removal efficiency of the plant at a wastewater temperature of 16°C. Assume 35% of BOD is removed by the primary treatment.

Given:

D = 70 ft, d = 6.5 ft   E = 35%   T = 16°C
BOD_{raw} = 200 mg/L   Q = 0.80 MGD Q_R = 0.4 + 1.0 = 1.4 MGD

Solution:

BOD loading

$$L = \frac{Q \times BOD_{PE}}{V_F} = \frac{0.80\,MG}{d} \times \frac{200 \times 8.34\,lb}{MG} \times 0.65 \times \frac{4}{\pi(70\,ft)^2 \times 6.5\,ft}$$
$$= 3.467 \times 10^{-2} = 0.0355\ lb/ft^3.d$$

Recirculation ratio and F factor

$$R = \frac{Q_R}{Q} = \frac{1.4}{0.80} = 1.75$$

$$F = \frac{1+R}{(1+0.1R)^2} = \frac{1+1.75}{(1+0.1\times1.75)^2} = 1.99 = 2.0$$

Removal efficiency

$$E_{20} = \frac{100\%}{\left(1+1.77\sqrt{\dfrac{BODL}{F}}\right)} = \frac{100\%}{\left(1+1.77\sqrt{3.467\times10^{-2}/2.0}\right)} = 81.1 = \underline{81\%}$$

Removal at 16°C operating temperature

$$E_T = E_{20}(1.035)^{T-20} = 81.1\%(1.035)^{(16-20)} = 70.6 = \underline{71\%}$$

EXAMPLE PROBLEM 35.12

Select the diameter of a trickling filter 7.5 ft deep to serve a population of 500 persons contributing wastewater @ 50 gal/c·d with BOD content of 280 mg/L. Assume 35% of BOD is removed in the primary clarification and design BOD loading is 30 lb/1000 ft³·d.

Given:

Pop. = 500 @ 50 gal/c·d   D = ?   d = 8.0 ft   PR = 35% BOD_{raw} = 280 mg/L

Solution:

BOD of the primary effluent

$$BOD_{PE} = BOD_{Raw}(1-PR) = \frac{280\,mg}{L}\left(1 - \frac{35\%}{100\%}\right) = 182\,mg/L$$

Daily flow rate

$$Q = \frac{50\,gal}{p.d} \times 500\,p = 25\,000\,gal/d$$

Volume of the filter

$$V_F = \frac{Q \times BOD_{PE}}{BODL} = \frac{25000\,gal}{d} \times \frac{182\,mg}{L} \times \frac{8.34\,lb/MG}{mg/L} \times \frac{1000\,ft^3.d}{30\,lb}$$

$$= 1264.9\,ft^3 = \underline{1260\,ft^3}$$

Diameter of the filter

$$D = \sqrt{\frac{4}{\pi} \times \frac{V_F}{d}} = \sqrt{\frac{4}{\pi} \times \frac{1264.9\,ft^3}{7.5\,ft}} = 14.65 = \underline{15\,ft}$$

PRACTICE PROBLEMS

PRACTICE PROBLEM 35.1

A trickling filter has a desired recirculation ratio of 1.4. If the flow to the filter is 4.4 MGD, what is the reciculated flow? (6.2MGD)

PRACTICE PROBLEM 35.2

A high-rate trickling filter receives a primary effluent flow of 4.5 ML/d. If the desired hydraulic loading rate is not to exceed 8.0 $m^3/m^2 \cdot d$, what should be the minimum diameter of the filter? (27 m)

PRACTICE PROBLEM 35.3

In a trickling filter wastewater treatment plant, trickling filters are operated in series to achieve a minimum of 85% BOD removal. Assuming 35% BOD removal in the primary clarification, determine the BOD removal efficiency of the plant. Also find the BOD of the plant effluent if the BOD of the raw wastewater is 190 mg/L. (90%, 19 mg/L)

PRACTICE PROBLEM 35.4

A trickling filter 85 ft in diameter, with a media depth of 5.0 ft, receives a flow of 1.2 MGD. If the BOD concentration of the primary effluent stream is 160 mg/L, what is the organic loading on the trickling filter? (56 lb/1000 ft^3.d)

PRACTICE PROBLEM 35.5

A flow of 4.3 ML/d is applied to a 15-m-diameter trickling filter filled with 1.2-m-deep stone media. The BOD of the wastewater after settling is 100 mg/L. Calculate the hydraulic and organic loading, assuming a recirculation ratio of 0.5. (20 $m^3/m^2 \cdot d$, 1100 g/m^3. d)

PRACTICE PROBLEM 35.6

A high-rate trickling filter is 65 ft in diameter, with an average media depth of 7.5 ft. The filter receives primary effluent with an average flow of 2.5 MGD, with an average BOD of 140 mg/L. Calculate hydraulic loading, return ratio and BOD loading rate. (750 gal/ft^2.d, 120 lb/1000 ft^3.d)

PRACTICE PROBLEM 35.7

In a trickling filter plant, the operating temperature during summer is 22°C. What is the increase in BOD compared to BOD removal at 20°C? (7%)

PRACTICE PROBLEM 35.8

A treatment plant receives a flow of 3.5 MGD. If the trickling filter effluent is recirculated at the rate of 4.5 MGD, what is the value of factor F in the BOD removal formula? (1.8)

PRACTICE PROBLEM 35.9

Rework Example Problem 35.9 if the direct recirculation is reduced from 3300 m³/d to 1500 m³/d. (69%)

PRACTICE PROBLEM 35.10

Repeat Example Problem 35.10 assuming a recirculation ratio of 2.0 and operating temperature of 16°C. (110 g/m³·d)

PRACTICE PROBLEM 35.11

A trickling filter is operated with organic loading of 30 lb/1000 ft²·d while maintaining a recirculation rate of 120%. What is the expected BOD removal if the operating temperature is 22°C? (87%)

PRACTICE PROBLEM 35.12

Select the diameter of a trickling filter 7.0 ft deep to serve a population of 500 persons contributing wastewater @ 35 gal/c·d with BOD content of 300 ppm. Assume 40% of BOD is removed in the primary clarification and design BOD loading is 30 lb/1000 ft³·d. (13 ft)

36 Rotating Biological Contactors

Rotating biological contactors (RBCs) are biological treatment systems similar to trickling filters. An RBC process typically consists of primary sedimentation and final clarification following the bio-disks. Recirculation through RBC units is not normally practised. As shown in Figure 36.1, an underflow from the final clarifier is returned to the primary clarifier. Waste sludge, similar in character to that from a trickling filter plant, is withdrawn from the primary clarifier for disposal. A series of four stages is normally installed in the treatment of domestic wastewater. To introduce nitrification, additional stages need to be added.

36.1 HYDRAULIC LOADING

Hydraulic loading in an RBC unit is based on the media surface area. The media discs are typically 3.6 m in diameter and are mounted on a horizontal shaft, generally 7.5 m long. A standard media shaft provides about 9300 m² of surface area for biomass to grow on. The surface area for the same shaft length also depends on the spacing between the two discs. If the discs are packed tightly, it will mean more disc surface. Media made with smaller spacing are known as high-density media. High-density media can be 20–80% more media surface than a standard shaft. Hydraulic loading is expressed as flow per unit disc surface area. Hydraulic loading on an RBC can range from 0.4 to 0.12 m³/m²·d.

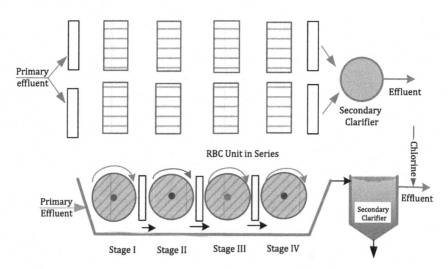

FIGURE 36.1 Rotating biological contactor plant.

 DOI: 10.1201/9781003468745-36

36.2 SOLUBLE BOD

Usually, the RBC unit is designed to remove soluble BOD. Total BOD in waste-water is contributed by organic solids (particulate BOD) and soluble and colloidal biodegradable matter. **Soluble BOD** can be thought as BOD of a filtered sample. Experience has shown that for a given wastewater BOD contributed by SS is propor-tional to the concentration of SS. The constant of proportionality is typically in the range of 0.5 to 0.7 for municipal wastewater. The k factor needs to be determined for a given operation. This can be done by observing total BOD, soluble BOD and suspended solids concentration data over a certain period of time. The k value is the ratio of the average particulate BOD to the average suspended solids concentration.

Particulate BOD

$$BOD_{part} = BOD_{tot} - BOD_{sol} = k \times SS$$

36.3 ORGANIC LOADING

Organic loading on RBC units is based on soluble BOD or total BOD. It is expressed as mass load of BOD in the primary effluent per unit disc surface.

Organic loading

$$BODLR = \frac{M_{BOD}}{A_S} = \frac{Q \times BOD_{PE}}{A_S}$$

A_S = *Total disc surface* Q = *Average flow* BOD = *BOD of settled wastewater*

Remember that the BOD loading rate in RBC is per unit surface as compared to per unit volume, as in the case of activated sludge and trickling filter units.

Key points:

- Typical loadings are 7.5 g/m²·d of soluble BOD or 15 g/m²·d of total BOD.
- In larger plants, RBC shafts are placed perpendicular to the direction of flow; thus, each shaft acts as one stage of BOD removal.
- The various stages of RBC simulate plug flow, with maximum loading on the first stage. Hence, the organic loading on the first stage is an important consideration.
- A loading of 60 g/m²·d of total BOD on the first stage should not be exceeded.
- For operating temperatures below 13°C, a temperature correction for addi-tional disc surface @ 15% for each 3°C below 13°C should be made.
- If operated properly, the biological growth on the first stage should be fairly uniform, thin and light brown in colour. The following stages should look similar, except with an additional gold or reddish tone.
- Overloading conditions are evidenced by grey or white biomass. DO in the first-stage effluent should not be allowed to fall below 0.5 mg/L, and DO of the final-stage effluent should preferably be more than 2.0 mg/L.

EXAMPLE PROBLEM 36.1

Based on hydraulic loading of 0.11 m³/m²·d, how many ML of flow can be treated by a standard shaft of surface area of 9300 m²?

Given:

$A_s = 9300 \text{ m}^2 \quad \text{HLR} = 0.11 \text{ m}^3/\text{m}^2 \cdot \text{d} \quad Q = ?$

Solution:

Design flow

$$Q_{Design} = A_S \times HLR = \frac{0.11 \, m^3}{m^2 . d} \times 9300 \, m^2 \times \frac{ML}{1000 \, m^3} = 0.99 = \underline{1.0 \, ML/d}$$

EXAMPLE PROBLEM 36.2

Over a period of 2 months, an average value of total BOD, soluble BOD and SS in the primary effluent respectively are 175 mg/L, 94 mg/L and 148 mg/L. What fraction of suspended solids contributes to BOD?

Given:

TBOD = 175 mg/L   SBOD = 94 mg/L   SS = 148 mg/L   k = ?

Solution:

Particulate BOD

$$BOD_{part} = BOD_{tot} - BOD_{sol} = 175mg/L - 94 \, mg/L = 81 \, mg/L$$

Fraction of SS contributing to BOD

$$k = \frac{BOD_{part}}{SS_{tot}} = \frac{81 \, mg/L}{148 \, mg/L} = 0.547 = \underline{0.55 \, or \, 55\%}$$

EXAMPLE PROBLEM 36.3

The wastewater entering an RBC plant has a BOD of 220 mg/L and SS content of 230 mg/L. If the k value is 0.50, what is the soluble BOD of the wastewater?

Given:

TBOD = 220 mg/L   SBOD = ?   SS = 230 mg/L   k = 0.50

Solution:

Particulate BOD

$$BOD_{part} = k \times SS = 0.50 \times 230 \, mg/L = 115 = 120 \, mg/L$$

Soluble BOD

$BOD_{sol} = BOD_{tot} - BOD_{part} = 220 - 115 = 105 = 110 \, mg/L$

EXAMPLE PROBLEM 36.4

An RBC unit receives a wastewater flow of 1.8 MGD with a BOD of 180 mg/L and SS concentration of 160 mg/L. Assuming a k value of 0.60, find the lbs of soluble BOD entering the RBC.

Given:

TBOD = 180 mg/L   Q = 1.8 MGD   SS = 160 mg/L   k = 0.60

Solution:

Soluble BOD

$BOD_{sol} = BOD_{tot} - BOD_{part} = 180 \, mg/L - 0.60 \times 160 \, mg/L = 84 \, mg/L$

Organic loading

$M_{BOD} = Q \times BOD = \dfrac{1.8 \, MG}{d} \times \dfrac{84 \, mg}{L} \times \dfrac{8.34 lb/MG}{mg/L} = 1261 = 1300 \, lb/d$

EXAMPLE PROBLEM 36.5

A treatment plant processes domestic wastewater by primary sedimentation RBCs and final clarification. Each RBC shaft has a length of 5.3 m with a 3.6-m-diameter disk for a nominal surface area of 5600 m². The installation has 16 shafts arranged with four rows of shafts of four stages each. The influent wastewater flow is 9.0 ML/d containing 180 mg/L of BOD. Assuming 30% BOD removal in the primary clarification, calculate the BOD loading on the first stage.

Given:

Disk surface area = 5600 m²/shaft   # of shafts = 16   # of stages = 4
PR_I = 30% BOD_{raw} = 180 mg/L

Solution:

Disk surface area

$A_S = \dfrac{5600 \, m^2}{shaft} \times \dfrac{4 \, shafts}{stage} = 21\,400 \, m^2/stage$

BOD load

$M_{BOD} = Q \times BOD_{PE} = \dfrac{9.0 \, ML}{d} \times \dfrac{180 \, kg}{ML} \times 0.7 = 1134 = 1100 \, kg/d$

BOD loading rate

$BODLR = \dfrac{M_{BOD}}{A_S} = \dfrac{1134 \, kg}{d} \times \dfrac{1}{21400 \, m^2} \times \dfrac{1000 \, g}{kg} = 50.6 = 51 \, g/m^2.d$

EXAMPLE PROBLEM 36.6

An RBC unit consists of 16 shafts with each shaft having a disc surface of 56,000 ft². The installation has 16 shafts arranged with four rows of shafts of four stages each. On average, a primary effluent flow of 2.5 MGD containing 150 mg/L of BOD and 120 mg/L of SS is treated. Assuming k = 0.50, calculate the soluble BOD loading on the RBC process.

Given:

Disc surface = 56000 ft²/shaft   shafts # = 16   Q = 2.5 MGD
BOD = 150 mg/L   SS = 120 mg/L   k = 0.50

Solution:

Disk surface area

$$A_S = \frac{56000\ ft^2}{shaft} \times 16\ shafts = 896\ 000\ ft^2$$

Particulate BOD

$$BOD_{part} = k \times SS = 0.50 \times 120\ mg/L = 60\ mg/L$$

Soluble BOD

$$BOD_{sol} = BOD_{tot} - BOD_{part} = 150\ mg/L - 60\ mg/L = 90\ mg/L$$

BOD loading rate

$$BODLR = \frac{Q \times BOD}{A_S} = \frac{2.5\ MG}{d} \times \frac{90\ mg}{L} \times \frac{8.34\ lb/MG}{mg/L} \times \frac{1}{896000\ ft^2}$$
$$= 2.09 \times 10^{-3} = 2.1\ lb/1000\ ft^2.d$$

EXAMPLE PROBLEM 36.7

How many RBC shafts (1 hm²/shaft) are required to treat a flow of 10 ML/d with a BOD concentration of 200 mg/L? Assume the primary removal of BOD is 35%. The design BOD loading rate is 15 g/m²·d.

Given:

Disc surface = 1 hm²/shaft   #of shafts = ?   Q = 10 ML/d
BOD_{raw} = 200 mg/L   BODL = 15 g/m²·d PR_I = 35% = 0.35

Solution:

BOD load on RBC

$$M_{BOD} = Q \times BOD = \frac{10\ ML}{d} \times \frac{200\ mg}{L} \times (1 - 0.35) = 1300\ kg/d$$

Disk surface area required

$$A_S = \frac{1300\ kg}{d} \times \frac{1000\ g}{kg} \times \frac{m^2.d}{15\ g} = 8.67 \times 10^4 = 87\ 000\ m^2$$

Number of RBC shafts

$$\# = 8.67 \times 10^4 \, m^2 \times \frac{shaft}{10000 \, m^2} = 8.67 = \underline{9.0 \, shafts}$$

EXAMPLE PROBLEM 36.8

A 16-ft-long RBC shaft is packed with 250 discs of 12 ft diameter and ¾ in spacing. Work out the total disc surface per shaft length. For a rotating speed of 1.5 rpm, determine the peripheral speed.

Given:

D = 12 ft   N = 1.5 rpm   250 discs/shaft

Solution:

Disk surface area

$$A_S = \frac{\pi(12\,ft)^2}{4} \times \frac{2 \, faces}{disc} \times \frac{250 \, disc}{shaft} = 5.65 \times 10^4 = \underline{57\,000 \, ft^2 \, /shaft}$$

Peripheral velocity v_p

$$v_p = \frac{1.5 \, rev}{min} \times \frac{\pi(12\,ft)}{rev} \times \frac{min}{60 \, s} = 0.942 = \underline{0.94 \, ft/s}$$

EXAMPLE PROBLEM 36.9

A treatment plant processes domestic wastewater by primary sedimentation, RBCs and final clarification. Each RBC shaft has a length of 5.3 m with a 3.6-m-diameter disk for a nominal surface area of 5600 m². The installation has 16 shafts arranged with four rows of shafts of four stages each. The influent wastewater flow is 9.0 ML/d containing 180 mg/L of BOD. Assuming 30% BOD removal in the primary clarification, calculate the BOD loading on the first stage.

Given:

A_S = 5600 m²/shaft   # of shafts = 16   # of stages = 4
PR_I = 30%   BOD_raw = 180 mg/L

Solution:

Disk surface area per stage

$$A_S = \frac{5600 \, m^2}{shaft} \times \frac{4 \, shafts}{stage} = 22\,400 \, m^2 \, /stage$$

BOD load on RBC

$$M_{BOD} = Q \times BOD_{PE} = \frac{9.0 \, ML}{d} \times \frac{180 \, kg}{ML} \times (1 - 0.30) = 1134 \, kg/d$$

BOD loading rate

$$BODLR = \frac{M_{BOD}}{A_S} = \frac{1134\ kg}{d} \times \frac{1}{22400\ m^2} \times \frac{1000\ g}{kg} = 50.6 = \underline{51\ g/m^2.d}$$

Hydraulic loading rate

$$HLR = \frac{Q}{A_S} = \frac{9000\ m^3}{d} \times \frac{1}{22400\ m^2} = 0.401 = \underline{0.40\ m^3/m^2.d}$$

EXAMPLE PROBLEM 36.10

A rotating biological contactor treats a flow of 2.4 MGD. The surface of the media is estimated to be 720,000 ft². The plant influent has a BOD of 220 mg/L and SS concentration of 240 mg/L with a k value of 0.50. Assume 35% removal of BOD and 45% removal of SS by primary clarification before wastewater enters the RBC process. Calculate the hydraulic loading and soluble BOD loading.

Given:

A_s = 720,000 ft²　　BOD_{raw} = 220 mg/L　　　SS_{raw} = 240 mg/L　　　Q = 2.4 MG/d
PR_{SS} = 45%　　　　PR_{BOD} = 35%　　k = 0.5

Solution:

Primary effluent BOD

$$BOD_{PE} = BOD_{RAW}\left(1 - \frac{PR}{100}\right) = \frac{220\ mg}{L}\left(1 - \frac{35}{100}\right) = 143 = 140\ mg/L$$

Primary effluent SS concentration

$$SS_{PE} = SS_{RAW}\left(1 - \frac{PR}{1000}\right) = \frac{240\ mg}{L}\left(1 - \frac{45}{100}\right) = 132 = 130\ mg/L$$

Soluble BOD

$$BOD_{sol} = BOD_{PE} - k \times SS_{PE} = 143\ mg/L - 0.5 \times 132\ mg/L = 77\ mg/L$$

Hydraulic loading

$$HLR = \frac{Q}{A_S} = \frac{2.4\ MG}{d} \times \frac{1}{720000\ ft^2} \times \frac{10^6\ gal}{MG} = 3.33 = \underline{3.3\ gal/ft^2.d}$$

Soluble BOD loading rate

$$BODLR = \frac{2.4\ MG}{d} = \frac{77\ mg}{L} \times \frac{8.34\ lb/MG}{mg/L} \times \frac{1}{720\ thousands} = \underline{2.1\ lb/1000\ ft^2.d}$$

PRACTICE PROBLEMS

PRACTICE PROBLEM 36.1

Based on hydraulic loading of 0.12 $m^3/m^2 \cdot d$, how many ML of daily flow can be treated by an RBC shaft with a surface area of 10,000 m^2? (1.2 ML/d)

PRACTICE PROBLEM 36.2

Over a period of 3 months, the average values of total BOD and soluble BOD in the primary effluent, respectively, are 190 mg/L and 90 mg/L. What fraction of BOD is contributed by suspended solids? (0.53)

PRACTICE PROBLEM 36.3

How many standard RBC shafts are required to treat of a flow of 1.5 ML/d while maintaining hydraulic loading of 8.5 $L/m^2 \cdot d$? (2 shafts)

PRACTICE PROBLEM 36.4

A sample of the primary effluent stream indicates TBOD of 152 mg/L and suspended solids concentration of 164 mg/L. Based on the K value of 0.55, estimate the soluble BOD in the primary effluent (62 mg/L).

PRACTICE PROBLEM 36.5

Calculate the BOD loading on an RBC shaft with a total surface area of 10,000 m^2 treating domestic wastewater containing 150 mg/L of BOD. The primary treatment preceding the RBCs removes 40% of the BOD, and average wastewater flow is 1100 m^3/d. By providing baffles, the total disc surface is divided into four equal stages. (40 $g/m^2 \cdot d$)

PRACTICE PROBLEM 36.6

An RBC installation has 4 trains of RBC shafts of 4 stages each. Each shaft has a total media surface of 9000 m^2. The average primary effluent flow is 12 ML/d, containing 135 mg/L of BOD and 110 mg/L of SS. Assuming k = 0.55, calculate the soluble BOD loading on the RBC unit. (6.2 $g/m^2 \cdot d$)

PRACTICE PROBLEM 36.7

How many RBC shafts (90,000 ft^2/shaft) are required to treat a flow of 4.5 MGD with a BOD concentration of 220 mg/L? Assume the primary residual of BOD is 30%. The design BOD loading rate is 4.0 $lb/1000\ ft^2 \cdot d$. (16 shafts)

PRACTICE PROBLEM 36.8

A 5-m-long RBC shaft is packed with 250 discs of 3.6 m diameter and 20 mm spacing. Work out the total disc surface per shaft length. For a rotating speed of 1.5 rpm, determine the peripheral speed in m/s. (5100 m^2, 0.28 m/s)

PRACTICE PROBLEM 36.9

A treatment plant processes domestic wastewater by primary sedimentation, RBCs and final clarification. Each RBC shaft has a length of 5.3 m with a 3.6-m-diameter disk for a nominal surface area of 5600 m². The installation has 16 shafts arranged with four rows of shafts of four stages each. The wastewater flow to the RBCs is 11 ML/d containing 150 mg/L of BOD. Find the hydraulic loading and BOD loading on the first stage. (0.49 m/d, 64 g/m².d)

PRACTICE PROBLEM 36.10

A secondary wastewater treatment plant consisting of RBCs processes municipal wastewater. Each RBC shaft has a nominal surface area of 56,000 ft². The installation has 16 shafts arranged with four rows of shafts of four stages each. The raw wastewater flow is 2.5 MGD containing 190 mg/L of BOD. Assuming 35% BOD removal in the primary clarification, calculate the BOD loading on the RBCs based on total disk area. (2.9 lb/1000 ft².d)

37 Stabilization Ponds

Stabilization ponds, also called lagoons or oxidation ponds, are generally employed as total treatment of wastewater in rural areas. Ponds are classified as facultative, aerobic and anaerobic according to the type of biological activity that takes place in them. A great many variations in ponds are possible due to differences in depth, operating conditions and loadings. For treating municipal wastewater, facultative ponds (Figure 37.1) are the most common type of stabilization pond. The upper zone of these ponds is aerobic, while the bottom layer is anaerobic. Algae growth is an important part of such ponds. During winter conditions, because of the snow cover, sunlight can reach the surface, and hence photosynthesis does not take place. Hence no effluent is released, and all incoming wastewater is stored until spring comes and algae establishes.

Operating depths range from 0.5 to 1.5 m, with 1.0 m of free board above the high-water level. In northern climates, the pond water level is controlled by controlling the discharge. Controlled discharge ponds provide detention time or storage time (no discharge) of up to 3 months or longer. Discharge in the winter is minimized or stopped, and influent wastewater is stored until spring.

37.1 BOD LOADING

BOD loading on stabilization ponds is expressed in terms of mass of BOD applied per unit surface area, $g/m^2 \cdot d$. It may range from as low as 2.0 $g/m^2 \cdot d$ in cooler climates to as high as 6.0 $g/m^2 \cdot d$ for warmer climates. These loadings are substantially less than volumetric BOD loads applied to aeration and filtration units.

37.2 HYDRAULIC LOADING

Hydraulic loading on ponds is expressed as surface overflow rate, that is, hydraulic flow per unit surface area. More commonly it is expressed as depth units and may vary from 10 mm to several millimetres per day depending on organic loading.

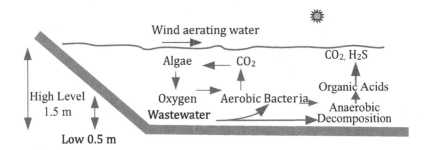

FIGURE 37.1 Facultative lagoon.

DOI: 10.1201/9781003468745-37 **489**

37.3 STORAGE TIME

Q is the influent rate without considering the water lost through evaporation and percolation. When working out the storage time (no discharge), similar calculations may be made, but it is important to consider water loss.

Storage time

$$
t_{sto} = \frac{V_{sto}}{Q_{net}} = \frac{A_S \times d_{sto}}{(Q_{in} - Q_{loss})} = \frac{d_{sto}}{(d_{in} - d_{loss})}
$$

Q_{loss} = *outflow or loss due to evaporation and percolation*
Q_{in} = *influent flow rate*
t_{sto} = *storage time*
d_{sto} = *depth of storage*
d_{in} = *hydraulic loading rate*
d_{loss} = *depth of water loss per day*

37.4 POPULATION LOADING

Loading calculated on a population-served basis is expressed simply as number of people served per unit of surface area of the pond, for example, persons/hectare. It may vary from 200–2000 persons per hectare. If there are significant industrial flows, an adjustment must be made to take into account the loading contributed by the industry. This is usually done by expressing the industrial contribution as BOD population equivalent, which is typically 90 g (0.2 lb) of BOD/d.

EXAMPLE PROBLEM 37.1

A facultative pond for a small town consists of a 10-acre primary cell and two smaller cells of 5 acres each. The average daily wastewater flow is 0.3 MGD containing a BOD of 200 mg/L. Calculate the BOD loading based on the area of primary cell.

Given:

Q = 0.30 MG/d   BOD = 200 mg/L   Primary cell Area = 10 acre

Solution:

BOD loading rate

$$
BODLR - \frac{Q \times BOD}{A_S} = \frac{0.3\,MG}{d} \times \frac{200\,mg}{L} \times \frac{8.34\,lb/MG}{mg/L} \times \frac{1}{10\,acre}
$$
$$
= 50.0 = \underline{50\,lb/acre.d}
$$

EXAMPLE PROBLEM 37.2

A facultative pond has an average length of 220 m with an average width of 140 m. The daily average flow rate to the pond is 1100 m³/d, and it is operated at a depth of 1.8 m. What is the detention time in days?

Given:

$Q = 1100$ m³/d $L = 220$ m $W = 140$ m $d = 1.8$ m

Solution:

Detention time

$$t_d = \frac{V}{Q} = \frac{L \times W \times d}{Q} = 220\ m \times 140\ m \times 1.8\ m \times \frac{d}{1100\ m^3} = 50.0 = \underline{50\ d}$$

EXAMPLE PROBLEM 37.3

Stabilization ponds for a town of 3600 people consist of two cells connected in series with a total area of 3.5 acres and average depth of 3.9 ft. The average daily waste-water flow is 0.32 MGD with a BOD content of 190 mg/L. Calculate detention time and hydraulic loading,

Given:

$A_S = 3.5$ acres $d = 3.9$ ft $Q = 0.32$ MG/d

Solution:

Detention time

$$t_d = 3.5\ acre \times 3.9\ ft \times \frac{d}{0.32 \times 10^6\ gal} \times \frac{43560\ ft^2}{acre} \times \frac{7.48\ gal}{ft^3} = 13.8 = \underline{14\ d}$$

Hydraulic loading rate

$$HLR = \frac{0.32 \times 10^6\ gal}{d} \times \frac{1}{3.5\ acre} \times \frac{acre}{43560\ ft^2} \times \frac{ft^3}{7.48\ gal} \times \frac{12\ in}{ft} = 3.36 = \underline{3.4\ in/d}$$

EXAMPLE PROBLEM 37.4

For the data of Example Problem 37.3, calculate population loading and BOD loading.

Given:

Population = 3600 persons $A_S = 3.5$ acres $Q = 0.32$ MG/d   BOD = 190 mg/L

Solution:

Population loading

$$Loading = \frac{Pop.}{A_S} = \frac{3600\ p}{3.5\ acre} = 1028 = \underline{1000\ p/acre}$$

BOD loading rate

$$BODLR = \frac{Q \times BOD}{A_S} = \frac{0.32\ MG}{d} \times \frac{190\ mg}{L} \times \frac{8.34\ lb/MG}{mg/L} \times \frac{1}{3.5\ acre}$$

$$= 144 = \underline{140\ lb/acre.d}$$

EXAMPLE PROBLEM 37.5

Stabilization ponds for a town consist of two cells connected in series, with each cell having an area of 1.5 ha and average depth of 1.2 m. The average daily wastewater flow is 1.2 ML/d, with a BOD control of 190 mg/L. Calculate detention time, hydraulic loading and BOD loading.

Given:

A_S = 3.0 ha   d = 1.2 m   Q = 12 ML/d = 1200 m³/d   BOD = 190 g/m³

Solution:

Detention time

$$t_d = 3.0\ ha \times 1.2\ m \times \frac{d}{1200\ m^3} \times \frac{10000\ m^2}{ha} = 30.0 = \underline{30\ d}$$

Hydraulic loading rate

$$HLR = \frac{1200\ m^3}{d} \times \frac{1}{3.0\ ha} \times \frac{ha}{10000\ m^2} \times \frac{1000\ mm}{m} = \underline{40\ mm/d}$$

BOD loading rate

$$BODLR = \frac{Q \times BOD}{A_S} = \frac{1200\ m^3}{d} \times \frac{190\ g}{m^3} \times \frac{1}{3.0\ ha} \times \frac{ha}{10000\ m^2} = \underline{7.6\ g/m^2.d}$$

EXAMPLE PROBLEM 37.6

Stabilization ponds for a town consist of two cells connected in series, and each cell has an area of 5.0 ha and average depth of 1.2 m. The average daily wastewater flow is 1.2 ML/d. The discharge is stopped in the early winter, and the water level is dropped to 0.6 m. Estimate the number of days of winter storage available between 0.6 m and 1.5 m water levels, assuming an evaporation and seepage loss of 2.5 mm/d.

Given:

$d_{sto} = 1.5 \text{ m} - 0.6 \text{ m} = 0.9 \text{ m}$ $A_S = 3.0 \text{ ha}$ $Q = 1200 \text{ m}^3/\text{d}$ $d_{loss} = 2.5 \text{ mm/d}$

Solution:

Water loss rate

$$Q_{loss} = \frac{2.5 \text{ mm}}{d} \times 10 \text{ ha} \times \frac{10 \text{ m}^3}{ha.mm} = 250.0 = 250 \text{ m}^3/d$$

$$Q_{net} = 1200 - 250 = 950 \text{ m}^3/d$$

Winter storage available

$$t_{sto} = \frac{V_{sto}}{Q_{net}} = \frac{A_S \times d_{sto}}{Q_{net}} = 0.9 \text{ m} \times 10 \text{ ha} \times \frac{10000 \text{ m}^2}{ha} \times \frac{d}{950 \text{ m}^3} = 94.7 = \underline{95 \text{ d}}$$

EXAMPLE PROBLEM 37.7

A sewage lagoon with discharge control has a total surface area of 22 ha. The average daily flow is 3200 m³/d. Calculate the minimum water depth for a storage time of 90 d, assuming that the difference between evaporation and seepage and precipitation is 1.0 mm/d.

Given:

$A_S = 22 \text{ ha} = 220\,000 \text{ m}^2$ $Q = 3200 \text{ m}^3/\text{d}$ $t_{sto} = 90 \text{ d}$ $d_{sto} = ?$

Solution:

Water loss rate

$$Q_{loss} = \frac{1.0 \text{ mm}}{d} \times \frac{m}{1000 \text{ mm}} \times 220000 \text{ m}^2 = 220.0 = \underline{220 \text{ m}^3/d}$$

Depth of storage

$$d_{sto} = \frac{t_{sto}(Q_{in} - Q_{loss})}{A_S} = 90 \text{ d} \times \frac{(3200 - 220)m^3}{d} \times \frac{1}{220000 \text{ m}^2} = \underline{1.2 \text{ m}}$$

EXAMPLE PROBLEM 37.8

Design a rectangular stabilization pond for treating wastewater from a subdivision of 6500 people contributing @ 130 L/c.d containing a BOD of 260 mg/L. Assume the width of the pond is 1/3rd of its length, and BOD loading of 13 g/m².d can be afforded. Find detention time if pond is operated by maintaining a water depth of 1.5 m.

Given:

Q = 130 L/c·d, BOD = 260 mg/L, L = 3 × W, BOD = 13 g/m².d, 6500 p, d = 1.5 m

Solution:

Daily flow rate

$$Q = \frac{130\,L}{p.\,d} \times 6500\,p \times \frac{m^3}{1000\,L} = 845 = \underline{850\,m^3/d}$$

BOD load on pond

$$BODL = \frac{845\,m^3}{d} \times \frac{260\,g}{m^3} \times \frac{kg}{1000\,g} = 219 = \underline{220\,kg/d}$$

Required pond area

$$A = \frac{BODL}{BODLR} = \frac{219.7\,kg}{d} \times \frac{m^2.d}{0.013\,kg} \times \frac{ha}{10000\,m^2} = 1.69 = \underline{1.7\,ha}$$

Width of pond

$$W = \sqrt{\frac{A}{3}} = \sqrt{\frac{16900\,m^2}{3}} = 75.05 = \underline{75\,m}$$

Select pond of size 225 m × 75 m × 2.5 m (1 m f.b.)

Detention time

$$t_d = \frac{V}{Q} = \frac{L \times W \times d}{Q} = 225 \times 75\,m \times 1.5\,m \times \frac{d}{845\,m^3} = 29.9 = \underline{30\,d}$$

EXAMPLE PROBLEM 37.9

A stabilization pond for a small town with a population of 2900 consists of a large cell (primary) of 12 acres and two smaller cells (secondary) of 6 acres each. The average daily flow is 0.25 MGD, with a BOD concentration of 220 mg/L. Calculate BOD loading based on area of the primary cell. Estimate the number of days of winter storage available between 2.0-ft and 5.0-ft water levels, assuming an evaporation and seepage loss of 0.11 in/d.

Given:

d_{sto} = (5 −2) ft = 3.0 ft A_S (primary) = 12 acres A_S (secondary) = 12 acres
Q = 0.25 MGD d_{loss} = 0.11 in/d

Solution:

Water loss rate

$$Q_{loss} = \frac{0.11\ in}{d} \times \frac{ft}{12\ in} \times 24\ acres \times \frac{43560\ ft^2}{acre} = 9583\ ft^2/d$$

Net flow rate

$$Q_{net} = \frac{0.25 \times 10^6\ gal}{d} \times \frac{ft^3}{7.48\ gal} - 9583 = 23839\ ft^3/d$$

Winter storage available

$$t_{sto} = \frac{V_{sto}}{Q_{net}} = \frac{A_S \times d_{sto}}{Q_{net}} = 3.0\ ft \times 24\ acres \times \frac{43560\ ft^2}{acre} \times \frac{d}{23839.4\ ft^3}$$

$$= 131.5 = \underline{130\ d}$$

EXAMPLE PROBLEM 37.10

A sewage lagoon with discharge control has a total surface area of 22 ha. The average daily flow is 3200 m³/d. Calculate the minimum water depth for a retention time of 90 d, assuming that the difference between evaporation and seepage and precipitation is 1.0 mm/d.

Given:

A_S = 22 ha = 220 000 m²   Q = 3200 m³/d t_{stor} = 90 d d_{sto} = ?

Solution:

Water loss rate

$$Q_{loss} = \frac{1.0\ mm}{d} \times \frac{m}{1000\ mm} \times 220000\ m^2 = \underline{220\ m^3/d}$$

Depth of storage

$$d_{sto} = \frac{t_{sto}(Q_{in} - Q_{loss})}{A_S} = 90\ d \times \frac{(3200 - 220)m^3}{d} \times \frac{1}{220000\ m^2} = \underline{1.2\ m}$$

- *No discharge from the lagoon is allowed until algae is re-established.*
- *In warmer climates, higher loadings can be afforded.*

PRACTICE PROBLEMS

PRACTICE PROBLEM 37.1

A facultative pond serves a population of 4500 people. The average dimensions of the pond are 200 m × 150 m, with an average water depth maintained at 1.0 m. The average daily wastewater flow is 1.5 ML/d, with a BOD concentration of 150 g/m^3. Calculate detention time, hydraulic loading, population loading and BOD loading. (20 d, 50 mm/d, 1500 person/ha, 7.5 g/m^2·d)

PRACTICE PROBLEM 37.2

Stabilization ponds with a total surface area of 6 ha receive a wastewater flow of 530 m^3/d. Calculate the winter storage available between 0.6 m and 1.5 m depth assuming a water loss of 0.40 mm/d during the winter months. (110 d)

PRACTICE PROBLEM 37.3

Stabilization ponds for a town of 5000 people consist of two cells connected in series. The area of the primary cell is 3.0 acres, and that of the secondary cell is 2.5 acres. The pond is operated by maintaining an average depth of 4.0 ft. The average daily wastewater flow is 0.50 MGD, with a BOD content of 150 mg/L. Calculate detention time, hydraulic loading and BOD loading. (14 d, 3.3 in/d,)

PRACTICE PROBLEM 37.4

For the data of Practice Problem 37.3, work out population loading and BOD loading. (910 p/acre, 110 lb/acre·d)

PRACTICE PROBLEM 37.5

A facultative pond for a small town consists of a 4-ha primary cell and two smaller cells of 2 ha each. The average daily wastewater flow is 1.0 MLD containing a BOD of 200 mg/L. Work out the BOD loading based on the area of the primary cell. (5.0 g/m^2·d)

PRACTICE PROBLEM 37.6

Stabilization ponds for a town consist of two cells connected in series, and each cell has an area of 5.5 ha and average depth of 1.1 m. The average daily wastewater flow is 1.1 ML/d. The discharge is stopped in the early winter, and the water level is dropped to 0.40 m. Estimate the number of days of winter storage available between 0.4-m and 1.4-m water levels, assuming an evaporation and seepage loss of 3.5 mm/d. (120 d)

PRACTICE PROBLEM 37.7

A stabilization pond for a town consists of two cells for series operation. The first cell has an area of 2.7 ha, and the second is smaller, with 1.5 ha. Water loss from the pond averages 2.1 mm/d, and the daily average influent wastewater is 300 m³/d. Calculate the minimum storage depth required to ensure zero discharge for 120 d. (0.61 m)

PRACTICE PROBLEM 37.8

Size a rectangular stabilization pond for treating wastewater from a subdivision of 5500 people contributing @ 150 L/c.d containing a BOD of 180 mg/L. Assume two cells connected in series, and the width of each pond is 1/3rd of its length; BOD loading of 28 g/m².d can be afforded. (each cell 90 m × 30 m)

PRACTICE PROBLEM 37.9

A stabilization pond for a small town of population 2200 consists of a primary cell of 6 acres and two smaller cells (secondary) of 4 acres each. The average daily flow is 80 gal/c.d containing 210 mg/L of BOD. Calculate BOD loading based on the area of the primary cell. Estimate the number of days of winter storage available between 2.0-ft and 5.0-ft water levels, assuming an evaporation and seepage loss of 0.15 in/d. (51 lb/acre·d, 72 d)

PRACTICE PROBLEM 37.10

A sewage lagoon with discharge control has a total surface area of 20 ha. The average daily flow is 3.2 ML/d. Calculate the minimum water depth for a retention time of 80 d, assuming that the difference between evaporation and seepage and precipitation is 1.2 mm/d. (1.2 m)

38 Sludge Processing

During processing of wastewater, solids removed as slurry are called sludge or, more appropriately, **biosolids**. Processing of sludges is a costly and difficult operation in wastewater treatment. In a secondary wastewater treatment plant, the solid slurry collected at the bottom of the primary clarifier is termed **primary** or **raw sludge**. Similarly, the sludge pumped from the bottom of a secondary clarifier is called **secondary sludge**. In some plants, secondary sludge is pumped back into the primary clarifier. Hence, sludge pumped out of the primary clarifier will be blended sludge.

38.1 GRAVITY THICKENER

Sludge thickening is important to reduce the volume of sludge by removing water. For example, when a sludge with 3% of solids (97% water) is thickened to 6% solids, the volume of sludge is reduced by a factor of two. This makes further processing or disposal easier and more efficient. Depending upon the characteristics of the sludge, two types of thickening are commonly used.

A gravity thickener is designed to further concentrate sludges before they are sent to additional sludge handling and treatment processes, such as digestion, conditioning and dewatering. By thickening the sludge, there is a reduced load on these subsequent processes (Figure 38.1).

A gravity thickener works on the same principle as a clarifier. The calculation of hydraulic loading is important in determining whether the process is underloaded or overloaded. Hydraulic loading for a gravity thickener is expressed as flow per unit surface area. The solids loading rate on the thickener is calculated in the same fashion as kg of solids entering a daily per square meter area.

The sludge detention time refers to the length of time the solids remain in the gravity thickener, which depends on the volume of the sludge blanket and pumping

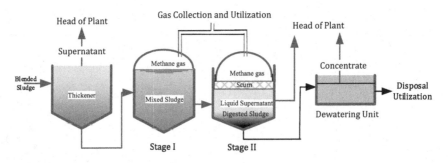

FIGURE 38.1 Sludge processing scheme.

 DOI: 10.1201/9781003468745-38

rate of sludge from the bottom of the thickener. **Sludge detention time** is the time sludge remains in the tank.

Sludge detention time

$$SDT = \frac{V_{Sl}}{Q_{Sl}} = \frac{A_S \times d_{Sl}}{Q_{Sl}}$$

The **efficiency** of a gravity thickener is a measure of the effectiveness in capturing the suspended solids form the influent sludge into the thickened sludge (underflow).

Concentration factor is another way of determining the effectiveness of a gravity thickener. This parameter indicates the factor by which sludge has been thickened. A concentration factor of three will mean that the thickened sludge is three times as concentrated as the influent sludge.

EXAMPLE PROBLEM 38.1

A primary sludge flow of 410 m³/d with a solid content of 3.5% is pumped into a 12.5-m-diameter gravity thickener. Calculate hydraulic loading rate and solid loading rate.

Given:

$Q = 410 \text{ m}^3/\text{d}$ $SS = 3.5\% = 35 \text{ kg/m}^3$ $D = 12.5 \text{ m}$

Solution:

Hydraulic loading rate

$$HLR = \frac{Q_{sl}}{A_S} = \frac{Q_{sl}}{0.785 D^2} = \frac{410\,m^3}{d} \times \frac{1.27}{(12.5\,m)^2} = 3.34 = \underline{3.3\,m^3/m^2.d}$$

Solids loading rate

$$SLR = \frac{Q_{sl} \times SS}{A_s} = \frac{410\,m^3}{d} \times \frac{35\,kg}{m^3} \times \frac{1.27}{(12.5\,m)^2} = 116 = \underline{120\,kg/m^2.d}$$

EXAMPLE PROBLEM 38.2

The sludge entering a gravity thickener contains 3.1% solids. The effluent from the thickener contains 1200 mg/L solids. What is the solids removal efficiency of the thickener? Also, calculate the concentration factor if the thickened sludge contains 7.4% of dry solids.

Given:

$SS_i = 3.1\%$ $SS_e = 1200 \text{ mg/L} = 0.12\%$ $SS_{thick} = 7.4\%$

Solution:

SS removal efficiency

$$PR = \frac{\left(SS_i - SS_e\right)}{SS_i} = \frac{\left(3.1 - 0.12\right)\%}{3.1\%} \times 100\% = 96.12 = \underline{96\%}$$

Concentration factor

$$CF = \frac{SS_{thick}}{SS_{feed}} = \frac{7.4\%}{3.1\%} = 2.38 = \underline{2.4\times}$$

EXAMPLE PROBLEM 38.3

Given the following data, determine the change in sludge blanket solids.

Parameter	Feed	Effluent	Thickened
Q, m³/d	630	240	290
SS, %	3.5	0.01	8.0
kg/m³	35	0.1	80

Solution:

Mass flow rate

$$M_{feed} = Q \times SS = \frac{630\,m^3}{d} \times \frac{35\,kg}{m^3} = 22050\,kg/d$$

Effluent SS load

$$M_{eff} = Q \times SS = \frac{240\,m^3}{d} \times \frac{0.10\,kg}{m^3} = 24\,kg/d$$

Thickened SS

$$M_{thick} = Q \times SS = \frac{290\,m^3}{d} \times \frac{80\,kg}{m^3} = 23200\,kg/d$$

Change in solids content
$$\Delta M = M_{in} - M_{out} = 22050 - 24 - 23200 = -1174 = \underline{-1200\,kg/d}$$

– Sign means sludge blanket will drop.

38.2   BLENDING OF SLUDGES

When sludges with different solid contents are mixed, the resulting sludge has a solid content somewhere between the sludge contents of the original sludges. The resulting value will be close to the solid content of the sludge with more volume. Based on the volume and mass balance, total solids and volume of mixed sludge will equal the sum of the values of the original sludges, as shown in Figure 38.2. It is assumed that the mass density of the sludges is the same irrespective of the solid content. This is valid for a majority of the situations where solid concentration is less than 10%, and density of sludge can be safely assumed to be equal to that of water, which is 1000 kg/m³.

$$SS_{mix} = \frac{\Sigma\left(SS_i \times Q_i\right)}{\Sigma Q_i}$$

EXAMPLE PROBLEM 38.4

Calculate the solids concentration in the sludge fed to a digester which consists of the mixing of a 4.0% primary sludge flowing at 6900 gal/d and 6.0% thickened secondary sludge flowing at 5000 gal/d.

Given:

Parameter	Primary Sludge, 1	Secondary Sludge, 2
Q, gal/d	6900	5000
SS, %	4.0	6.0

Solution:

SS load in the blended sludge

$$SS_{mix} = \frac{\Sigma\left(SS_i \times Q_i\right)}{\Sigma Q_i} = \frac{6900 \times 4.0\% + 5000 \times 6.0\%}{6900 + 5000} = 4.84 = \underline{4.8\%}$$

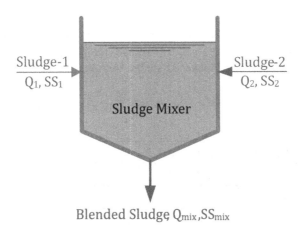

FIGURE 38.2 Blending of sludges.

EXAMPLE PROBLEM 38.5

A flow of 1.5 MGD with SS = 240 mg/L and BOD of 220 enters a trickling filter plant. Removal by primary clarification is 50% for SS and 30% for BOD. Compute the mass of solids generated as primary sludge and secondary sludge, assuming 25% of incoming BOD is converted to microbial mass. Primary and secondary sludges are combined in a holding tank. Assuming the blended sludge has a solids concentration of 4.0%, estimate the quantity of sludge generated.

Given:

$$Q = 1.5\text{ MGD} \qquad BOD = 220\text{ mg/L} \qquad SS = 240\text{ mg/L} \qquad K = 0.25 \qquad SS_{sl} = 4.0\%$$

Solution:

Mass of dry solids in primary sludge

$$M_{ss}\left(I\right) = Q \times SS_{rem} = \frac{1.5\,MG}{d} \times \frac{240 \times 8.34\ lb}{MG} \times \frac{50\%}{100\%} = 1501\ lb/d$$

Mass of dry solids in secondary sludge

$$M_{ss}\left(II\right) = Q \times BOD \times K = \frac{1.5\ MG}{d} \times \frac{220 \times 0.70 \times 8.34\ lb}{MG} \times \frac{25\%}{100\%} = 481.6\ lb/d$$

Volume of total sludge

$$V_{sl} = \frac{M_{ss}}{SS_{sl}} = \frac{\left(1501 + 481.6\right)lb}{d} \times \frac{100\%}{4.0\%} \times \frac{gal}{8.34\,lb} = 5943 = \underline{5900\ gal/d}$$

EXAMPLE PROBLEM 38.6

For the data of Example Problem 38.5, secondary sludge is thickened before pumping it to an anaerobic digester. During thickening, 95% of SS are captured, and the sludge is concentrated to 5.0% solids. Estimate the quantities and SS content of mixture of the primary and thickened sludge. Assume the SS content of primary sludge is 3.5%.

Given:

Parameter	Primary, 1	Thickened, 2	Mixed, 3
M_{ss}, lb/d	1501	95% × 481.6	1958.5
SS, %	3.5	5.0	?

Solution:

Quantity of primary sludge

$$Q(1) = \frac{M_{ss}}{SS_{sl}} = \frac{1501\ lb}{d} \times \frac{gal}{8.34\ lb} \times \frac{100\%}{3.5\%} = 5142.1\ gal/d$$

Quantity of thickened sludge

$$Q(2) = \frac{M_{ss}}{SS_{sl}} = \frac{0.95 \times 481.6\ lb}{d} \times \frac{gal}{8.34\ lb} \times \frac{100\%}{5.0\%} = 1097.1\ gal/d$$

Mass of solids in the mixed sludge

$$M_{ss}(3) = \frac{1501\ lb}{d} + 0.95 \times 481.6\ lb/d = 1958.5\ lb/d$$

Quantity of mixed sludge

$$Q(3) = Q(1) + Q(2) = 5142.1\ gal/d + 1097.1\ gal/d$$
$$= 6239.2\ gal/d$$

SS concentration of mixed sludge

$$SS(3) = \frac{\Sigma(SS_i \times Q_i)}{\Sigma Q_i} = \frac{5142 \times 3.5\% + 1097 \times 5.0\%}{5142 + 1097} = 3.76 = \underline{3.8\%}\quad or$$

$$SS(3) = \frac{M_{SS}(3)}{Q(3)} = \frac{1958.5\ lb}{d} \times \frac{d}{6239.2\ gal} \times \frac{gal}{8.34\ lb} \times 100\% = 3.76 = 3.8\%$$

EXAMPLE PROBLEM 38.7

For the data of Example Problem 38.6, work out the tonnage of wet sludge produced at this plant.

Given:

Parameter	Primary, 1	Thickened, 2	Mixed, 3
M_{ss}, lb/d	1501	95% × 481.6	1958.5
SS, %	3.5	5.0	3.76
M_{sl}, t/d	?	?	?

Solution:

Tonnage of mixed sludge

$$M_{sl}(3) = \frac{M_{ss}}{SS_{sl}(3)} = \frac{1958.5\,lb}{d} \times \frac{100\%}{3.76\%} \times \frac{t}{2000\,lb} = 26.0 = 26\,t\,/\,d$$

EXAMPLE PROBLEM 38.8

Daily wastewater flow to a sewage treatment plant averages 5.8 ML/d. On average, primary treatment removes 150 mg/L of solids and produces raw sludge containing 3.2% solids. Waste sludge from the activated sludge has a consistency of 1.1% solids. Waste sludge is produced as a result of 130 mg/L of BOD removal, and each unit of BOD is converted to 0.5 unit of solids. Estimate the biosolids production in wet metric tons/d.

Given:

Q = 1.45 MGD $SS_r(1)$ = 150 mg/L $SS_{sl}(1)$ = 3.2% $SS_{sl}(2)$ = 1.1%
$BOD_r(2)$ = 130 mg/L   K = 0.5

Solution:

Quantity of primary sludge

$$Q(1) = \frac{Q \times SS_r}{SS_{sl}(1)} = \frac{5.8\,ML}{d} \times \frac{150\,kg}{ML} \times \frac{100\%}{3.2\%} \times \frac{t}{1000\,kg} = 27.2\,t/d$$

Quantity of secondary sludge

$$Q(2) = \frac{Q \times K \times BOD_r}{SS_{sl}(2)} = \frac{5.8\,ML}{d} \times \frac{0.5 \times 130\,kg}{ML} \times \frac{100\%}{1.1\%} \times \frac{t}{1000\,kg}$$
$$= 34.3\,t/d$$

Quantity of mixed sludge
Q(3) = Q(1) + Q(2) = 27.2 + 34.3 = 61.5 = <u>62 t/d</u>

38.3   SLUDGE VOLUME PUMPED

The quantity of sludge pumped per day, both in terms of mass and volume, is an important variable that a plant operator needs to determine. In smaller plants using positive displacement pumps, the volume of raw sludge is determined by the volume displaced by the pump during each revolution. Positive displacement pumps are equipped with a counter at the end of the shaft and are rarely operated faster than 200 L/min.

Sludge pumping rate

$$Volume \times rpm = \frac{\pi D^2 LN}{4} \quad N = rpm$$

Converted to appropriate units, this is the maximum value that can be pumped for a set stroke length. Actual volume will be less due to slow or incomplete valve closures and slippage. You may calibrate the pump by recording the number of strokes required to fill an empty tank of known volume. In case of centrifugal pumps, it would be necessary to determine the volume pumped within the system. A pump can be calibrated by determining how long it took to pump a given volume of sludge by observing the pumping time to raise a fixed depth of sludge in the digester.

EXAMPLE PROBLEM 38.9

A sludge pump has a bore of 25 cm and a stroke of 10 cm. If the pump operates at 40 rpm, how many litres of sludge are pumped each minute? Account for the slippage by reducing the stroke volume by 10%.

Given:

$D = 25$ cm $= 0.25$ m $L = 10$ cm $= 0.10$ m $N = 40$ strokes/min   Slippage $= 10\%$

Solution:

Theoretical pump rate

$$Q = ALN = \frac{\pi (0.25\,m)^2 \times 0.10\,m}{4} \times \frac{40}{min} \times \frac{1000\,L}{m^3} = 196.3\,L/min$$

Actual pumping rate

$$Q = \frac{90\%}{100\%} \times 196.3 = 176.7 = 180\,L/min$$

EXAMPLE PROBLEM 38.10

A digester 12 m in diameter has a side water depth of 6.0 m. If the digester seed sludge is 20% of the digester capacity, how many m³ of seed sludge will be required?

Given:

$D = 12$ m $H = 6.0$ m   Seed sludge volume $= 20\%$ of digester capacity

Solution:

Capacity of the digester

$$V = AH = \frac{\pi(12\,m)^2}{4} \times 6.0\,m = 678.24 = \underline{680\,m^3}$$

Volume of seed sludge required

$$Seed = \frac{20\%}{100\%} \times 678.2 = 135.6 = \underline{140\,m^3}$$

38.4 VOLATILE SOLIDS

Volatile solids are the component of total solids in sludge which goes through biological degradation. During digestion, a fraction of solids is destroyed and converted to gases. The fraction of solids destroyed is expressed as the mass rate of solids destroyed per unit digester volume. Volatile solids destroyed are also expressed as the percentage of volatile solids reduced.

38.4.1 VOLATILE SOLIDS REDUCTION IN DIGESTION

During anaerobic digestion of sludge, a fraction of **volatile** solids is converted to carbon dioxide and methane (Figure 38.3). This reduction in volatile solids can be found by the following expression.

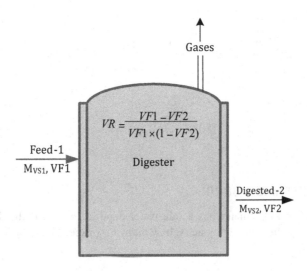

FIGURE 38.3 Volatile solids reduction.

Volatile reduction

$$VR = \frac{(VF_1 - VF_2)}{VF_1 - VF_1 \times VF_2}$$

VF = volatile fraction
VF_1 = volatile fraction of feed sludge
VF_2 = volatile fraction of digested sludge

EXAMPLE PROBLEM 38.11

What is the percent reduction in volatile matter in a primary digester if the volatile content of the raw sludge is 69% and the volatile content of the digested sludge is 51%?

Given:

$VF_1 = 69\% = 0.69$ $\qquad$ $VF_2 = 51\% = 0.51$

Solution:

Volatile solids reduction

$$VR = \frac{(VF_1 - VF_2)}{VF_1 - VF_1 \times VF_2} = \frac{0.69 - 0.51}{0.69 - 0.69 \times 0.51} = 0.532 = \underline{53\%}$$

EXAMPLE PROBLEM 38.12

The volatile content of the feed sludge is 72%. If it is desired to have 50% reduction of volatile solids during anaerobic digestion, what must be the volatile fraction in the digested sludge?

Given:

$VF_1 = 72\% = 0.72$ $\qquad$ $VF_2 = ?$ $\qquad$ $VR = 50\% = 0.50$

Solution:

Volatile solids reduction

$$VR = \frac{(VF_1 - VF_2)}{VF_1 - VF_1 \times VF_2} \quad or$$

$$VR(VF_1 - VF_2 \times VF_1) = (VF_1 - VF_2) \quad reaaranging$$

Volatile fraction

$$VF_2 = \frac{\left(VF_1 - VR \times VF_1\right)}{1 - VR \times VF_1} = \frac{0.72 - 0.5 \times 0.72}{1 - 0.5 \times 0.72} = 0.56 = 56\%$$

38.5 VOLATILE SOLIDS DESTROYED

This is another way of expressing volatile solids destroyed during digestion. It can be found by knowing the feed rate of sludge, volatile fraction, reduction and volume of the digester. It is illustrated in the following example problem.

EXAMPLE PROBLEM 38.13

A digester 30 ft in diameter and with 25 ft of water depth is fed daily at the rate of 1100 gal of feed sludge containing 6.5% total solids. The volatile fraction in the feed sludge is 68%, and that in the digested sludge is 50%. Find the percent reduction in volatile matter and express it as volatile solids destroyed per unit volume of the digester.

Given:

$VF_1 = 68\% = 0.68$ $VF_2 = 5\% = 0.5$ $Q_1 = 1100$ gal/d $SS_1 = 6.5\%$

Solution:

Digester capacity

$$V_{dig} = \frac{\pi D^2}{4} \times d = \frac{\pi \left(32\,ft\right)^2}{4} \times 25\,ft = 20106\,ft^3 = \underline{20000\,ft^3}$$

Volatile solids reduction

$$VR = \frac{\left(VF_1 - VF_2\right)}{VF_1 - VF_1 \times VF_2} = \frac{0.68 - 0.50}{0.68 - 0.68 \times 0.50} = 0.529 = \underline{53\%}$$

Volatile solids destroyed

$$M_{VS(des)} = \frac{1100\,gal}{d} \times \frac{8.34\,lb}{gal} \times 0.065 \times 0.68 \times 0.529 = 3.30 \times 10^4 = 214.5 = \underline{214\,lb/d}$$

Volatile solids destroyed loading

$$VSL = \frac{214.5\,lb}{d} \times \frac{1}{20106\,ft^3} = 0.0106 = \underline{0.011\,lb/ft^3.d}$$

38.6   DIGESTER SOLIDS MASS BALANCE

The principle of mass balance dictates what enters a process or a combination of processes in series, must have the same mass in the effluent streams. If you can account for 90% of this solid material leaving your plant in the form of sludge, liquid (effluent) or gas (digester gas), then you have control of your plant. This accounting process provides a good check on your metering devices, sampling procedures and analytical techniques. The mass entering the digester is sludge which is composed of solids and water. During the digestion process, part of the volatile solids is converted to gases such as methane and carbon dioxide, and the rest leaves the digester as supernatant and digested sludge.

38.7   ANAEROBIC DIGESTER CAPACITY

The determination of digester tank volume is a critical step in the design of an anaerobic system. The digester volume must be sufficient to prevent the process from failing under all accepted conditions. Process failure is defined as the accumulation of volatile acids that result in a decrease in pH, what is usually called pickling of the sludge. When the **volatile acids/alkalinity** ratio becomes greater than 0.5, acid-forming bacteria takes over, the second stage virtually stops and no methane is produced. Once the digester turns sour, it usually takes several days to return to normal operation after corrective actions are taken.

The digester capacity must also be large enough to ensure that raw sludge is adequately stabilized and there is room for storage if needed. The relationship between volume of sludge reduction and detention time is shown in Figure 38.4. During digestion, the reduction in volume is assumed to be linear. Hence, in a single-stage digester, the volume of the digester equals the area of the trapezoid and area of the rectangle, as seen in Figure 38.4.

Single tank digester capacity

$$V = \frac{(Q_1 + Q_2)}{2} \times T_1 + Q_2 \times T_2$$

Q_1 = *pump rate of feed sludge*
Q_2 = *rate of digested sludge*
T_1 = *digestion period*
T_2 = *sstorage period*

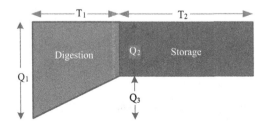

FIGURE 38.4   Anaerobic digester capacity.

EXAMPLE PROBLEM 38.14

What is the hydraulic digestion time for a 49-ft-diameter digester with a level of 9.2 ft and sludge flow of 8200 gallons per day?

Given:

$D = 49$ ft $d = 9.2$ ft $Q = 8200$ gal/d $T_d = ?$

Solution:

Detention time

$$t_d = \frac{V}{Q} = \frac{\pi (49\,ft)^2}{4} \times 9.2\,ft \times \frac{d}{8200\,gal} \times \frac{7.48\,gal}{ft^3} = 15.8 = \underline{16\,d}$$

EXAMPLE PROBLEM 38.15

Work out the digester capacity required for a low-rate anaerobic digester to serve a population of 50,000 people. Assume the solids contribution in the production of sludge is 90 g/c·d, with a solid content of sludge of 4.0% and volatile fraction of 70%. Assume a digestion period of 25 d, storage period of 20 d and volatile reduction of 50% during digestion, and the solid content of digested sludge is 8.0%.

Given:

$SS_1 = 4.0\%$ $VF_1 = 70\%$ $T_1 = 25$ d.
$SS_2 = 8.0\%$ $VF_2 = 50\% \times 70\% = 35\%$ $T_2 = 20$ d

Solution:

Feed sludge (1)

$$M_{SS} = \frac{90\,g}{p.d} \times 50000\,p \times \frac{kg}{1000\,g} = 4500\,kg/d$$

$$V_1 = \frac{M_{SS}}{SS} = \frac{4500\,kg}{d} \times \frac{m^3}{40\,kg} = 112.5 = 110\,m^3/d$$

Digested sludge (2)

$$M_{SS} = \frac{4500\,kg}{d} \times (0.30 + 0.70 \times 0.50) = 2925 = 2900\,kg/d$$

$$V_2 = \frac{M_{SS}}{SS} = \frac{2925\,kg}{d} \times \frac{m^3}{80\,kg} = 36.56 = 37\,m^3/d$$

Digester capacity

$$V = \frac{(112.5 + 36.56)\,m^3}{2d} \times 25\,d + \frac{36.56\,m^3}{d} \times 20\,d = 2594 = \underline{2600\,m^3}$$

38.8   OPERATION OF ANAEROBIC DIGESTERS

During the operation of an anaerobic sludge digester, it is important to monitor volatile solids, alkalinity, gas production and composition of gas. Operating a digester by monitoring pH is not recommended since damage is already done before the pH changes.

38.8.1   VOLATILE ACID/ALKALINITY RATIO

As long as the volatile acid/alkalinity ratio remains low, the anaerobic sludge process will occur optimally. However, if this ratio starts to increase, it is the first indication of the digestion process becoming sour. If correct action is not taken immediately, it may lead to the failure of the digestion process. Since it is a biological process, it takes time to bring it back to normal operation. Sometimes caustic is added to bring up the pH of a soured sludge. A jar test is carried out to determine the optimum dosage. Details about volatile acid and alkalinity tests can be found in Chapter 40 on laboratory tests.

EXAMPLE PROBLEM 38.16

Jar tests indicate that a caustic dose of 17 mg/L is required to raise the pH of sludge to the neutral range. If the digester is 50 ft in diameter with a side water depth of 12 ft, find how much caustic is required to adjust the pH of the sludge.

Given:

C = 17 mg/L = 17 ppm   D = 50 ft   d = 12 ft

Solution:

Volume of sludge

$$V_{sl} = \frac{\pi D^2 d}{4} = \frac{\pi (15\,ft)^2}{4} \times 12\,ft = 23561 = \underline{24000\,ft^3}$$

Quantity of caustic required

$$m_{cau} = C \times V = \frac{17\,lb}{10^6\,lb} \times \frac{62.4\,lb}{ft^3} \times 23561\,ft^3 = 24.99 = \underline{25\,lb}$$

38.8.2 ESTIMATING GAS PRODUCTION

As a result of volatile solids reduction during anaerobic digestion, carbon dioxide and methane gas is produced. During operation of a anaerobic digester, gas production and its composition are monitored and recorded. The carbon dioxide content is usually tested once or twice a week. Any change in production of CO_2 in the digester gas is an indicator of the health of the digester. During optimum operation, digester gas will have a CO_2 content of 30–35%. Before CO_2 content begins climbing, the volatile solids to alkalinity ratio starts to increase. If the CO_2 content of the digester gases exceeds 40%, it indicates the digestion process is poor and needs attention. Gas production should range between 440–750 L per kg of volatile solids destroyed during digestion.

EXAMPLE PROBLEM 38.17

An anaerobic sludge digester is fed at the rate of 4100 gal/d of raw sludge with solid content of 6.5%, of which 69% are volatile. What is the percent reduction in volatile matter if the volatile content of the digested sludge is 51%? Estimate the gas production, assuming 15 ft³ of gas per lb of volatile solids destroyed.

Given:

$Q_1 = 4100$ gal/d $SS_1 = 6.5\%$ $VF_1 = 69\% = 0.69$ $VF_2 = 51\% = 0.51$
Gas = 15 ft³/lb VS destroyed

Solution:

Fraction of volatile solids reduced/destroyed

$$VR = \frac{\left(VF_1 - VF_2\right)}{VF_1 - VF_1 \times VF_2} = \frac{0.69 - 0.51}{0.69 - 0.69 \times 0.51} = 0.532 = \underline{53\%}$$

Mass of volatile solids fed

$$M_{VS1} = \frac{4100\,gal}{d} \times \frac{8.34\,lb}{gal} \times 0.065 \times 0.69 = 1533.6\,lb/d$$

Mass of volatile solids destroyed

$$M_{VSR} = \frac{1533.6\,lb}{d} \times 0.532 = 815.8 = \underline{820\,lb/d}$$

Gas production

$$Q_{gas} = \frac{815.8\,lb}{d} \times \frac{15\,ft^3}{lb} = 12238 = \underline{12\,000\,ft^3/d}$$

38.9 DEWATERING OF SLUDGES

Digested sludge or blended sludge usually contains more than 90% water. It would be expensive to carry this watery sludge for disposal. Hence, sludge is dewatered before disposal. Depending on the method of **dewatering**, sludge water content can be as low as 20–30%, thus reducing the volume many times. In warmer climates, the most common practice is to spread the sludge on **drying** beds and let it dry in the open. When space required for dry beds is not available, sludge is conditioned, followed by mechanical dewatering. Mechanical dewatering methods include vacuum filters, filter presses or centrifugation.

38.9.1 SLUDGE DRYING BEDS

This method can be used in all places where adequate land is available, and dried sludge can be used for soil conditioning. Sludge drying beds are affected by weather, sludge characteristics, system design including depth of sludge layer and frequency of scraping the beds after drying.

38.9.1.1 Area of Beds

The land area needed for dewatering and drying the sludge is dependent on the volume of sludge. The cycle time between two successive dryings depends on the characteristics of sludge, including water content, drainage ability, rate of evaporation and acceptable water content in the dried sludge.

38.9.1.2 Drying Beds

A sludge drying bed usually consists of a bottom layer of gravel of uniform size, which is laid over a bed of clean sand. Graded gravel is placed around the underdrain in layers up to 30 cm with a minimum of 15 cm above the top of under drains. The top layer should be at least 3.0 cm thick and consist of gravel of 3–6 mm in size. Gravel should be covered with a 20–30-cm-thick layer of clean sand of effective size of 0.50–0.75 mm and uniformity coefficient less than 4.0. The finished surface should be level. Open-jointed under drainpipes of diameter 100–150 mm are laid with spacing of 6.0 m or less. The pipes are laid on a grade of 1%. The beds are about 15m × 30m in plan and are surrounded by about 1.0-m-high brick walls above the sand surface.

EXAMPLE PROBLEM 38.18

Find the number of beds (30 m × 15 m) required for drying digested sludge, discussed in Example Problem 38.15. The population served is 50,000, and the average production of digested sludge is 37 m^3/d. Assume 25 cm of digested sludge is spread every couple of weeks. Though drying in summer is done over 10 d, a drying cycle of 14 days is assumed to compensate for wet weather.

Given:

$Q_{sl} = 37 \ m^3/d$ $t = 14 \ d$ $d = 25 \ cm/cycle$ Area of each bed = 30 m × 15 m

Solution:

Area of drying beds

$$Area, A = \frac{37\,m^3}{d} \times \frac{cycle}{0.25\,m} \times \frac{14\,d}{cycle} = 2072\,m^2$$

Number of drying beds

$$\# = 2072\,m^2 \times \frac{bed}{30\,m \times 15\,m} = 4.6 = 5\,beds\,say$$

Making 100% allowance for space for storage, repairs and resting of beds, it is suggested to have 10 beds each 30 m × 15 m.

38.10   MASS VOLUME RELATIONSHIP

The solids concentration of sludges varies depending on point of production, removal of solids and specific gravity of sludge. As mentioned earlier, for sludge with solid concentration of less than 10%, it is safe to assume the SG of the sludge is close to that of water. Solids in sludge consist of volatile and non-volatile or fixed solids. Non-volatile solids, being inorganic in nature, are heavier. The specific gravity of total solids in sludge can be found as follows:

Specific gravity of TSS

$$\frac{1}{SG_{TSS}} = \frac{VSS_f}{SG_{VSS}} + \frac{FSS_f}{SG_{Fss}}$$

Sub f = decimal fraction   TSS = total suspended solids
VSS = volatile suspended solids   FSS = fixed suspended solids

Wet sludge or slurry consists of dry solids and water. Knowing the specific gravity of solids in sludge, the SG of sludge slurry can be found.

Specific gravity of sludge

$$\frac{1}{SG_{SL}} = \frac{SS_f}{SG_{SS}} + \frac{WC_f}{1} \quad or \quad SG_{SL} = \frac{SG_{SS}}{\left(SS_f + SG_{SS} \times WC_f\right)}$$

Sub f = decimal fraction   SL = sludge slurry
SS = dry solids   WC = water content

EXAMPLE PROBLEM 38.19

Determine the volume of wet sludge before and after digestion produced for every 1000 lb for dry solids as feed sludge with the characteristics shown in the following table:

Parameter	Feed Sludge, 1	Digested Sludge, 2
SS, %	5.5	11
VF, %	65	60% destroyed
SG_{FS}	2.5	2.5
SG_{VS}	1.0	1.0

Given:

$$SS_1 = 5.5\% \qquad SS_2 = 11\% \qquad VF_1 = 65\%$$

Solution:

Specific gravity of dry solids in feed sludge

$$\frac{1}{SG_{TSS}} = \frac{VSS_f}{SG_{VSS}} + \frac{FSS_f}{SG_{FSS}} = \frac{0.65}{1} + \frac{0.35}{2.5} = 0.79$$

$$SG_{TSS} = \frac{1}{0.79} = 1.265$$

Specific gravity of wet sludge (1)

$$SG_1 = \frac{SG_{SS}}{\left(SS_f + SG_{SS} \times WC_f\right)} = \frac{1.265}{\left(0.055 + 1.265 \times 0.945\right)} = 1.0116$$

Volume of feed sludge

$$V_1 = \frac{M_{SS}}{SS} = 1000\,lb \times \frac{gal}{8.34\,lb \times 0.055 \times 1.012} = 2154 = \underline{2150\,gal}$$

Mass of solids in the digested sludge (2)
$M_{ss} = fixed + volatile = 1000\ lb \times (0.35 + 0.65 \times 0.40) = 610\ lb$

Volatile fraction of digested sludge solids

$$VF = \frac{M_{SS}}{M_{TSS}} = \frac{0.65 \times 0.4 \times 1000\,lb}{610\,lb} = 0.436 = 44\%$$

Specific gravity of sludge solids in digested solids (2)

$$\frac{1}{SG_{TSS}} = \frac{VSS_f}{SG_{VSS}} + \frac{FSS_f}{SG_{FSS}} = \frac{0.44}{1} + \frac{0.56}{2.5} = 0.718$$

$$SG_{TSS} = \frac{1}{0.66} = 1.51$$

Specific gravity of digested sludge (2)

$$SG_2 = \frac{SG_{SS}}{\left(SS_f + SG_{SS} \times WC_f\right)} = \frac{1.51}{\left(0.11 + 1.51 \times 0.89\right)}$$

$$= 1.0387 = 1.039$$

Volume of digested sludge

$$V_2 = \frac{M_{SS}}{SS} = 610\,lb \times \frac{gal}{0.11 \times 1.0387 \times 8.34\,lb}$$

$$= 640.14 = \underline{640\,gal}$$

EXAMPLE PROBLEM 38.20

Determine the volume of wet sludge before and after digestion produced for every 1000 kg for dry solids as feed sludge with the characteristics shown in the following table:

Parameter	Feed Sludge, 1	Digested Sludge, 2
SS, %	6.5	12
VF, %	70	50% destroyed
SG_{FS}	2.5	2.5
SG_{VS}	1.0	1.0

Given:

$SS_1 = 6.5\%$ $SS_2 = 12\%$ $VF_1 = 70\%$

Solution:

Specific gravity of sludge solids in the feed sludge (1)

$$\frac{1}{SG_{TSS}} = \frac{VSS_f}{SG_{VSS}} + \frac{FSS_f}{SG_{FSS}} = \frac{0.70}{1} + \frac{0.30}{2.5} = 0.82$$

$$SG_{TSS} = \frac{1}{0.82} = 1.219 = 1.22$$

Specific gravity of wet feed sludge (1)

$$SG_1 = \frac{SG_{SS}}{\left(SS_f + SG_{SS} \times WC_f\right)} = \frac{1.22}{\left(0.065 + 1.22 \times 0.935\right)} = 1.011$$

Volume of feed sludge (1)

$$V_1 = \frac{M_{SS}}{SS} = 1000\,kg \times \frac{m^3}{1000\,kg \times 0.065 \times 1.011} = 15.20 = \underline{15\,m^3}$$

Mass of solids in the digested sludge (2)

$M_{ss} = fixed + volatile = 1000\ kg \times (0.30 + 0.70 \times 0.50) = 650\ kg$

Volatile fraction of digested sludge solids

$$VF = \frac{M_{VSS}}{M_{TSS}} = \frac{0.70 \times 0.50 \times 1000\ kg}{7545\ kg} = 0.538 = 54\%$$

Specific gravity of digested sludge solids (2)

$$\frac{1}{SG_{TSS}} = \frac{VSS_f}{SG_{VSS}} + \frac{FSS_f}{SG_{Fss}} = \frac{0.54}{1} + \frac{0.46}{2.5} = 0.72$$

$$SG_{TSS} = \frac{1}{0.72} = 1.38$$

Specific gravity of digested sludge (2)

$$SG_2 = \frac{SG_{SS}}{\left(SS_f + SG_{SS} \times WC_f\right)} = \frac{1.38}{\left(0.12 + 1.38 \times 0.88\right)} = 1.034$$

Volume of digested sludge (2)

$$V_2 = \frac{M_{SS}}{SS} = 650\ kg \times \frac{m^3}{0.12 \times 1.034 \times 1000\ kg} = 5.23 = \underline{5.2\ m^3}$$

PRACTICE PROBLEMS

PRACTICE PROBLEM 38.1

The flow of primary sludge with solid content of 4.5% to a 10.5-m-diameter gravity thickener is 250 m³/d. What is the solid loading rate in kg/m² · d? (130 kg/m² · d)

PRACTICE PROBLEM 38.2

Calculate the solid removal efficiency and concentration factor if the influent sludge to a gravity thickener is 2.8% solids. The solid content of overflow (effluent) and underflow (thickened sludge) respectively are 0.6% and 5.8%. (79%, 2.1×)

PRACTICE PROBLEM 38.3

For the data of the previous problem, calculate the change in solids contained in a sludge blanket if the amount of sludge withdrawn is 220 m³/d. (Increase of 4400 kg/d)

PRACTICE PROBLEM 38.4

A primary sludge flow of 27 m³/d containing 3.8% solids is mixed with a thickened secondary sludge flow of 154 m³/d containing 5.0% solids. What is the percent of solids of the combined sludge flow? (4.2%)

PRACTICE PROBLEM 38.5

A conventional activated sludge plant treats 7.0 MGD of municipal wastewater with BOD content of 240 mg/L and SS of 200 mg/L. Estimate the quantities of the primary and secondary sludges. Assume 50% SS removal and 35% BOD removal in the primary, a raw sludge solids content of 4.0%, 50% of BOD converted to new cells and waste sludge solids concentration of 1.5%. (17500 gal/d, 36400 gal/d)

PRACTICE PROBLEM 38.6

For the data of Practice Problem 38.5, secondary sludge is thickened before pumping it to anaerobic digester. During thickening, 98% of SS are captured and the sludge is concentrated to 5.0% solids. Estimate the quantities and SS content of the blend of primary and thickened sludge. (25000 gal/d, 4.3%)

PRACTICE PROBLEM 38.7

For the conventional plant discussed in Practice Problem 38.6, estimate the tonnage of blended sludge produced daily. (100 t/d)

PRACTICE PROBLEM 38.8

The average daily flow to a wastewater treatment plant is 6.5 ML/d. On average, primary treatment removes 120 mg/L of suspended solids and produces raw sludge containing

2.8% solids. Waste sludge from activated sludge contains 0.9% solids. Waste sludge is produced as result of 120 mg/L of BOD removal, and each unit of BOD is converted to 0.6 unit of solids. Estimate the biosolids production in wet metric tons/d. (2600 t)

PRACTICE PROBLEM 38.9

For the data of Example Problem 38.8, calculate the volume of sludge pumped if the length is reduced to 7.5 cm. (130 L/min)

PRACTICE PROBLEM 38.10

A digester 10 m in diameter has a side water depth of 5.0 m. If the seed sludge to be used is 25% of the tank capacity, how many cubic metres of seed sludge will be required? (98 m^3)

PRACTICE PROBLEM 38.11

Calculate the percent reduction in volatile acids if the fraction of volatiles entering the digester is 70% and that leaving the digester is 45%. (33%)

PRACTICE PROBLEM 38.12

The volatile content of the sludge fed to the anaerobic digester is 75%. If it is desired to have a minimum 50% reduction of volatile solids during anaerobic digestion, what must be the volatile fraction in the digested sludge? (60%)

PRACTICE PROBLEM 38.13

A digester 35 ft in diameter and with 25 ft of water depth is fed daily at the rate of 160 ft^3/d of feed sludge with solid content of 6.0%. The fraction of volatile solids in the feed sludge is 70%, and that in the digested sludge is 55%. Find the percent reduction in volatile matter and express it as volatile solids destroyed per unit volume of the digester. (0.008 lb/ft^3.d)

PRACTICE PROBLEM 38.14

What is the hydraulic digestion time for a 52-ft-diameter digester with a level of 9.5 ft and sludge feed rate of 6.0 gpm? (18 d)

PRACTICE PROBLEM 38.15

Size an anaerobic digester to treat a blend of primary and secondary sludge. The feed sludge daily production is 250 m^3/d. The solid content of the feed sludge and digested sludge, respectively, are 5.0% and 9.5%. Based on the prevailing temperatures, a digestion period of 25 d is considered, and a minimum storage period of 40 d is required. Assume that the density of wet sludge for solid concentrations less than 10% is the same as that of water. Make other assumptions as appropriate. (70% volatile, 60% reduction, 7200 m^3)

PRACTICE PROBLEM 38.16

One unit of lime is required to neutralize one unit of sour sludge. The digester sludge contains 2100 mg/L of volatile acids. If the digester is 35 ft in diameter with a side water depth of 15 ft, how much lime is required to adjust the pH? (1900 lb)

PRACTICE PROBLEM 38.17

An anaerobic sludge digester is fed at the rate of 3200 gal/d of raw sludge with a solid content of 6.5%, of which 71% are volatile. What is the percent reduction in volatile matter if the volatile content of the digested sludge is 52%? Estimate the gas production, assuming 12 ft^3 of gas per lb of volatile solids destroyed. (8200 ft^3/d)

PRACTICE PROBLEM 38.18

Find the number of drying beds required for digested sludge. The population served is 65,000, and the average production of digested sludge is 0.2 gal/c.d. Assume 1.0 ft of digested sludge is spread every couple of weeks. Though drying in summer is done over 12 d, a drying cycle of 14 days is assumed to compensate for wet weather. (8 beds each 100 ft × 30 ft)

PRACTICE PROBLEM 38.19

Determine the volume of wet sludge after digestion produced for every 1000 lb of dry solids as feed sludge with the characteristics shown in the following table: (710 gal)

Parameter	Feed Sludge, 1	Digested Sludge, 2
SS, %	5.5	11
VF, %	65	50% destroyed
SGFS	2.5	2.5
SGVS	1.0	1.0

PRACTICE PROBLEM 38.20

Determine the volume of wet sludge before and after digestion produced for every 100 kg for dry solids as feed sludge with the characteristics shown in the following table: (1.8 m^3, 0.65 m^3)

Parameter	Feed Sludge, 1	Digested Sludge, 2
SS, %	5.5	11
VF, %	65	40% destroyed
SGFS	2.5	2.5
SGVS	1.0	1.0

39 Head Works Loadings

The amount of pollutant that can be allowed to enter a plant depends upon the process removal efficiency and the following limiting factors:

- Pollutant discharge limitations.
- Water quality standard or criteria.
- Process inhibitions, for example, activated sludge or digestion.
- Sludge disposal standards or criteria.

The allowable head works loading is worked out in reverse mathematical order, as illustrated in Figure 39.1. Starting with the effluent quality, calculate the influent quality. Based on the maximum concentration permitted, calculate back the influent mass loading of the pollutant of concern allowable at the head end.

Influent loading

$$M_i = Q_i \times C_i = \frac{Q_e \times C_e}{(1 - PR/100)}$$

39.1 DISCHARGE LIMITATION BY PERMIT

Knowing the maximum concentration permitted in the final effluent and removal efficiency, back calculation of the allowable concentration of the pollutant of concern in the influent stream can be worked out. Mathematical relationship is shown here:

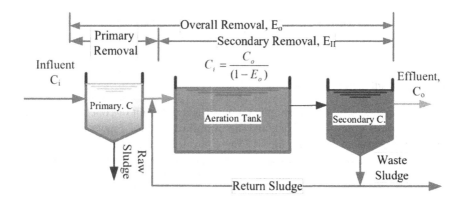

FIGURE 39.1 Back calculation for head works loading.

DOI: 10.1201/9781003468745-39

Allowable concentration

$$C_i = \frac{C_p}{(1 - PR/100)}$$

PR = *overall removal of the specific contaminant*
C_p = *concentration of the pollutant in the final effluent*

39.2 ACTIVATED SLUDGE PROCESS INHIBITION

To prevent the inhibition of the activated sludge process, the maximum loading of the pollutant of concern allowed into the aeration tank (primary effluent) should not exceed the process inhibiting concentration or loading.

Allowed influent

$$C_i = \frac{C_{i/AS}}{(1 - PR_1/100)}$$

$C_{I/AS}$ = *concentration inhibiting activated sludge process*
PR_I = *primary percent removal*

39.3 ANAEROBIC SLUDGE PROCESS INHIBITION

Sludge is the by-product of the wastewater treatment process. Whatever is removed in this process becomes sludge.

Maximum mass loading in sludge

$$M_{sl} = Q_{sl} \times C_{i/an} \frac{Q_e \times C_e}{(1 - PR/100)} \quad M_i = \frac{Q_{sl} \times C_{i/an}}{(1 - PR/100)} \quad C_i = \frac{Q_{sl} \times C_{i/an}}{Q_i (1 - PR/100)}$$

M_i = *maximum allowable pollutant* C_i = *allowable influent concentration*
Q_{sl} = *volume of sludge pumped* $CI_{/an}$ = *inhibiting concentration of the pollutant*

EXAMPLE PROBLEM 39.1

In a municipal sewage treatment plant, copper is a pollutant of concern. The daily average flow of water is 40 ML/d, and the overall removal of copper in the plant is 83%, with the primary accounting for 25% removal. If the discharge criteria require copper concentration not to exceed 0.10 mg/L in the final effluent, calculate the maximum allowable head works loading.

Given:

$Q = 40$ ML/d $E_I = 25\%$ $PR = 83\%$ $C_{e\,(max)} = 0.10$ mg/L

Solution:

Maximum concentration of Cu in the plant influent

$$C_i = \frac{C_p}{(1-PR/100)} = \frac{0.10\,mg/L}{(1-0.83)} = 0.588 = \underline{0.59\,mg/L}$$

Allowable head works loading of Cu

$$M_i = Q_i \times C_i = \frac{40\,ML}{d} \times \frac{0.588\,kg}{ML} = 23.5 = \underline{24\,kg/d}$$

EXAMPLE PROBLEM 39.2

Primary effluent with copper more than 1.0 mg/L inhibits the growth of microorganisms in the activated sludge process. For the municipal plant referred to in Example Problem 39.1, what is the maximum amount of copper that can be allowed to enter the plant without inhibiting the activated sludge process?

Given:

$Q = 40$ ML/d $PR_I = 25\%$ C_e (Primary) $= C_{I/AS} = 1.0$ mg/L

Solution:

Allowable head works loading of Cu without inhibiting ASP

$$M_i = \frac{Q_i \times C_{i/as}}{(1-PR/100)} = \frac{40\,ML}{d} \times \frac{1.0\,kg}{ML} \times \frac{1}{(1-0.25)} = 53.3 = \underline{53\,kg/d}$$

EXAMPLE PROBLEM 39.3

In the case of the municipal plant discussed earlier, the average daily production of sludge is 160 kL/d. If a concentration of copper more than 5.0 mg/L inhibits the sludge digestion process, what maximum loading at the head works can be allowed?

Given:

$Q_{SL} = 160$ kL/d $C_{I/AN} = 1.5$ mg/L (g/kL) $PR = 83\%$

Solution:

Allowable loading of Cu in anaerobic sludge

$$M_i = Q_{sl} \times C_{i/an} = \frac{160\,kL}{d} \times \frac{5.0\,g}{kL} \times \frac{kg}{1000\,g} = 0.80\,kg/d$$

Allowable head works loading of Cu

$$M_i = \frac{M_{sl}}{PR/100} = \frac{0.80\,kg}{L} \times \frac{1}{0.83} = 0.963 = \underline{0.96\,kg/L}$$

Obviously, head works loading based on this criterion is more stringent.

EXAMPLE PROBLEM 39.4

The average concentrations of phosphorus in the influent and effluent of 16 ML/d plant are 5.6 mg/l and 4.1 mg/L, respectively. The phosphorus removal in the primary treatment is on average 10%. What is the phosphorus removal in the second stage of treatment?

Given:

Primary treatment: $C_i = 5.6$ mg/L $PR_I = 10\%$
Secondary treatment: $C_e = 4.1$ mg/L $PR_{II} = ?$

Solution:

Phosphorus content in primary effluent

$$C_{PE} = \frac{C_{PI}}{\left(1 - PR/100_I\right)} = \frac{5.6\,mg/L}{\left(1 - 0.10\right)} = 5.04 = 5.0\,mg/L$$

Note primary effluent is secondary influent.

Percent removal secondary

$$PR_{II} = \left(1 - \frac{C_e}{C_i}\right) = \left(1 - \frac{4.1\,mg/L}{5.04\,mg/L}\right) = 0.189 = \underline{19\%}$$

EXAMPLE PROBLEM 39.5

On average, a sewage treatment plant discharges 28 ML/d of treated wastewater into a river that enters a lake a short distance downstream. To preserve the quality of lake water, the permissible phosphorus loading from the plant is limited to 30 kg/d. The phosphorus removal at the plant is on average 40%. Calculate the allowable phosphorus level in the raw wastewater.

Given:

M_e = 30 kg/d PR_o = 40% Q_i = Q_e = 28 ML/d C_i = ?

Solution:

Mass load of phosphorus allowed in raw wastewater

$$M_i = \frac{M_e}{\left(1 - PR_o/100\right)} = \frac{30\,kg}{d} \times \frac{1}{\left(1 - 0.40\right)} = 50\,kg/d$$

Allowable phosphorus concentration in plant influent

$$C_i = \frac{M_i}{Q_i} = \frac{50\,kg}{d} \times \frac{d}{28\,ML} = 1.78 = \underline{1.8\,mg/L}$$

EXAMPLE PROBLEM 39.6

The daily average flow from a STP is 22 ML/d. The plant effluent is discharged into a river. On average, phosphorus in the raw wastewater is 4.8 mg/L, and overall removal is 30%. What is the phosphorus level in the plant effluent and loading into the river?

Given:

M_e = 30 kg/d PR_o = 40% Q_i = Q_e = 28 ML/d C_i = ?

Solution:

Concentration of phosphorus in the plant effluent

$$C_e = C_i \times \left(1 - \frac{PR_O}{100}\right) = \frac{4.8\,mg}{L} \times \left(1 - 0.30\right) = 3.36 = 3.4\,mg/L$$

Phosphorus loading into the river water

$$M_e = Q \times C_e = \frac{22\,ML}{d} \times \frac{3.36\,kg}{ML} = 73.9 = \underline{74\ kg/d}$$

PRACTICE PROBLEMS

PRACTICE PROBLEM 39.1

If the pollutant of concern is chromium and the discharge limit is 0.05 mg/L, calculate the allowable head works loading, assuming overall removal of 7.5% by the plant. (8 kg/d)

PRACTICE PROBLEM 39.2

For the same plant, if the primary removal for chromium is typically 35% and the inhibiting concentration is 0.5 mg/L, calculate the allowable head works loading. (31 kg/d)

PRACTICE PROBLEM 39.3

Work out the allowable concentration of chromium if the inhibiting concentration of chromium for the anaerobic digestion process is 1.5 mg/L. As discussed previously, the overall removal of chromium is 75%, and daily sludge production is 160 kL/d. (8 µg/L)

PRACTICE PROBLEM 39.4

It is desired to produce a final effluent with a phosphorus level not exceeding 1.0 mg/L. It is intended to enhance phosphorus removal by the addition of alum before the final clarification. What should be the minimum removal by the secondary treatment to prolduce an effluent containing 1.0 mg/l of phosphorus? Also, calculate the overall phosphorus removal in this case. (80%, 82%)

PRACTICE PROBLEM 39.5

Refer to Example Problem 39.5, the permissible phosphorus level in the plant effluent is 1.0 mg/L. To meet regulations, modifications of the secondary process are carried out. Assuming 10% phosphorus removal in the primary, what should the minimum removal of the advanced biological process be to meet the effluent criteria? (77%)

PRACTICE PROBLEM 39.6

The average concentration of phosphorus in the plant influent is 6.5 mg/L. What should be the minimum removal of phosphorus in the plant to meet the criteria? (84%)

40 Laboratory Tests

Lab tests are required for two main purposes: to meet the regulatory requirements and to monitor and control plant processes. To meet some of the regulatory requirements, it may be required to use a credited lab. Some of the key tests performed in wastewater treatment labs are discussed.

40.1 SETTLEABLE SOLIDS (IMHOFF CONE)

The settleable solids measure the volume of settleable solids in 1 L of sample that will settle to the bottom of an Imhoff cone (Figure 40.1) during a specific time period, usually 1 hour. This test indicates the volume of solids that are removed by sedimentation tanks and clarification ponds. When conducted on influent and effluent samples of a sedimentation tank, this test allows us to determine the percent removal of settleable solids. The Imhoff cone test may not indicate the true performance of the clarifier.

EXAMPLE PROBLEM 40.1

Samples were collected from the influent and effluent of a primary clarifier. After 1 hour of settling, the volume of settleable solids in influent and effluent were recorded to be 12.0 mL/L and 0.2 mL/L, respectively. Calculate the percent removal of settleable solids.

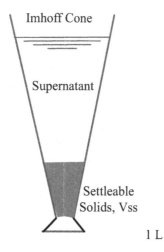

FIGURE 40.1 Settleable solids: an Imhoff cone.

DOI: 10.1201/9781003468745-40

Given:

V_{SS} (i) = 12.0 mL/L V_{SS} (e) = 0.2 mL/L

Solution:

Percent removal of settleable solids

$$PR = \left(1 - \frac{V_e}{V_i}\right) = \left(1 - \frac{0.2}{12}\right) = 0.98 = \underline{98\%}$$

EXAMPLE PROBLEM 40.2

For the data of Example Problem 40.1, estimate the volume of primary sludge if the daily wastewater flow is 4.5 MGD.

Given:

V_{ss} (r) = 12.0–0.2 = 11.8 mL/L   Q = 4.5 MG/d Q_{SI} = ?

Solution:

Volume of sludge

$$Q_{sl} = V_{ss}(r) \times Q = \frac{11.8\,mL}{L} \times \frac{L}{1000\,mL} \times \frac{4.5 \times 10^6\,gal}{d} = 53100 = \underline{53000\,gal/d}$$

40.2   SUSPENDED SOLIDS (NON-FILTERABLE)

The suspended solids test is designed to determine the concentration of suspended solids. It thus measures the strength of a given wastewater sample. The aliquot size for this test is based on the concentration of suspended solids in a given sample. For example, in the case of raw wastewater (SS = 100–400 mg/L), an aliquot of 50–100 mL is recommended. However, for precision, a large aliquot is required for the secondary effluent.

A dry filter disc is weighed (A). The selected aliquot is filtered, and the solids are retained on the filter disc. After 2 to 3 hours of drying in the oven (102°C), the dry solids and filter disc are weighed (B). To determine the volatile fraction, the solids are ignited in the muffle furnace (550°C) for about half an hour. After burning the residue, the ash left on the fibreglass disc represents the non-volatile (fixed) solids, which are primarily inorganic in nature (C).

Suspended solids

$$SS = \frac{m_{SS}}{V} = \frac{(B-A)}{V} \qquad VF = \frac{m_{VS}}{m_{SS}} = \frac{(B-C)}{(B-A)} \qquad VSS = VF \times SS$$

> *Typically, 40–60% of the total suspended solids are removed by gravity settling.*

EXAMPLE PROBLEM 40.3

A suspended solids test carried on 50 mL of a sample of plant influent yielded the following data. Calculate the concentration of suspended solids and fraction of volatiles.

Weight of the filter disc = 0.2335 g   Filter + dry solids = 0.2531 g
Filter + ash = 0.2360 g

Given:

V = 50 mL   A = 0.2335 g   B = 0.2531 g   C = 0.2375 g

Solution:

SS content in the plant influent (raw wastewater)

$$SS = \frac{(B-A)}{V} = \frac{(0.2531 - 0.2335)g}{50\,mL} \times \frac{1000\,mL}{L} \times \frac{1000\,mg}{g} = 392 = \underline{390\,mg/L}$$

Volatile fraction of SS

$$SS = \frac{(B-C)}{(B-A)} = \frac{(0.2531 - 0.2375)g}{(0.2531 - 0.2335)g} \times 100\% = 79.5 = \underline{79.5\%}$$

40.3   TOTAL SLUDGE SOLIDS

Total solids are the combined amounts of suspended and dissolved matter in the sample. This test is commonly used for wastewater sludges or where the solids concentration is greater than 1000 mg/L. Just like suspended solids, total solids are composed of two components—volatile and fixed solids. Volatile solids represent the organic fraction of solids.

The total solids test is similar to the suspended solids test, except that the sample is not filtered because of the excessive solids, and rather than volume, the sample is weighed. For the same reason, concentration of total solids is expressed as mass to mass in % as compared to mass per unit volume. The four weighings done as part of the test are as follows:

Dish (tare) = A   Dish + Sample = B   Dish + Dry solids = C   Dish + Ash = D

$$TS = \frac{m_{TS}}{m_{SL}} = \frac{(B-C)}{(B-A)} \qquad VF = \frac{m_{VTS}}{m_{SL}} = \frac{(C-D)}{(B-C)} \qquad TVS = \frac{(C-D)}{(B-A)} = VF \times TS$$

EXAMPLE PROBLEM 40.4

Given the following weighings from a solids test on a sludge sample, calculate the concentration of total solids and volatile fraction.

Given:

Dish (tare) $= 20.31$ g   Dish + Sample $= 70.31$ g
Dish + Dry Solids $= 22.81$ g   Dish + Ash $= 20.93$ g

Solution:

Total solids in the sludge

$$TS = \frac{m_{TS}}{m_{SL}} = \frac{(B-C)}{(B-A)} = \frac{(70.31-22.81)g}{(70.31-22.31)g} = 0.050 = \underline{5.0\%}$$

Volatile fraction of TS

$$VF = \frac{m_{VTS}}{m_{SL}} = \frac{(C-D)}{(B-C)} = \frac{(22.81-20.93)g}{(70.31-22.81)g} = 0.76 = \underline{76\%}$$

> *Total solids in sludges closely represent the concentration of suspended solids.*

40.4 BIOCHEMICAL OXYGEN DEMAND TEST

The biochemical oxygen demand (BOD) of a given sample is determined by measuring the DO depletion in a given sample. This depletion occurs as a result of the oxygen used by the microorganisms as they break down (biodegrade) the organic matter in the wastewater. The BOD test can't be conducted on a full-strength sample of wastewater, as there would not be enough oxygen available to last for 5 days. Therefore, the test is conducted on a diluted sample. Depending on the percent dilution, the DO depletion is also modified by the same factor to calculate the BOD of the sample.

BOD of sample

$$BOD = \frac{(D_1 - D_2)}{P} \times 100 = (D_1 - D_2) \times DF \qquad P = \frac{Voume\ of\ sample\ aliquot}{Voume\ of\ diluted\ aliquot} \times 100$$

$P = percent\ dilution$ $DF = dilution\ factor = 100/P$

Selecting a proper dilution is very important for a valid BOD test. Dilution should be such that the percentage of depletion over a 5-day period is in the range of 30% to 70% and the final DO reading is not below 2.0 mg/L. Achieving this will entail

DO depletion around 4.0 mg/L. To select a proper dilution, you should know the approximate value of the BOD of the given sample. Based on the expected BOD, percent dilution can be selected as follows:

$$P = 400/(Expected\ BOD)$$

EXAMPLE PROBLEM 40.5

A BOD test was run off a raw wastewater sample by selecting 2.0% dilution (20 mL of sample diluted to 1 L). Initial and final DO readings were observed to be 8.20 mg/L and 4.15 mg/L, respectively. Calculate the BOD of the raw wastewater.

Given:

$D_1 = 8.20$ mg/L $D_2 = 4.15$ mg/L $P = 2.0\%$

Solution:

Dilution factor

$$DF = \frac{100}{P} = \frac{100\%}{2.0\%} = 50$$

BOD of raw wastewater

$$BOD = (D_1 - D_2) \times DF = (8.20 - 4.15)\,mg/L \times 50 = 202. = \underline{200\ mg/L}$$

40.5   BOD TEST WITH SEED

When a sample contains very few microorganisms as a result, of chlorination, extreme pH, industrial wastewater, etc., microorganism (seed) must be added to the sample for proper BOD test. For such situations, we also need to run a BOD test on the seed (blank). Depending on the portion of seed used in the sample test compared to seed in the blank, DO depletion due to the seed is calculated. Calculation of BOD in this case is made based on the DO depletions in the sample test and those in the blank test, as shown:

BOD of seeded sample

$$BOD = \frac{(D_1 - D_2) - f(B_1 - B_2)}{P/100} \qquad f = \frac{Portion\ of\ seed\ in\ sample}{Portion\ of\ seed\ in\ blank}$$

B_1, B_2 = *initial and final DO readings in the blank test*
$D_1 - D_2$ = *initial and final DO readings in the seeded sample test*

If the seed is added to directly to the dilution water, $f = 1 - P/100$. For example, if P is 90%, then the corresponding f is 0.10. Thus, to apply seed correction, you need to subtract 10% of the BOD exerted by the seed.

EXAMPLE PROBLEM 40.6

A sample of final effluent is dechlorinated, and a BOD test is carried out, resulting in the following data. Determine the BOD of the plant effluent.

Given:

	Dilution, mL/L		DO, mg/L	
Test	Sample	Seed	Initial	Final
Sample	400	2.0	8.45	4.45
Blank	0.0	10	8.80	5.40

Solution:

BOD concentration of plant effluent

$$BOD = (D_1 - D_2) - f(B_1 - B_2) \times F$$

$$= \left[(8.45 - 4.45) - \frac{2}{10}(8.80 - 5.40)\right] \frac{mg}{L} \times \frac{1000 \ mL}{400 \ mL} = 8.30 = \underline{8.3 \ mg/L}$$

40.6   COD TEST

COD is the chemical oxygen demand of a given water sample. It is the amount of oxygen required for oxidation to carbon dioxide and water without any regard for biodegradability of the waste. Therefore, it is difficult to predict the effects of an effluent on the DO in receiving waters and the treatability of particular wastewater by biological processes. The COD test is also used to measure the strength of the waste that is too toxic for the BOD test.

The COD test method oxidizes organics substances in the wastewater sample using potassium dichromate (PDC) in 50% sulphuric acid solution. Silver sulphate is used as a catalyst, and mercuric sulphate is added to remove chloride interference. The dichromate is titrated with standard ferrous ammonium sulphate (FAS), using ferroin as an indicator. The advantage of the COD test is that it is quick (3 to 4 hours) and is a good measure of the strength of a given wastewater.

40.6.1   STANDARDIZATION OF FAS SOLUTION

$$N_1 = N_2 \times \frac{V_2}{V_1}$$

1 = FAS solution   2 = PDC solution

40.6.2 SAMPLE TEST

Normality of the sample

$$N_1 = N_2 \times \frac{V_2}{V_1} = N_1 \times \frac{(A-B)}{V_2} \quad or \quad COD = N_2 \times \frac{8000\,mg}{eq}$$

A = mL FAS used for blank   B = mL FAS used for sample
V = mL sample 1, 2 = FAS, Sample

EXAMPLE PROBLEM 40.7

In standardization of FAS solution, 11 mL of FAS were used for 10 mL of 0.25 N PDC. 20 mL of the sample was taken and digested with 10 mL of PDC. The same test was conducted with a blank (no sample, only dilution water). The digested solution was titrated with FAS solution. The volume of FAS used for the blank and sample were observed to be 10.0 mL and 3.0 mL, respectively. What is the COD of the sample?

Given:

$A = 10.0\,mL$	$B = 3.0\,mL$	$V = 20\,mL$
$V_1 = 11\,mL$	$V_2 = 10\,mL$	$N_2 = 0.25\,N$

Solution:

Normality of the sample

$$N_1 = N_2 \times \frac{V_2}{V_1} = \frac{0.25\,eq}{L} \times \frac{10\,mL}{11\,mL} = 0.227 = \underline{0.23\,eq/L}$$

COD of the sample

$$COD = N_1 \times \frac{(A-B)}{V_2} = \frac{0.227\,eq}{L} \times \frac{(10.0-3.0)mL}{20\,mL} \times \frac{8000\,mg}{eq} = 636.6 = \underline{640\,mg/L}$$

40.7 SETTLEABILITY TEST

The settleability test is used to study the settling characteristics and density of the activated sludge. A 1-L sample of mixed liquor is poured into a graduated cylinder, and the volume occupied by sludge solids is observed. A suspended solids test is also conducted on the same sample. This enables us to calculate sludge volume index (SVI) or sludge density index (SDI). Sludge volume index is defined as the volume of sludge occupied by 1 gram of solids.

Sludge settleability indices

$$\boxed{SVI(mL/g) = \frac{V_{SS}}{MLSS} \qquad SDI(\%) = \frac{100}{SVI}}$$

Looking closely, SVI indirectly illustrates the density of the sludge. As concentration is the mass of solids contained in a given volume of sludge, SVI is essentially the inverse of the solid concentration in the sludge. Hence, lower values of SVI indicate a quick-settling denser sludge. Sludge density index is the inverse of SVI expressed as percent solids. SDI is the concentration of solids in the sludge after 30 minutes of settling. The settleability test is a useful test to indicate changes in sludge characteristics. The proper SVI range for a given facility is determined at the time when removal of BOD and SS is best.

EXAMPLE PROBLEM 40.8

The MLSS concentration in an aeration tank is 2400 mg/L. If the volume of settled sludge after 30 minutes of settling is observed to be 220 mL/L, calculate the SVI and SDI.

Given:

MLSS = 2400 mg/L V_{SS} = 240 mL/L

Solution:

Sludge volume index

$$SVI = \frac{V_{SS}}{MLSS} = \frac{240\,mL}{L} \times \frac{L}{2400\,mg} \times \frac{1000\,mg}{g} = 100.0 = \underline{100\,mL/g}$$

Sludge density index

$$SDI = \frac{1}{SVI} = \frac{g}{100\,mL} \times \frac{mL}{1.0\,g} \times 100\% = 1.00 = \underline{1.0\%}$$

EXAMPLE PROBLEM 40.9

A 30-minute settleability test indicates 18% of settleable solids. Determine SVI and SDI given an MLSS concentration of 1500 mg/L.

Given:

MLSS = 1500 mg/L V_{SS} = 18% = 180 mL/L

Solution:

Sludge volume index

$$SVI = \frac{V_{SS}}{MLSS} = \frac{180\,mL}{L} \times \frac{L}{1500\,mg} \times \frac{1000\,mg}{g} = 120.0 = \underline{120\,mL/g}$$

Sludge density index

$$SDI = \frac{1}{SVI} = \frac{g}{120\,mL} \times \frac{mL}{1.0\,g} \times 100\% = 0.833 = \underline{0.83\%}$$

EXAMPLE PROBLEM 40.10

A settleability test was performed on a sample of mixed liquor, and the following data were observed:

MLSS = 2850 mg/L V_{SS} = 688 mL/2L   Sample size = 2.0 L

Determine the sludge volume index

Given:

MLSS = 2800 mg/L V_{SS} = 688 mL/2L   SVI = ?

Solution:

Sludge volume index

$$SVI = \frac{V_{SS}}{MLSS} = \frac{688\,mL}{2.0\,L} \times \frac{L}{2850\,mg} \times \frac{1000\,mg}{g} = 120.7 = \underline{120\,mL/g}$$

Sludge density index

$$SDI = \frac{1}{SVI} = \frac{g}{120.7\,mL} \times \frac{mL}{1.0\,g} \times 100\% = 0.828 = \underline{0.83\%}$$

40.8   OXYGEN UPTAKE RATE TEST

Oxygen uptake rate with regard to the activated sludge process is the rate at which the microorganisms in the aeration tank utilize oxygen. This essentially represents the respiration rate of microorganisms, which depends on the BOD loading characterized by the F/M ratio. The oxygen uptake rate (OUR) is commonly expressed as mg/L and can be calculated by observing the drop in dissolved oxygen in a given sample of mixed liquor. Based on the observed data, the average rate of oxygen depletion (uptake rate) is calculated and expressed as mg/L ·h. The DO drop rate should be uniform over the chosen time interval.

Oxygen uptake rate and specific uptake rate (SUR)

$$OUR = \frac{\Delta DO}{\Delta t} = \frac{(DO_1 - DO_1)}{t_2 - t_1} \qquad SUR = \frac{OUR}{MLSS}$$

Respiration rate or uptake rate varies according to the solids concentration in the aeration tank (MLSS). As a result, the uptake rate is frequently expressed as the specific uptake rate, which is expressed as mg of O_2 utilized by 1 gram of mixed liquor solids in 1 hour. The most common units of expression are mg/g.h. SUR is more useful when comparing the uptake rate of different plants. Oxygen uptake rates are lowest for extended aeration systems, typically about 10 mg/L.h due to relatively low loading, and can be as high as 100 mg/L.h for high-rate systems. For conventional activated sludge processes, OUR is typically around 30 mg/L.h.

EXAMPLE PROBLEM 40.11

The dissolved oxygen drop in a sample of mixed liquor is observed over a period of 10 min. During this period, DO drops from initial value of 7.8 mg/L to 3.0 mg/L. Calculate the oxygen uptake rate and specific uptake rate knowing MLSS is 2400 mg/L.

Given:

DO$_2$ (10 min) = 3.0 mg/L   DO$_1$ (initial) = 7.8 mg/L   MLSS = 2400 mg/L Δt = 10 min

Solution:

Oxygen uptake rate

$$OUR = \frac{(DO_1 - DO_2)}{t_2 - t_1} = \frac{(7.8 - 3.0) mg/L}{10 min} \times \frac{60 min}{h} = 28.8 = \underline{29\, mg/L.h}$$

Specific uptake rate

$$SUR = \frac{OUR}{MLSS} = \frac{28.8\, mg}{L.h} \times \frac{L}{2.4\, g} = 12.0 = \underline{12\, mg/g.h}$$

- *A typical range of SVI is 80–120 mL/g.*
- *The SVI range for best removals may vary from plant to plant.*
- *OUR decreases with a decrease in the F/M ratio.*

40.9 VOLATILE ACID ALKALINITY (RAPID METHOD)

The titration method allows getting the information rapidly on volatile acid and alkalinity in one single test. Though it is not a standard method, it is an acceptable method for digester control. A given volume of sample is titrated to a pH of 4.0 with an appropriate strength of sulphuric acid. The volume of the titrant used allows us to calculate the total alkalinity of the sample. Titration with acid is continued until the pH level reaches 3.3. Next, titrate the sample back to a pH of 4.0 using a appropriate-strength (0.05–0.1N) sodium hydroxide (NaOH). The volume of the NaOH used allows us to calculate volatile acid alkalinity and hence volatile acid concentration.

40.9.1 TITRATION WITH H_2SO_4

Total alkalinity

$$\frac{mL\ of\ acid}{mL\ of\ sample} \times N\ of\ acid$$

40.9.2 TITRATION WITH NaOH

Volatile acid alkalinity

$$\frac{mL\ of\ NaOH}{mL\ of\ sample} \times N\ of\ NaOH$$

HCO_3 alkalinity = total Alk − Volatile acid Alk
When < 180 mg/L voltile acids = voltile acid alk
When > 180 mg/L voltile acids = 1.5 × voltile acid alk

EXAMPLE PROBLEM 40.12

The standard method using silicic acid is used to determine volatile acid concentration in a given sample of digested sludge. A 10-mL aliquot of acidified sample (supernatant) is passed through the silicic acid column. A volume of 2.3 mL of 0.02 N NaOH is used to titrate the sample and 0.5 mL for blank titration. Calculate the volatile acid concentration in mg/L as acetic acid.

Given:

	Titrant = 1	Sample = 2
Volume, mL	2.3–0.50 = 1.8	10.0
Normality, N	0.02	?

Solution:

Normality of the sludge sample

$$N_2 = N_1 \times \frac{V_1}{V_2} = \frac{0.02\,eq}{L} \times \frac{1.8\,mL}{10\,mL} \times \frac{1000\,meq}{eq} = 3.60 = \underline{3.6\,meq/L}$$

Volatile acid concentration

$$C_2 = \frac{3.6\,meq}{L} \times \frac{60\,mg\,of\,acetic\,acid}{meq} = 216 = \underline{220\,mg/L}$$

EXAMPLE PROBLEM 40.13

A 50-mL aliquot of a centrifuged digested sludge sample (supernatant) is titrated with 0.1 N sulphuric acid. A volume of 25.5 mL of titrant is used to reach the end point. Calculate alkalinity as mg/L of $CaCO_3$.

Given:

Parameter	Titrant = 1	Sample = 2
Volume, mL	25.5	50.0
Normality, N	0.10	?

Solution:

Non-volatile acid concentration

$$N_2 = N_1 \times \frac{V_1}{V_2} = \frac{0.1\,eq}{L} \times \frac{25.5\,mL}{50\,mL} \times \frac{1000\,meq}{eq} \times \frac{50\,mg}{meq} - 2550 = \underline{2600\,mg/L}$$

EXAMPLE PROBLEM 40.14

In the titration method for determining volatile acids and alkalinity, a 50-mL sample from an operating digester is used. Calculate the total alkalinity for every mL of the titrant (0.05 N H_2SO_4) to lower the pH to 4.0.

Given:

Parameter	Titrant = 1	Sample = 2
Volume, mL	1.0	50.0
Normality, N	0.05	?

Solution:

Alkalinity of sludge sample

$$\frac{N_2}{V_1} = \frac{N_1}{V_2} = \frac{0.05\,eq}{L} \times \frac{1}{50\,mL} \times \frac{1000\,meq}{eq} \times \frac{50\,mg}{meq} = 50.0 = \underline{50\,mg/L.\,mL}$$

EXAMPLE PROBLEM 40.15

In a titration method for determining alkalinity, a 50-mL sample from an operating digester required 12.4 mL of 0.05 N H_2SO_4 to lower the pH to 4.0. Calculate the total alkalinity.

Given:

Parameter	Titrant = 1	Sample = 2
Volume, mL	12.4	50.0
Normality, N	0.05	?

Solution:

Alkalinity of the sludge sample

$$N_2 = N_1 \times \frac{V_1}{V_2} = \frac{0.05\,eq}{L} \times \frac{12.4\,mL}{50\,mL} \times \frac{1000\,meq}{eq} \times \frac{50\,mg}{meq} = 620.0 = \underline{620\,mg/L}$$

PRACTICE PROBLEMS

PRACTICE PROBLEM 40.1

The settleable solids of the raw wastewater are 14 mL/L. If the settleable solids of the primary clarifier effluent are 0.5 mL/L, calculate the percentage removed by the primary clarification. (96%)

PRACTICE PROBLEM 40.2

Samples were collected from the influent and effluent of a primary clarifier. After 1 hour of settling, the volume of settleable solids in the influent and effluent were recorded to be 14 mL/L and 2.0 mL/L, respectively. Estimate the volume of raw sludge produced when the daily wastewater flow rate is 2.5 MGD. (30,000 gal/d)

PRACTICE PROBLEM 40.3

A suspended solids test carried on 100 mL of a sample of plant influent yielded the following data. Calculate the concentration of suspended solids and fraction of volatile. (296 mg/L, 68%)

Weight of the filter disc = 0.2335 g   Filter + dry solids = 0.2631 g   Filter + ash = 0.2430 g

PRACTICE PROBLEM 40.4

Given the following weighing from a total solids test on a sludge sample, calculate the concentration of total solids. (6.5%)

Crucible mass = 19.31 g   Crucible + wet solids = 68.34   Crucible + dry solids = 22.49 g

PRACTICE PROBLEM 40.5

A BOD test run on the primary effluent from the same plant using a 4% dilution yielded initial and final DO readings of 7.90 mg/L and 3.60 mg/L, respectively. Determine the BOD. (110 mg/L)

PRACTICE PROBLEM 40.6

A dechlorinated sample of secondary effluent was seeded with 0.4% of stale sewage, and a seed control (blank) was incubated at 3.0% concentration. The DO depletions in the sample using 25% dilution were found to be 3.8 mg/L, and that in the blank test was 4.9 mg/L. Calculate the BOD of the sample. (13 mg/L)

PRACTICE PROBLEM 40.7

In standardization of FAS solution, normality is found to be 0.25 N. A volume of 20 mL of the digested solution water was titrated with FAS solution. The volumes of FAS used for the blank and sample were observed to be 12.0 mL and 2.5 mL, respectively. What is the COD of the sample? (950 mg/L)

PRACTICE PROBLEM 40.8

The MLSS concentration in the aeration tank is 1870 mg/L. The settleability test indicates 175 mL settled in the 1-L graduated cylinder. Calculate SVI and SDI. (94 mL/g, 1.1%)

PRACTICE PROBLEM 40.9

A settleability test performed on a sample of mixed liquor indicates the volume of settleable solids to be 150 mL/L. Suspended solids made on the sample yield an MLSS of 1800 mg/L. Calculate SVI and SDI. (83 mL/g, 1.2%)

PRACTICE PROBLEM 40.10

A settleability test was performed on a sample of mixed liquor, and the following data were observed: MLSS = 2550 mg/L V_{SS} = 770 mL/2L Sample size = 2.0 L. Determine the SVI and SDI. (150 mL/g. 0.66%)

PRACTICE PROBLEM 40.11

Data from an uptake rate test is graphed, and a straight line is fitted. The slope of the line is read to be 0.21 mg/L ·min. A suspended test run on the same sample indicated an MLSS of 2200 mg/L, of which 71% is volatile. Calculate OUR and SUR based on MLVSS. (13 mg/L.h, 8.1 mg/g·h)

PRACTICE PROBLEM 40.12

A standard volatile acid test on a digesting sludge sample yielded the following data: Volume of sample = 15 mL; Volume of 0.01 N NaOH = 6.6 mL for titration and 1.1 mL for the blank. What is the concentration of volatile acids in the digesting sludge in mg/L of acetic acid? (220 mg/L)

PRACTICE PROBLEM 40.13

In an alkalinity test of 50 mL of supernatant of a digested sludge after centrifugation, 27.3 mL of 0.1 N H_2SO_4 are used. What is the alkalinity of the sludge? (2700 mg/L)

PRACTICE PROBLEM 40.14

Work out the alkalinity for every mL of 0.02 N titrant used when 100 mL of the digested sludge sample is used. (10 mg/L ·mL)

PRACTICE PROBLEM 40.15

In the same test, 3.5 mL of 0.05 NaOH were required to raise the pH back to 4.0. Calculate the volatile acid to alkalinity ratio. (0.42)

Index

Printed in the United States
by Baker & Taylor Publisher Services